Nonlinear Modeling Analysis and Predistortion Algorithm Research of Radio Frequency Power Amplifiers

Nonlinear Modeling Analysis and Predistortion Algorithm Research of Radio Frequency Power Amplifiers

Jingchang Nan and Mingming Gao

CRC Press is an imprint of the
Taylor & Francis Group, an **informa** business

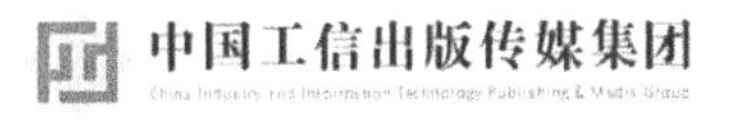

First edition published 2022

by CRC Press
6000 Broken Sound Parkway NW, Suite 300, Boca Raton, FL 33487-2742

and by CRC Press
2 Park Square, Milton Park, Abingdon, Oxon, OX14 4RN

CRC Press is an imprint of Taylor & Francis Group, LLC

Library of Congress Cataloging-in-Publication Data
Names: Nan, Jingchang, 1971- author. | Gao, Mingming, 1980- author.
Title: Nonlinear modeling analysis and predistortion algorithm research of radio frequency power amplifiers / Jingchang Nan, Mingming Gao.
Other titles: She pin gong fang fei xian xing jian mo fen xi yu yu shi zhen suan fa yan jiu. English
Description: First edition. | Boca Raton : CRC Press, 2021. | Translation of: She pin gong fang fei xian xing jian mo fen xi yu yu shi zhen suan fa yan jiu. | Includes bibliographical references. | Summary: "This book is a summary of a series of achievements made by the authors and colleagues in the areas of radio frequency power amplifier modeling (including neural Volterra series modeling, neural network modeling, X-parameter modeling), nonlinear analysis methods, and power amplifier predistortion technology over the past 10 years. Blending theory with analysis, this book will provide researchers and RF/microwave engineering students with a valuable resource"--Provided by publisher.
Identifiers: LCCN 2021000440 (print) | LCCN 2021000441 (ebook) | ISBN 9781032010267 (hbk) | ISBN 9781003176855 (ebk)
Subjects: LCSH: Amplifiers, Radio frequency--Mathematical models. | Microwave amplifiers--Mathematical models. | Power amplifiers--Mathematical models. | Electric distortion--Mathematical models.
Classification: LCC TK6565.A55 N3613 2021 (print) | LCC TK6565.A55 (ebook) | DDC 621.3841/2--dc23
LC record available at https://lccn.loc.gov/2021000440
LC ebook record available at https://lccn.loc.gov/2021000441

ISBN: 978-1-032-01026-7 (hbk)
ISBN: 978-1-032-01032-8 (pbk)
ISBN: 978-1-003-17685-5 (ebk)

Typeset in Minion
by MPS Limited, Dehradun

Contents

Preface

Since 2009, the authors' research group has been approved by the National Natural Science Foundation of China for two projects, which are "Research on Nonlinearity and Behavioral Model of Wideband RF Power Amplifier" (Grant No. 60971048) and "Research on Nonlinear Modeling and Analysis Methods of Radio Frequency and Microwave Circuit and System" (Grant No. 61372058). The research group has also been approved by the Start-up Grant of the Doctoral Program of Liaoning Province for the project "Research on CAD Simulation Technology of RF and Microwave Circuit and RF Card Reader" (Grant No. 20091033), as well as the Excellent Talents Support Program of Higher Education Institutions in Liaoning Province for the project "Radio Frequency Circuit and System" (Grant No. LR2013012). Furthermore, the research group also participated in the provincial teaching quality project "Pilot Program for the Reform of Undergraduate Engineering Talents Training Scheme in General Colleges of Liaoning Province (Major: Communication Engineering)". The research group applied for the project of Liaoning Provincial Key Laboratory of Higher Education Institutions, and was approved in 2018. Supported by these projects, the research group has made certain achievements in the design of radio frequency and microwave modules (power amplifiers, power dividers, filters, antennas, etc.), module behavioral modeling and reverse modeling, radio frequency and microwave nonlinear analysis methods and power amplifier predistortion. These achievements have been published as papers in various journals, and have also been highlighted in annual reports and concluding reports. All along, we planned to summarize and classify these achievements systematically, with each achievement refined into innovation points and collectively presented in a book. We hope to enable readers to learn more deeply or investigate more systematically in a certain direction, acquire valuable knowledge and experience, benefit from the book and, at the same time, feel, experience and share the hardship and happiness of the research group in the research process. But this plan has not been carried out because of a tight working schedule. As the team grows over time, both postgraduates and undergraduates need systematic materials. Therefore, we are determined to fulfill the longstanding plan.

We finally prepared two books, *Design and Implementation of High Efficiency Wideband Radio Frequency Module with Dual Frequency and High Linearity* and *Nonlinear Modeling Analysis and Predistortion Algorithm Research of Radio Frequency Power Amplifier*, based on the published academic papers and the annual reports or concluding reports generated for various funds.

This book is a summary of a series of achievements made by Professor Nan Jingchang, Associate Professor Gao Mingming and the postgraduates supervised in the areas of radio frequency power amplifier modeling (including neural Volterra series modeling, neural network modeling, X-parameter modeling), nonlinear analysis methods, and power amplifier predistortion technology in the past 10 years. Volterra series modeling of power amplifier, power amplifier modeling based on neural network, power amplifier modeling with X-parameters, modeling of other power amplifiers, nonlinear circuit analysis methods, and predistortion algorithms and applications are described in this book. To increase readability, the book adds the basic descriptions of research status, power amplifier nonlinearity, power amplifier model, nonlinear analysis and predistortion theory. This book comprehensively summarizes the power amplifier models, modeling methods and related algorithms, harmonic balance method and related algorithms for power amplifier nonlinear analysis, predistortion algorithms, techniques and applications. This book has proposed accurate and reliable models and related analysis methods for system simulation analysis and predistortion system construction.

Professor Nan Jingchang is responsible for classifying, editing and summarizing Chapters 5 to 9 of this book, while Associate Professor Gao Mingming is responsible for Chapters 1 to 4 and Chapter 10. We thank Zhao Jingmei, Ren Jianwei, Huang Lina, Qu Yun, Li Houru, Li Shiyu, Xu Jing, Wang Ruina, Zhou Dan, Sang Baixing, Liu Yue, Sun Danping, Tian Na, Wang Zhenxia, Cong Rijing, Zhang Nan, Lu Yanan, Zhang Yunxue, Cui Hongyan and Zhan Suli for their contributions to this book. During the preparation process of this book, full-time postgraduate students Cui Hongyan, Wang Ziqi, Zhan Suli, Sheng Shuangshuang, Hu Tingting,

Liu Yinling, Wu Yue, Tao Chengjian and Zang Jing devoted a lot of their time and effort to this book. They determined the format, proofread the book, and revised figures and charts. Other teachers of the research team have provided specific guidance in postgraduate training, publication of academic papers and achievements, and have made many valuable comments during the preparation of this book. In addition, I have received steadfast support and abiding understanding from my family in the process of leading the RF and microwave research team to engage in this research. Here, I would like to express my thanks to all of you for your painstaking efforts for the research group. Furthermore, I would also like to thank the readers. Your support is the unremitting driving force for our continuous research. I hope we will develop together and make contributions to the development of the radio frequency and microwave industry for our country.

This book is specially funded by the General Program of National Natural Science Foundation of China "Research on Nonlinear Modeling and Analysis Methods of Radio Frequency and Microwave Circuit and System" (Grant No. 61372058) and the Liaoning Distinguished Professor project "Design and Research of Reconstructable Multimode Multib and Radio Frequency Power Amplifier and Passive Module" (Grant No. 551710007004). I would like to express my thanks here.

Nan Jingchang, Huludao

CHAPTER 1

Overview of Research Status

The rapid development of wireless communications and microwave technology as well as the continuous innovation of digital communication technology has highlighted the design and research of power amplifier (PA), an important component of radio frequency (RF) front-end, in the RF/microwave field. The first-generation mobile communication network is an analog system, with its modulation scheme adopting a constant envelope method. The second-generation mobile communication network, such as the Global System for Mobile Communication (GSM) widely used in China, uses Gaussian Filtered Minimum Shift Keying (GMSK) modulation scheme, which is also a constant envelope modulation. The superiority of constant envelope modulation technology lies in that it reduces the impact of nonlinearity on signals and requires less linearity of the transmitter PA. Therefore, the PA operates near the saturation region, improving the working efficiency. However, with the continuous development of society and the continuous expansion of communication system capacity, spectrum resources are becoming scarce. In this context, constant envelope modulation technology can no longer meet the needs of modern wireless mobile communications. A PA often working in the high-power state causes nonlinear distortion. The design quality of PA directly affects the performance of the entire wireless communication system. To meet the new communication technology standards and the requirements of high efficiency, low cost and miniaturization, the structure of PA has become more and more complex. In addition, the rapid development of wideband communication services has been demanding faster data transmission and higher spectrum utilization, resulting in more crowded frequency bands used for the communication systems. Digital modulation schemes, such as Wide Code Division Multiple Access (WCDMA), Orthogonal Frequency Division Multiplexing (OFDM), Quadrature Amplitude Modulation (QAM) and Quadrature Phase Shift Keying (QPSK), have been widely used for modern wireless communication systems to modulate signals for an improvement in spectrum utilization. These new modulation schemes feature wide spectrum, nonconstant envelope and high peak-to-average ratio. Consequently, modulated signals after passing the RF PA will experience serious distortion and the resulting in-band distortion will cause an increased bit error rate (BER), while out-of-band distortion generated by intermodulation signals will lead to varying degrees of interference to adjacent channels. Plus, the input signals with different envelopes will produce memory effects resulting in distortion. Therefore, improving the linearity of RF PA is one of the urgent problems to be solved.

PA behavioral modeling is of great significance in the simulation analysis and design of communication systems. In a communication system, setting up a hardware circuit for actual measurements has the drawbacks of high complexity, high cost, low efficiency, etc. Therefore, an accurate and effective RF PA behavioral model for system-level analysis and simulation is established instead, which accurately evaluates the impact of RF PA on the communication system with a shortened R&D cycle and reduced development cost. The circuit design of a modern communication system focuses on low cost, high efficiency and modularity, which makes component modeling far more difficult, setting higher requirements for PA modeling technology in the RF module design. Various PA models have been proposed, such as Volterra series, memory polynomials, neural networks, etc. Furthermore, people are still pursuing to improve the accuracy and simplify the complexity of existing models, as well as to explore new models that can comprehensively describe the characteristics of PA. This type of research has been a highlight in modeling technology development.

Power back-off is a traditional method used to improve the linearity of RF PA and eliminate the impact of nonlinearity, but the back-off of the operating point reduces the power efficiency of PA and produces heat dissipation. New linearization technologies include feed-forward linearization [1], negative feedback [2], digital predistortion [3], linear amplification with nonlinear components (LINC), etc. The performance comparison of these technologies is given in Table 1.1.

At present, digital predistortion is considered as the most effective and widely used linearization technology. The premise of digital predistortion is to build a PA model with high accuracy, simple structure and easy measurement, and to develop more effective optimization algorithms. Consequently, the PA behavioral model has been one of the highlights in the industry and research work in recent years.

As indicated by Table 1.1, predistortion technology has gained a wide application prospect owing to its obvious superiorities in terms of high accuracy, low cost, and absence of the stability problem. Wireless mobile

TABLE 1.1 Performance Comparison of Various Linearization Technologies

Classification	Advantage	Disadvantage	System Bandwidth	Remarks
Negative feedback technology	High accuracy, low cost	Poor stability	Narrow band	Rarely used
Feed-forward technology	Fast speed, high linearity, absence of the stability problem	Relatively complex structure, high cost, not high efficiency, poor adaptive capability	Wide band	Widely used
Envelope elimination and restoration technology	Relatively high efficiency	Difficult time delay calibration	Wide band	Rarely used
LINC technology	Relatively high efficiency	Difficult accurate matching	Narrow band	Rarely used
Predistortion technology	High accuracy, low cost, absence of the stability problem	Slow adjustment	Wide band	The most promising outlook

communication systems are migrating to higher frequency bands, faster data rates and broader bandwidths, which causes the problem of stronger nonlinearity and memory effects of PA. It is extremely challenging but quite practical to investigate the behavioral model of wideband PA with strong nonlinearity and memory effects.

1.1 RESEARCH STATUS AND DEVELOPMENT OF THE BEHAVIORAL MODEL FOR PA

As one of the important parts of RF front-end for wireless communications, the RF PA has been under development in pursuit of linearization, integration and broadband. Continuous technical improvements have enabled a more extensive research on RF PA modeling all over the world.

As a method to characterize the real PA circuit, the RF PA model's characteristics should be consistent with those of the real PA. RF PA models can be categorized into physical models based on circuit simulation and behavioral models based on system-level simulation. For the physical model, an equivalent model is established by using the relevant method according to the specific design of PA circuit, which can describe the physical relationship between the components of each part inside the PA. Although the simulation result is accurate, there is a cost in simulation time and description of the internal structure. The behavioral model is used to describe the behavior characteristics of PA. It extracts each parameter of the model by means of the preset structure and algorithm based on the input-output relationship of PA, and then establishes a PA model that meets the requirements without the need to obtain the circuit structure inside the PA. Therefore, the behavioral model is also called "black box model".

The behavioral model of PA is categorized into the memoryless behavioral model and the memory behavioral model, regarding the presence of memory characteristics of PA. The research on PA behavioral modeling first proposed the memoryless model. The so-called "memoryless model" means that the PA output depends only on the input at the current moment, independent of the inputs at previous moments. For a narrowband communication system, the signal bandwidth is far less than that of the PA itself, and therefore the impact of memory effects may be ignored in a PA nonlinearity analysis. That is, the characteristics of PA can be described by the memoryless PA model in a narrowband communication system. However, memory effects are significant in a wideband communication system since the PA is affected by self-heating of internal active devices and DC bias, resulting in the memoryless model unable to accurately express the PA nonlinearity. Therefore, the memory behavioral model has been proposed, and memory PA modeling and linearization technology have been highlighted. In the analysis and design of microwave circuits, the methods for analyzing nonlinear RF/microwave circuits include power series method, Volterra series method, harmonic balance method, neural network method, etc.

Regarding the research on behavioral modeling technology of small-signal RF PA, the memoryless PA models include Saleh model [4], power series model, memoryless polynomial model, etc., and the memory PA models include Volterra series model [5] (also called Volterra model), memory polynomial model [6], Wiener model [7], neural network model [8], etc.

The power series method is usually selected to analyze the weakly nonlinear memoryless circuit system. With the method, system modeling is achieved by using a filter (or another frequency sensitive component) followed by a memoryless, wideband transfer nonlinear "component". It is quite convenient to use the power series method to quantitatively analyze the second harmonic, second-order or third-order intercept point, 1 dB compression point, 1 dB gain and other indicators of the amplifier. But note that the method is only suitable for weakly nonlinear circuits. A circuit containing any memory component such as a capacitor cannot be expressed as a power series, because a power series ignores the fluctuations of a nonlinear reactance and can only serve as an approximation.

New algorithms have been developed to address the drawbacks of the existing power series method. The idea of new algorithms is to combine the generalized power series method with the Newton method, partial Newton method and secant iteration method. The frequency-dependent complex power series varying with the time delay can simply exhibit memory effects, with only a few terms in the series expanded. For a typical circuit comprising only a few components, complex coefficients may not be required, and the time delay may be included outside the phase shift of the input signal.

The Volterra series method addresses the drawbacks of the power series method and does not require memory of circuits. It is a general method to analyze practical realizable nonlinear systems or circuits, having no difference from the power series method in computations of small-signal, memoryless nonlinear circuits. The Volterra series method uses a circuit transfer function indicating the excitation-response relationship, considers the interaction between input frequencies, and expresses the system as the sum of infinite Volterra series (third-, fourth- or fifth-order expansion generally used in practical applications).

The Volterra series method is one of the methods to describe nonlinear systems. It can explain the physical meaning of a nonlinear system more clearly and accurately. However, its disadvantage lies in that the number of model coefficients grows exponentially with the increase of the nonlinearity and memory depth of the system, resulting in an extremely complicated identification process of model coefficients. In most cases, the Volterra series method cannot meet the requirements of model accuracy and complexity at the same time, and is usually used only in PA modeling with weak nonlinearity. To solve these problems of the Volterra series model, some simplified Volterra series models are often selected for PA modeling in recent years. The most frequently used models are Wiener model, Hammerstein model, memory polynomial (MP) model, generalized memory polynomial (GMP) model, etc. Compared with the Volterra series model, the above simplified models greatly reduce the computational complexity and better adapt to modern wireless communication systems.

The MP model is composed of the diagonal terms of Volterra series model, which is more widely applied than the Volterra series model. Based on the diagonal terms of Volterra series model, the GMP model additionally includes the cross terms between the signal and its lagging envelope and leading envelope, thereby delivering higher accuracy. The GMP model is suitable for modeling strongly nonlinear PAs, but it requires quite a number of coefficients. The PA nonlinearity rises with the increase of the input signal power. Meanwhile, the application of wideband PA makes the PA exhibit more obvious memory effects. In this situation, the traditional PA behavioral model can no longer accurately describe the nonlinearity and memory effects of PA. For PAs with strong nonlinearity and memory effects, Oualid Hammi [9] has proposed a Twin Nonlinear Two-Box (TNTB) model, which consists of the Look UP Table (LUT) and MP model. The TNTB model is categorized into forward TNTB model, reverse TNTB model and parallel TNTB model, regarding the specific construction method. The parallel TNTB model can achieve the same accuracy as the memory polynomial by using only 50% of the coefficients. Mayada Younes proposed an accurate and low-complexity Parallel-LUT-MP- EMP (PLUME) model and experimentally validated it on the Doherty PA in the multi-carrier WCDMA signal context. The PLUME model is composed of three nonlinear models in a parallel structure: LUT, MP and Envelope Memory Polynomial (EMP). The experimental results show that the PLUME model has higher accuracy compared with the MP model and the parallel TNTB model, and that the PLUME model achieves the same accuracy as the GMP model while reducing 45% of coefficients.

The Hammerstein model separates the nonlinearity from memory effects of PA, equating the PA into a cascade of a memoryless nonlinear system (static nonlinearity) with a memory linear system (linear filter). The electrical memory effects caused by the inconstant frequency response near the transmitter carrier can be compensated by the linear filter, but the electrical memory effects caused by the impedance changes of the transmitter bias circuit and harmonics cannot be compensated. To improve the accuracy of Hammerstein model, Liu Taijun et al. proposed an improved Hammerstein model. A second-order linear filter is added to the second branch of the original Hammerstein model in parallel, thus improving the performance of the model.

Neural network is a new method applied to behavioral modeling in recent years, by which the convergence and fitting degree can precisely approach the objective. It has become a cutting-edge topic in nonlinear circuit research. The output and input data can be obtained by means of measurements on the device under test (DUT). The neural network model is trained with some data, and the accuracy of the model is validated with the remaining data. Once the model is validated to be accurate, it can provide strong theoretical support for the design.

This method has been widely used to solve various problems in RF circuits. Representative neural network models include multilayer perceptual network, back propagation (BP) network, radial basis function (RBF) network, knowledge-based neural network, bidirectional associative memory, Hopfield model, etc. These models can be used to enforce function approximation, data clustering, pattern recognition, optimized computation, etc.

At present, domestic and foreign scholars are carrying out the research on RF/microwave circuit modeling based on neural networks. Professor Q. J. Zhang, as the founder and leader of neural network modeling in the RF/microwave field, proposed and validated the feasibility of neural networks applied to microwave circuit simulation and statistical design. He has proposed the use of dynamic neural networks in behavioral modeling of RF circuits and engaged in the cutting-edge research on neuro-modeling and space mapping optimization, which are two major technologies in the development history of RF/microwave computer-aided design. There are also many research groups from domestic universities that have been carrying out similar researches in this field, such as Tsinghua University, Shanghai Jiao Tong University, Xidian University, Hangzhou Dianzi University, etc. The research results have been widely published in international authoritative journals, such as *IEEE Transactions on Microwave Theory and Techniques, IEEE Transactions on Computer-Aided Design of Integrated Circuits and Systems.*

A variety of neural network structures have been used for PA modeling and some achievements have been made, basically including BP neural network, RBF neural network [10], Elman neural network, real-valued time-delay neural network, etc. Various structures are also emerging as the combinations of neural networks with other modeling theories and technologies, such as the combination of support vector machine and neural network, combination of wavelet theory and neural network [11], and combination of fuzzy theory and neural network [12]. Each model can well describe the PA characteristics and can be used in behavior-level simulation. Furthermore, various new learning algorithms are emerging to improve the accuracy of neural network model and accelerate the convergence process. Such algorithms include gradient descent method and its improved algorithm, least square method and its improved algorithm, quasi-Newton method, genetic algorithm, particle swarm optimization (PSO) algorithm, etc.

An effective way to improve neural network technology is to enable it to effectively exploit the prior knowledge in the learning process. A fuzzy system just possesses this advantage, which can express fuzzy or qualitative knowledge. But its disadvantage lies in the absence of self-learning and adaptive capability. Fuzzy neural network is a combination of fuzzy system and neural network. It combines the advantages of the neural network's learning mechanism and the fuzzy system's verbal reasoning capability, and therefore it is applied to nonlinear system modeling. However, these models suffer the disadvantages of complex network structure, low model accuracy and slow convergence. Therefore, researchers are pursuing various methods that can simplify the model structure, improve the model accuracy and speed up the convergence process.

Neural network modeling is a behavioral modeling method with fast speed, high accuracy, favorable universality, and short development cycle. Plus, it is easy to enable modeling automation. However, certain deficiencies in the neural network make its modeling unable to describe the circuit characteristics of RF/microwave devices. In addition, the neural network has a relatively complex structure, which is difficult to be implemented by hardware. This makes it still at the theoretical research stage. Meanwhile, the application of neural network technology also depends on the development of digital integration technology and fast optimization algorithms.

Generally, modeling technology of large-signal RF PA needs to consider the actual working conditions of devices. The PA works in the nonlinear region under the excitation of large signals. In this case, the strong nonlinearity and memory effects have to be considered. Therefore, large-signal modeling is relatively difficult. This problem is resolved by large-signal X-parameters, which can more accurately describe the characteristics of a device itself in the modeling process, providing precise theoretical support for the system-level design.

Communication systems are demanding higher power, higher efficiency, lower power consumption and more complex modulation schemes. RF PAs long working under the excitation of large signals may produce serious nonlinear distortion and multiple harmonic components. This will significantly degrade the performance of communication systems, as evidenced in gain compression, amplitude distortion and phase distortion. Therefore, accurately characterizing the RF PA can not only precisely realize the RF PA model as per design requirements, but also shorten the development cycle, improve the success rate of products and save costs. RF PA modeling technology is also tending to the research on large signal nonlinear modeling.

With the continuous development of large signal measurement techniques, the theoretical research and application of RF/microwave circuits in the case of large signal nonlinearity has been highlighted in the field of modern communications. In June 2000, a symposium named "Going Beyond S-Parameters" was held in Boston (Massachusetts, USA), by the Automatic RF Technology Group of the IEEE Microwave Theory and Techniques Society (MTT-S). New challenges of RF networks were discussed on the symposium. Participants discussed seven topics: from large-signal measurements to modeling, from standard to model validation, load traction, deployment and noise solutions, modeling and simulation technique, large-signal characterization and calibration technique, large-signal measurement technique, large-signal modeling and signal simulation. In December 2000, Dr. Marc and Dr. Jan, former employees of HP's Electronic Measurement Division, jointly published a famous paper entitled "Large-Signal Network Analysis: Going Beyond S-Parameters" [13], which defined large-signal network analysis for the first time and proposed a scheme for the measurements and modeling of an RF network in the large-signal context. With vector measurement technology, a variety of black-box modeling methods based on frequency- and time-domain measurements have been proposed at home and abroad. Professor Wang Jiali from Xidian University with his laboratory team proposed the nonlinear scattering function method for device modeling based on frequency-domain measurements [14–16]. This method is based on test data and uses various neural network models to fit the coefficients of the scattering function, thereby characterizing the nonlinearity of the device. In 2008, Agilent put forward the concept of large-signal X-parameter, developed the world's first nonlinear network analyzer (PNA-X series), and proposed a large-signal

X-parameter model. The X-parameter model can replace the S-parameter model to characterize the linear and nonlinear behaviors of RF devices, and enable X-parameter technology to be compatible with the commercial software Advanced Design System (ADS). The X-parameter model can obtain accurate data under actual working conditions and automatically extract various nonlinear characteristics of the DUT, such as gain compression, harmonic amplitude and phase/spectral gain. The X-parameter model, as a nonlinear analysis method for RF large signals that integrates measurements, modeling, simulation and design, has been widely investigated in the industry and used in the RF/microwave nonlinear circuit design.

In 2008, scientists first proposed a nonlinear behavioral model for the GSM mobile phone PA characterized by X-parameters [17]. In 2009, Dr. Jan Verspecht, in cooperation with scientists from Agilent and the University of Leeds in the UK, proposed the theory of a dynamic X-parameter model for characterizing, modeling and simulating RF devices in the envelope domain [18], considering the impact of strong nonlinearity and long-term dynamic memory effects. This proposal won the annual Best Oral Presentation Award at the IEEE MTT International Symposium that year. In 2010, Jan Verspecht and Dr. D. E. Root [19] collaborated again and were invited to deliver a presentation on the latest achievements of the dynamic X-parameter model at the 76th IEEE ARFTG Conference. In 2011, Professor Lin Maoliu of Harbin Institute of Technology and his team proposed a new method for X-parameter measurements of RF PA [20], and independently developed relevant software for X-parameter measurements and analysis, supporting measurement data preprocessing, X-parameter modeling and visualization of measurements and X-parameters. In 2012, Dr. D. E. Root published "Future Device Modeling Trends" [21], acknowledging the significance of X-parameters in future microwave device modeling by means of extensive analysis and investigation.

As an emerging technology, the X-parameter model is immature and still has some deficiencies. Regarding the description of memory effects, scholars have put forward the dynamic X-parameter theory, and the associated research work is also in full swing. However, the theory still needs to be deeply investigated and applied to practical measurements. Several literatures have provided specific schemes for the extraction method of X-parameters, but further research work is required on how to accurately measure the extracted data, improve the model accuracy and the efficiency.

In summary, under the influence of large signal nonlinearity, RF/microwave modeling technology will evolve to resolve the nonlinear problem, fully consider the impact of memory effects, well fit the characteristics of RF devices, improve the model accuracy, and accelerate the convergence process. There is no doubt that these efforts will make modeling technology achieve a leap-forward breakthrough, and further promote the progress of science and technology.

1.2 RESEARCH STATUS AND DEVELOPMENT OF PREDISTORTION TECHNOLOGY

In the 1920s, H.S. Black of Bell Laboratories proposed feedback technology and negative feedback technology. Applications of these technologies greatly reduced the nonlinear distortion of PA, serving as a perfect start for the research on linearization technology of PA. Later, D. C. Cox improved negative feedback technology and derived Cartesian negative feedback. Then in the early 1970s, D. C. Cox proposed a technology called LINC by applying the phase-shifting technique. The linearization technology of PA continues to improve and develop.

In the 1980s, the development of linearization technology took a critical step, namely the emergence of predistortion, which was initially analog predistortion. In the late 1980s, applications of analog predistortion in mobile communication systems enabled rapid development. With the development of digital signal processing (DSP) technology, predistortion has also been achieved in the digital domain, which is referred to as Digital Pre-Distortion (DPD). DPD can be applied to baseband, intermediate frequency (IF) and RF parts of communication systems, and can use well-functioning adaptive tracking compensation to adapt to the changes in PA characteristics caused by environmental temperature, humidity, power supply drift, device aging and other factors, which is called adaptive predistortion. Because baseband predistortion technology is easy to control and implement, adaptive digital baseband predistortion has become the focus of current investigations.

In recent years, some foreign chip manufacturers have developed different DPD chips. For example, ISL5239 chip released by Intersil is a DPD chip completely used for PA linearization. Xilinx proposed a predistortion scheme based on a polynomial method integrating software and hardware. Altera introduced a reference design based on LUT and polynomial digital predistortions. In addition, TI, Ericsson, PMC, and Optichron of the United States have also performed the research work on predistortion technology and released the predistortion chips with their own features. In the future development, the DPD module will have higher integration level, lower cost, and better performance. Its application fields will no longer be limited to telecommunications, but expanded to aviation, aerospace radar, etc.

In China, some major universities have also studied predistortion. For example, Tsinghua University has investigated the LUT predistortion method for compensating nonlinear distortion of memoryless PAs. Southeast University has also focused on the LUT research; the research group simulated the real PA and then used the inversion method to extract the compensation parameters stored in the predistortion LUT, and also proposed a new index method for LUT entry addresses. Beijing University of Posts and Telecommunications has basically focused on the memory polynomial digital predistortion. Enterprises and institutions such as Huawei, ZTE and No.29 Research Institute of CETC have also performed DPD research work. These researches focus on reducing the complexity of algorithms, constructing an easy-to-implement structure, and how to make the system more flexible and stable.

Earlier digital predistortion methods usually adopt a direct learning structure to obtain the predistorter coefficients (polynomial coefficients or LUT values). This type of learning structure is simple but it is difficult to be enforced in practical engineering because the adaptive algorithms may not converge. In contrast, the indirect learning structure has gained popularity because it can use classical adaptive algorithms. Since the indirect learning structure is less resistant to noises, G. Montoro et al. proposed a learning structure based on model identification, which improved the linearization level for the indirect learning structure.

In recent years, many well-known foreign companies and scientific research institutions have actively participated in the research of digital predistortion technology. Many PA behavioral models, digital predistortion algorithms and learning structures have been proposed. At present, digital predistortion related products have been commercially available. Ericsson and some other companies have applied digital predistortion technology to base station systems. Domestic companies such as Huawei, Potevio and Comba as well as universities such as Tsinghua University, Beijing University of Posts and Telecommunications and Ningbo University, are also actively performing investigations on digital predistortion and have obtained series of research achievements. However, the existing digital predistortion technology cannot meet the requirements of wireless communication systems with broader bandwidths, so further research is required to improve the linearization capability of digital predistortion technology and reduce the complexity of implementation in engineering.

REFERENCES

1. KATZ A, WOOD J, CHOKOLA D. The Evolution of PA Linearization: From Classic Feedforward and Feedback through Analog and Digital Predistortion[J]. IEEE Microwave Magazine, 2016, 17(2):32–40.
2. CHEN W, ZHANG S, LIU Y J, et al. Efficient Pruning Technique of Memory Polynomial Models Suitable for PA Behavioral Modeling and Digital Predistortion[J]. IEEE Transactions on Microwave Theory and Technique, 2014, 62(10):2290–2299.
3. 王伟, 曹民, 朱娟娟, 等. 功率放大器及预失真模型的设计[J]. 通信技术, 2014, 47(1):111–114.
   WANG W, CAO M, ZHU J, et al. Designs of Power Amplifier and Predistorter Models[J]. Communications Technology, 2014, 47(1): 111–114.
4. SALEH A. Frequency-Independent and Frequency-dependent Nonlinear Models of TWTA Amplifier[J]. IEEE Transactions on Communication, 1981, 29(11):1715–1720.
5. MKADEM F, CLAUDE FARES M, BOUMAIZA S. Complexity-reduced Volterra Series Model for Power Amplifier Digital Predistortion[J]. Analog Integrated Circuits Signal Processing, 2014, 79:331–343.
6. KIM J, KONSTANTINOU K. Digital Predistortion of Wideband Signals Based on Power Amplifier Model with Memory[J]. IET Electronics Letters, 2001, 37(23):1417–1418.

7. LI J, QI X, FENG L. Correlation Analysis Method Based SISO Neuro-Fuzzy Wiener Model[J]. Journal of Process Control, 2017, 58:100–105.
8. LI J, NAN J, ZHAO J. Study and Simulation of RF Power Amplifier Behavioral Model Based on RBF Neural Network. International Conference on Microwave & Millimeter Wave Technology, Chengdu, 2010[C]. Piscataway:IEEE, 1465–1467.
9. HAMMI O, SHARAWI M S, GHANNOUCHI F M. Generalized Twin-Nonlinear Two-Box Digital Predistorter for GaN Based LTE Doherty Power Amplifiers with Strong Memory Effects. IEEE International Wireless Symposium 2013, Beijing, 2013[C]. Piscataway:IEEE, 1–4.
10. LIU T. Behavioral Modeling and Digital Predistortion Linearization for Wideband RF Power Amplifiers with Neural Networks. International Conference on Microwave & Millimeter Wave Technology, Beijing, 2016[C]. Piscataway:IEEE, 316.
11. 闫星星. WCDMA 系统中基于小波网络的数字预失真技术研究[D]. 沈阳:东北大学, 2013.
    YAN X. Research on Digital Predistortion Technology Based on Wavelet Network in WCDMA System[D]. Shenyang: Northeastern University, 2013.
12. CHIH LIN Y. An Application of Artificial Neural Networks to Forecast Wining Prices: A Comparison of Back-Propogation Networks and Adaptive Network Based Fuzzy Inference System[D]. Taiwan, China: Yuan Ze University, 2010.
13. VERSPECHT J, VERBEYST F, VANDEN BOSSCHE M. Network Analysis Beyond S-Parameters: Characterizing and Modeling Component Behaviour under Modulated Large-Signal Operating Conditions. 56th ARFTG Conference Digest, Boulder, 2000[C]. AZ: USA, 1–4.
14. 倪峰, 王家礼. 基于非线性散射函数的微波电路 CAD 技术[J]. 微波学报, 2008, 24(S1):176–179.
    NI F, WANG J. Microwave Circuit CAD Technology Based on Nonlinear Scattering Function[J]. Journal of Microwaves, 2008, 24(S1):176–179.
15. 孙璐. 基于散射函数的一种微波非线性电路建模新方法[D]. 西安:西安电子科技大学, 2010.
    SUN L. A New Modeling Method of Microwave Nonlinear Circuit Based on Scattering Function[D]. Xi'an: Xidian University, 2010.
16. 孙璐, 王家礼, 陈晓龙. 两种基于测量的微波非线性电路频域黑箱模型[J]. 电子测量与仪器学报, 2009, 23(10):19–24.
    SUN L, WANG J, CHEN X. Two Frequency Domain Black-Box Models of Microwave Nonlinear Circuit Based on Measurement[J]. Journal of Electronic Measurement and Instrument, 2009, 23(10): 19–24.
17. H ORN J M, VERSPECHT J, GUNYAN D, et al. X-Parameter Measurement and Simulation of a GSM Handset Amplifier. Microwave Integrated Circuit Conference, EUMIC2008. European. Amsterdam, Netherlands, 2008[C]. Piscataway: IEEE, 135–138.

18. VERSPECHT J. Extension of X-Parameters to Include Long-term Dynamic Memory Effects. 2009 IEEE MTT-S International Microwave Symposium Digest, Boston, MA, 2009[C]. Piscataway: IEEE, 741–744.
19. VERSPECHT J, HORN J, ROOT D E. A Simplified Extension of X-Parameters to Describe Memory Effects for Wideband Modulated Signals. Microwave Measurements Conference, USA, 2010[C]. Piscataway: IEEE, 1–6.
20. 辛栋栋. 基于 NVNA 的射频功率放大器的X参数测量提取仿真研究[D]. 哈尔滨:哈尔滨工业大学, 2011.
XIN D. Study on the X-Parameter Extraction and Simulation of RF Power Amplifiers Based on NVNA Measurements[D]. Harbin: Harbin Institute of Technology, 2011.
21. ROOT D E. Future Device Modeling Trends[J]. Microwave Magazine IEEE, 2012, 13(7):45–59.

CHAPTER 2

Nonlinear Characteristics of Power Amplifier

Power amplifier (PA) is considered an important part of a communication system. In applications, a PA usually works near its saturation point and in a strong nonlinear state for an improvement in the working efficiency. In addition, a PA also exhibits a certain degree of memory effects. In the communication process, the nonlinearity and memory effects of PA will generate spectrum spreading in the output signals, resulting in adjacent-channel interference. Consequently, the bit error rate (BER) of the whole system increases, seriously degrading the communication system's performance. To address this issue, an investigation on the nonlinearity and memory effects of PA is quite necessary.

2.1 NONLINEARITY OF POWER AMPLIFIER

The nonlinear feature of PA usually causes harmonic distortion, intermodulation distortion, amplitude distortion (AM/AM) and phase distortion (AM/PM). Harmonic distortion and intermodulation distortion generate new frequency components, so these two are also called frequency distortion. A brief analysis of these distortions is provided in the following subsections so that readers are able to further grasp the principle of PA nonlinearity.

2.1.1 Harmonic Distortion

Harmonic distortion [1] is the presence of new frequency components at the output of a device that are not present in the input signal, and these new frequency components are integer multiples of the input frequency. Harmonic distortion is caused by the inability of the nonlinear system to comply with the principle of homogeneity and superposition when a signal is transmitted by a PA.

In general, the input–output relationship in a nonlinear model can be expressed in the form of power series

$$y(t) = \sum_{n=1}^{\infty} a_n x^n(t) \tag{2.1}$$

where $x(t)$ is the input signal, $y(t)$ is the output signal and a_n is the coefficient of each term.

Assume that the PA input signal is $x(t) = A\cos\omega t$. Generally, a weak nonlinear system can be precisely characterized by taking the first three terms of Equation (2.1). After passing through the nonlinear system, the output is

$$\begin{aligned} y(t) = {} & \frac{1}{2}a_2A^2 + \left(a_1 + \frac{3}{4}a_3A^2\right)A\cos\omega t + \frac{1}{2}a_2A^2\cos 2\omega t \\ & + \frac{1}{4}a_3A^3\cos 3\omega t \end{aligned} \tag{2.2}$$

As indicated in Equation (2.2), when the input is a single tone signal, the output signal components of the PA contain a DC component and harmonic components 2ω and 3ω, in addition to the fundamental frequency ω. It also indicates that the odd term of power series determines odd harmonic components, and the even term of power series determines the DC component and even harmonic components. For a narrow-band system, harmonic frequencies are far from the signal frequency, so they can be easily filtered out by using a low-pass filter. However, for a wideband system, harmonic components are likely to fall within the signal passband range, which are difficult to filter using a filter. In this case, the influence of harmonic distortion cannot be ignored.

In Equation (2.2), the fundamental coefficient a_3 is a negative value, and $3a_3A^3/4$ decreases rapidly with the increase of the amplitude of the input signal A, resulting in gain compression. When the real PA gain becomes 1 dB lower compared with the ideal linear gain, the corresponding output powerpoint is defined as the output power 1 dB compression point (P_{1dB}), as shown in Figure 2.1.

2.1.2 Intermodulation Distortion

Intermodulation distortion [2] refers to a distortion caused by intermodulation components generated by two or more input signals of different frequencies passing through a nonlinear PA. The output signal contains additional combined frequency components, in addition to harmonic components. These additional components are referred to as intermodulation components. The intermodulation components falling within the working range of the signal may cause serious interference to the system.

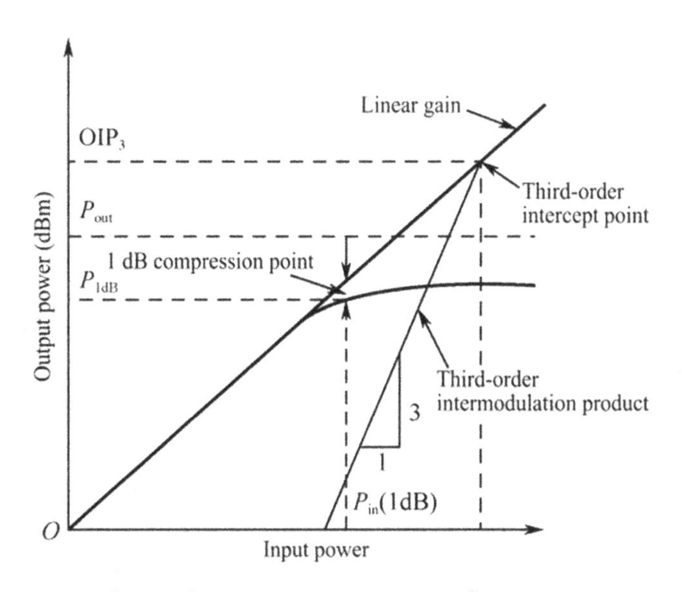

FIGURE 2.1 Nonlinear characteristic parameters of PA.

Assume that the input signal is an equal-amplitude dual-tone signal. The amplitude is A and the frequencies are ω_1 and ω_2, respectively. It can be expressed as

$$x(t) = A(\cos \omega_1 t + \cos \omega_2 t). \tag{2.3}$$

Substituting Equation (2.3) into Equation (2.1) and taking the first three terms yields the following output signal:

$$\begin{aligned} y(t) = {} & a_1 A(\cos \omega_1 t + \cos \omega_2 t) + a_2 A^2(\cos^2 \omega_1 t + \cos^2 \omega_2 t \\ & + 2\cos \omega_1 t \cos \omega_2 t) + a_3 A^3(\cos^3 \omega_1 t + \cos^3 \omega_2 t \\ & + 3\cos \omega_1 t \cos^2 \omega_2 t + 3\cos^2 \omega_1 t \cos \omega_2 t) \end{aligned} \tag{2.4}$$

Further extending Equation (2.4), we have

$$\begin{aligned} y(t) = {} & a_2 A^2 + a_2 A^2 \cos(\omega_1 - \omega_2)t + \left(a_1 A + \frac{9}{4}a_3 A^3\right)\cos \omega_1 t \\ & + \left(a_1 A + \frac{9}{4}a_3 A^3\right)\cos \omega_2 t + \frac{3}{4}a_3 A^3 \cos(2\omega_1 - \omega_2) \\ & + \frac{3}{4}a_3 A^3 \cos(2\omega_2 - \omega_1) + a_2 A^2 \cos(\omega_1 + \omega_2)t + \frac{1}{2}a_2 A^2 \cos 2\omega t \\ & + \frac{1}{2}a_2 A^2 \cos 2\omega_2 t + \frac{3}{4}a_3 A^3 \cos(2\omega_1 + \omega_2)t \\ & + \frac{3}{4}a_3 A^3 \cos(2\omega_2 + \omega_1)t + \frac{1}{4}a_3 A^3 \cos 3\omega_1 t + \frac{1}{4}a_3 A^3 \cos 3\omega_2 t. \end{aligned} \tag{2.5}$$

As indicated by (2.5), when the input signal is a dual-tone signal, the output signal $y(t)$ contains second-order intermodulation components at frequency $\omega_1 \pm \omega_2$ and third-order intermodulation components at frequencies $2\omega_1 \pm \omega_2$ and $2\omega_2 \pm \omega_1$, in addition to the DC component, fundamental component and harmonic components. The signal spectrum is shown in Figure 2.2.

As shown in Figure 2.2, the components at frequencies $\omega_1 \pm \omega_2$, $2\omega_1$, $2\omega_2$, $2\omega_1 + \omega_2$, $2\omega_2 + \omega_1$, $3\omega_1$ and $3\omega_2$ are far from the fundamental frequency and generally fall outside the passband, which may interfere with

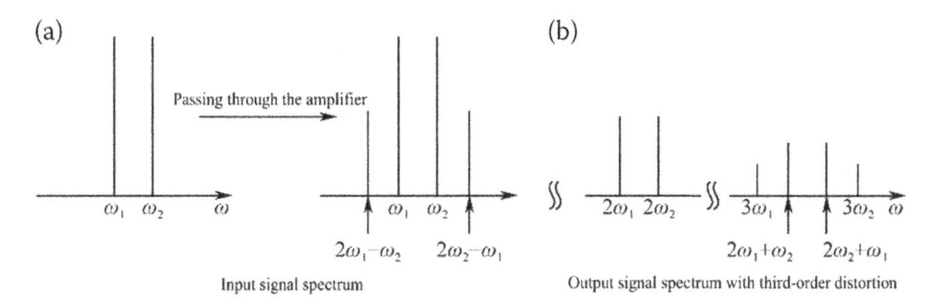

FIGURE 2.2 Input and output spectrum.

signals in other frequency bands. These frequency components can be removed by using appropriate filters. However, the third-order intermodulation components at frequencies $2\omega_1 - \omega_2$ and $2\omega_2 - \omega_1$ are very close to the fundamental frequency and always fall within the passband, which are difficult to filter out. Therefore, such intermodulation components should be specially considered in the system design. The fifth-order intermodulation distortion at frequencies $3\omega_1 - 2\omega_2$ and $3\omega_2 - 2\omega_1$ is also close to the fundamental frequency but has a very small power value, so the third-order intermodulation distortion is generally considered in the system design.

With the increasing input power of PA, the third-order intermodulation component of PA and the output extension line of the fundamental frequency component in the ideal state will intersect at one point. This point is known as the third-order intercept point and is denoted as IP_3. See Figure 2.1. The input power corresponding to IP_3 is IIP_3, and the corresponding output power is called the cutoff power (OIP_3). IP_3 can reflect the nonlinear characteristics of PA. When the output power is constant, a greater IP_3 value indicates better linearity of PA, and a farther distance between IP_3 and the 1dB compression point indicates a smaller influence of intermodulation components on the whole system.

2.1.3 AM/AM and AM/PM Distortion

The nonlinearity of PA is basically indicated by amplitude and phase. When the input signal passes through the nonlinear PA, the amplitude and phase of the output signal will vary with the amplitude of the input signal. AM/AM distortion is characterized as the changes in the amplitude of output signal caused by the changes in the amplitude of input signal [3], as

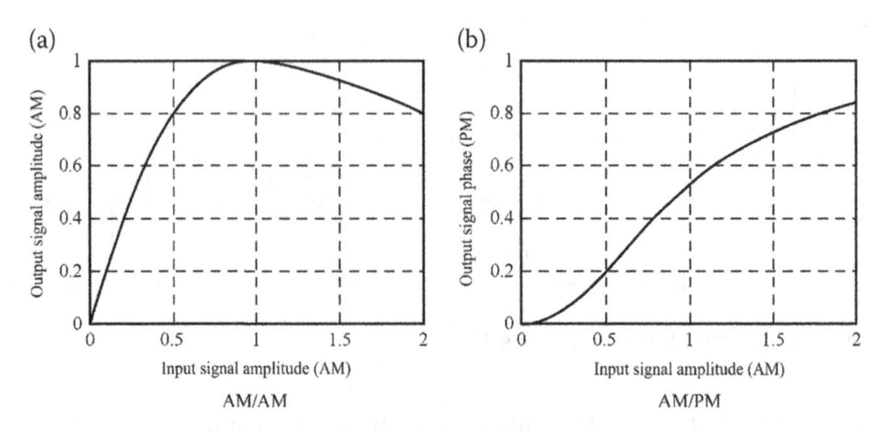

FIGURE 2.3 AM/AM and AM/PM characteristic curves of PA.

shown in Figure 2.3 (a). AM/PM distortion is characterized as the changes in the phase of output signal caused by the changes in the amplitude of input signal [4], as shown in Figure 2.3 (b). Assume the PA input is

$$v_i = r(t)\cos(\omega_0 t + \theta(t)) \tag{2.6}$$

According to AM/AM and AM/PM characteristics, the output of PA can be described as

$$v_o = G(r(t))\cos(\omega_0 t + \theta(t) + \Phi(r(t))) \tag{2.7}$$

where $r(t)$ and $\theta(t)$ are the amplitude and phase of the input signal, respectively; $G(\cdot)$ and $\Phi(\cdot)$ are the amplitude and phase characteristic functions of the PA output, respectively.

Due to amplitude or phase distortion, the PA experiences gain compression, power saturation and phase shift of the output signal. In a narrowband system, the weak nonlinear distortion of PA is primarily AM/AM distortion. However, in a wideband system such as an OFDM system, both AM/AM and AM/PM distortions need to be considered for a full representation of the PA nonlinearity, since the PA has strong memory effects.

2.2 MEMORY EFFECTS OF POWER AMPLIFIER

Any PA has a certain degree of memory effects in applications. For a narrowband communication system, memory effects of PA can be ignored in system analysis, since the signal bandwidth is relatively small and memory effects of PA are pretty weak. In contrast, memory effects of PA become significant with the increase of input signal bandwidth in a wideband communication system, and therefore the impact of memory effects on the system cannot be ignored.

Memory effects [5] of PA are evidenced in the following ways: (1) The PA output depends both on the current input and on the previous inputs; (2) AM/AM and AM/PM characteristics of PA depend both on the amplitude of input signal and on the frequency of signal envelope; (3) ACPR asymmetry is observed in the continuous spectrum signal test; (4) Third-order intermodulation point asymmetry is observed in the dual-tone test.

Figure 2.4 shows the AM/AM and AM/PM characteristic curves of PA with memory effects. As shown in the figure, with the presence of memory effects, the AM/AM and AM/PM characteristics of PA are no longer clear curves but appear as a "divergent" state, which is called hysteresis phenomenon. Memory effects of PA are a kind of dynamic characteristic, which cannot be described by an existing memoryless model.

2.2.1 Causes of Memory Effects

The so-called "memoryless circuit" refers to a circuit that does not contain any energy-storing component. The output of the memoryless circuit at any time depends only on the current input. In applications, a circuit always contains energy-storing component(s), so it is inevitable for a circuit to come with memory effects. There are many factors that

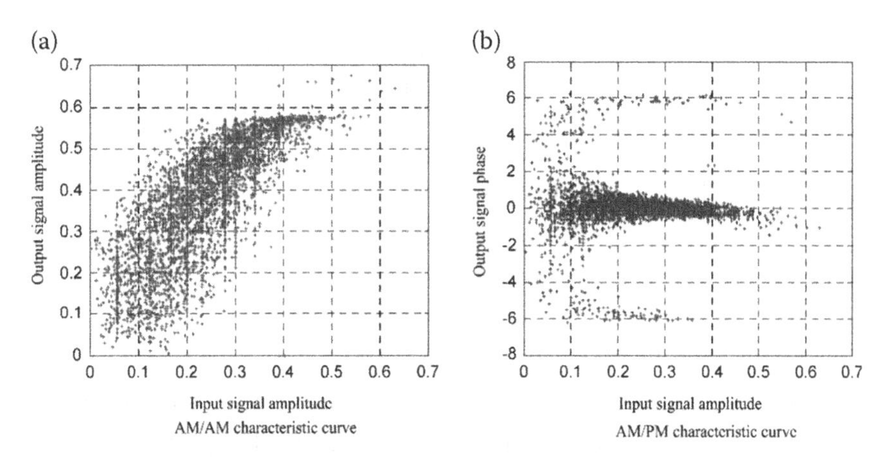

FIGURE 2.4 AM/AM and AM/PM characteristic curves of PA with memory effects.

affect the memory effects of PA, including the self-heating of active devices, power supply characteristics, bias network and mismatching of even harmonics. Figure 2.5 shows the circuit diagram where the positions prone to memory effects of PA are indicated.

There are usually two types of memory effects in PA, which are electrical memory effects and thermal memory effects.

The electrical memory effects on the node impedance network are caused by frequency changes of the input signal, such as the matching circuit impedance and the bias network impedance. The envelope frequency changes play a major role in the electrical memory effects. When the envelope frequency reaches tens of Hertz, the PA will produce strong electrical memory effects. If the node impedance contains an energy-storing component with a large time constant, electrical memory effects are inevitably produced. Therefore, design optimization in the matching and bias circuits is required to avoid the electrical memory effects as much as possible.

The thermal memory effects are basically affected by the power dissipation of active devices such as transistors. Changes in the thermal impedance determine the influence of power dissipation on the temperature changes. The temperature changes due to power dissipation generated by active devices cannot disappear or be reflected instantly, resulting in an increase in the junction temperature of transistors. Consequently, the values of transistor parameters vary.

The aforementioned analysis indicates that memory effects will affect the linearization of PA, so memory effects should be considered in the establishment of the PA behavioral model. One of the major purposes of predistortion linearization technology is to enable the predistorter to fully compensate memory effects of PA, thereby minimizing memory effects.

2.2.2 Methods to Eliminate Memory Effects

The methods to eliminate memory effects are provided from two aspects, based on the previous analysis on the causes of memory effects.

1. Elimination of electrical memory effects

Spectrum resources are limited in wireless communications, so a variety of modulation schemes have been proposed to improve the spectrum utilization, resulting in changes in the instantaneous signal bandwidth. Regarding

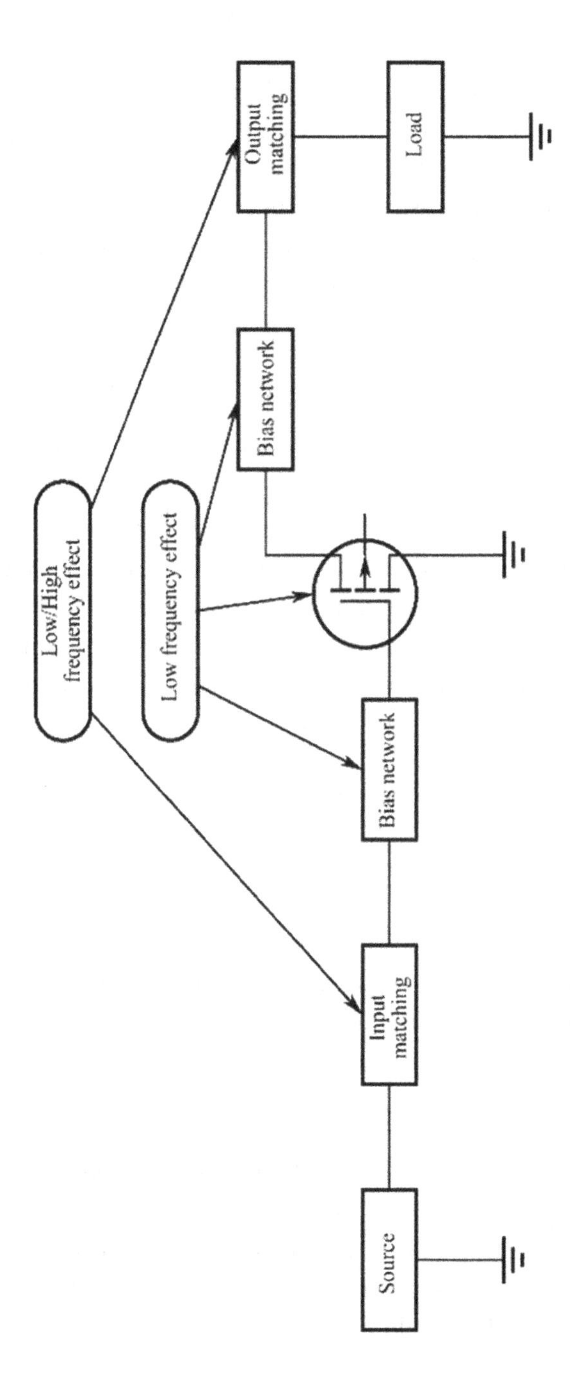

FIGURE 2.5 Positions prone to memory effects.

the envelope frequency, if the input and output impedances are nonconstant, different envelope signals will produce different impedance characteristics, causing the electrical memory effects. At different envelope frequencies, active loads can be used to make the node impedance constant. Specifically, add a signal source working at the same frequency as the envelope, and adjust its amplitude and phase for node impedance matching; alternatively, connect a capacitor in parallel at the drain and gate of the transistor, and change the capacitor value so that the signal is short-circuited within the bandwidth range. In this way, the electrical memory effects are eliminated.

2. Elimination of thermal memory effects

Power dissipation of PA is an important factor in the generation of thermal memory effects. Therefore, an appropriate heat dissipation part can be designed in the circuit to accelerate heat diffusion. However, in the high power state, the PA generates massive heat. Simply using the heat dissipation method cannot completely suppress the thermal memory effects. In this situation, add the temperature compensation module and the heat dissipation module to the predistortion system to eliminate the thermal memory effects.

2.3 IMPACT OF POWER AMPLIFIER NONLINEARITY ON COMMUNICATION SYSTEMS

In complex modulation schemes such as OFDM, IP_3, P_{1dB} and other PA, nonlinear indicators can no longer comprehensively measure the nonlinearity and linearization performance of PA. Therefore, nonlinear indicators specific to wideband signals are required, such as adjacent channel power ratio (ACPR) and error vector magnitude (EVM), in addition to the existing indicators.

2.3.1 ACPR

The adjacent channel power [6] (ACP) is the mean power of the two channels immediately adjacent to the main channel. The ACPR is the ratio of the main channel's mean power to the adjacent channel's mean power, as shown in Figure 2.6. It is a measurement of how well the signal extends to adjacent channels. The nonlinearity of PA results in the adjacent channel power, causing the main channel signal to generate adjacent-channel

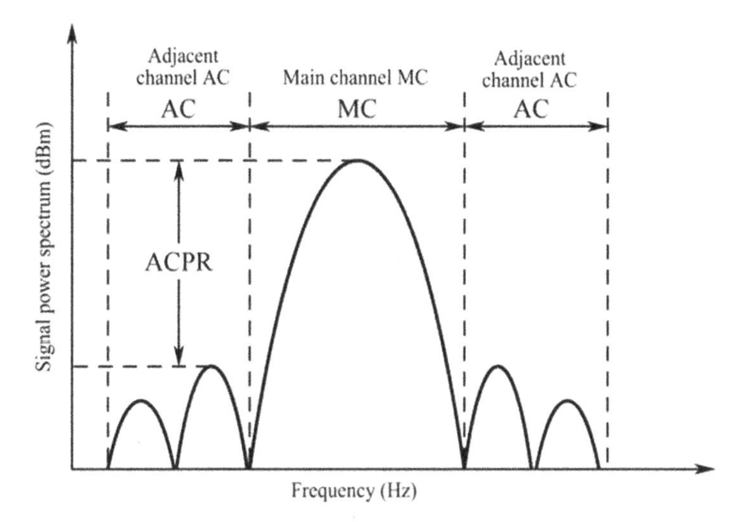

FIGURE 2.6 Adjacent channel power ratio.

spectrum spreading. In addition, the adjacent channel power will also be generated if the spectrum of the signal itself is wider than the channel spectrum.

2.3.2 EVM

EVM [7] is defined as the square root ratio of the error vector signal's mean power to the reference signal's mean power; that is, the root mean square (RMS) ratio. This ratio is expressed as a percentage. EVM is a measurement of the degree to which the signal constellation point deviates from the ideal position. The deviation of the signal constellation point may cause misjudgments. See Figure 2.7. Assuming the measurement signal is $\boldsymbol{S}$ and the reference signal is $\boldsymbol{R}$, the EVM is expressed as

$$\mathrm{EVM} = \frac{\mathrm{RMS}(|\boldsymbol{E}|)}{\mathrm{RMS}(|\boldsymbol{R}|)} = \frac{\mathrm{RMS}(|\boldsymbol{S} - \boldsymbol{R}|)}{\mathrm{RMS}(|\boldsymbol{R}|)} \times 100\,\%. \tag{2.8}$$

In Figure 2.7, the left-hand S_I and S_Q indicate the real axis and imaginary axis of the constellation diagram, respectively; the right-hand I is the in-phase component and Q is the orthogonal component.

ACPR and EVM are important indicators in the measurement of the linearization techniques of PA, which respectively indicate the out-of-

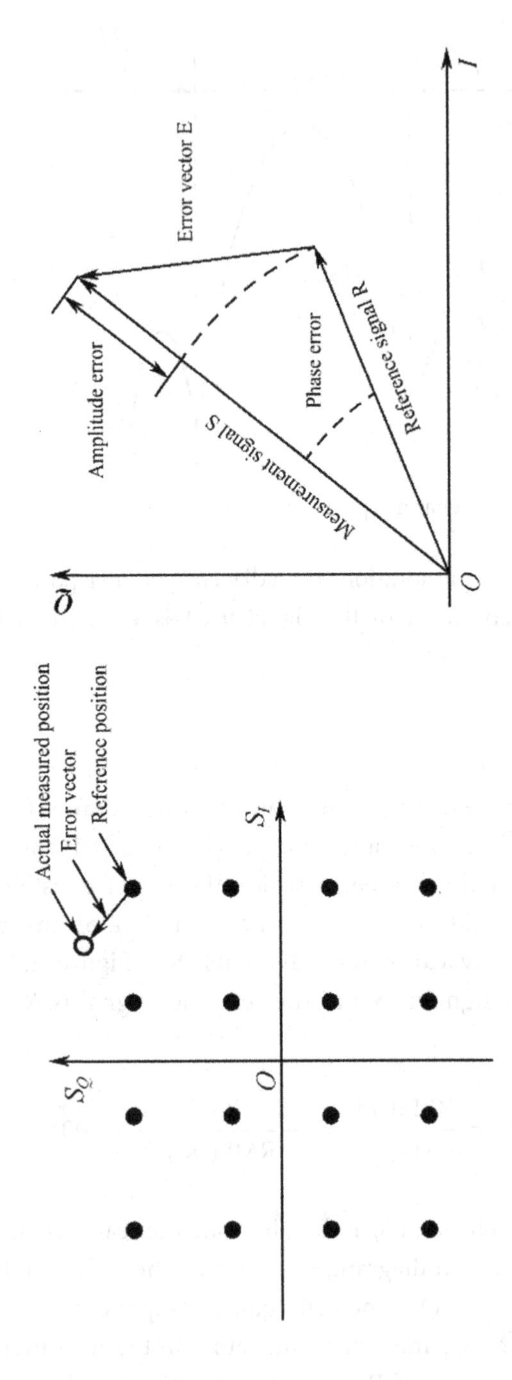

FIGURE 2.7 Schematic diagram of error vector magnitude.

band and in-band performance in PA transmission. Actually, ACPR is inversely proportional to EVM. That is, if the linearization techniques compensate the nonlinear distortion of PA and optimize the linearity of the entire communication system, then both in-band and out-of-band transmission performance will be improved.

REFERENCES

1. 王亮. 污水处理厂内变频器引起的谐波失真分析和处理措施[J]. 数据通信, 2018(04):50–54.
   WANG L. Analysis of and Countermeasures for Harmonic Distortion Caused by Frequency Converter in Sewage Treatment Plant[J]. Data Communications, 2018, (04):50–54.
2. 陈丹, 柯熙政, 张璐. 湍流信道下激光器互调失真特性[J]. 光子学报, 2016, 45(02):99–103.
   CHEN D, KE X, ZHANG L. Laser Intermodulation Distortion and Characteristic under the Turbulence Channel[J]. Acta Photonica Sinica, 2016, 45(02):99–103.
3. 曹学龙. 基于模拟预失真的高效率功率放大器的设计与实现[D]. 北京: 北京邮电大学, 2015.
   CAO X. The Design and Implementation of a Highly Efficient Power Amplifier Based on Analog Predistortion[D]. Beijing: Beijing University of Posts and Telecommunications, 2015.
4. 申倩伟. 基于预失真和前馈的功率放大器线性度优化[D]. 天津:天津大学, 2016.
   SHEN Q. The Optimization of Power Amplifier's Linearity Based on Pre-distortion and Fee-Forward[D]. Tianjin: Tianjin University, 2016.
5. 南敬昌, 樊爽, 高明明. 基于负载牵引和记忆效应的 X 参数的功放建模[J]. 计算机应用, 2018, 38(10):2982–2989.
   NAN J, PAN S, GAO M. Power Amplifier Modeling of X-Parameter Based on Load-Pulling and Memory Effect[J]. Journal of Computer Applications, 2018, 38(10):2982–2989.
6. 张作龙, 熊祥正. 一种平衡 AB 类功率放大器的设计[J]. 科学技术与工程, 2012, 12(32):8703–8706.
   ZHANG Z, XIONG X. A Design of Class AB Balanced Power Amplifier[J]. Science Technology and Engineering, 2012, 12(32):8703–8706.
7. 杨凡, 曾孝平, 毛海伟, 等. κ-μ 阴影衰落信道下非数据辅助的误差矢量幅度性能分析[J]. 通信学报, 2018, 39(05):177–188.
   YANG F, ZENG X, MAO H, et al. Nontata-Aided Error Vector Magnitude Performance Analysis over κ-μ Shadowed Fading Channel[J]. Journal on Communications, 2018, 39(05):177–188.

CHAPTER 3

Power Amplifier Behavioral Model and Nonlinear Analysis Basis

3.1 MEMORYLESS BEHAVIORAL MODEL

Memory effects are not significant in narrow-band systems. You can ignore the memory effects of PA when modeling a narrow-band PA. The existing memoryless behavioral models [1] can be categorized into Saleh model, Rapp model and power series model.

The Saleh model uses only four parameters and is a common static memoryless behavioral model. The model is originally applied to behavioral modeling of traveling-wave tube PAs. The mathematical expression of Saleh model is

$$A\left[r(n)\right] = \frac{a_A r(n)}{1 + b_A \left[r(n)\right]^2} \tag{3.1}$$

$$\varphi\left[r(n)\right] = \frac{a_\varphi r(n)^2}{1 + b_\varphi \left[r(n)\right]^2} \tag{3.2}$$

$$y(n) = A[r(n)]e^{[j\phi(n)+\varphi[r(n)]]} \tag{3.3}$$

where $r(n)$ and $\phi(n)$ are the amplitude and phase of the input signal, respectively; a_A and b_A are the fitted parameters based on the AM/AM characteristic curve of the actually measured PA; a_φ and b_φ are the fitted parameters based on the AM/PM characteristic curve of the actually measured PA; and $y(n)$ is the output of Saleh model.

The Rapp model can accurately describe the characteristics of solid-state PAs and is also a widely used memoryless behavioral model. The mathematical expression of Rapp model is

$$A[r(n)] = \frac{r(n)}{[1 + (r(n)/V_{sat})^{2p}]^{(1/2p)}} \tag{3.4}$$

$$\varphi[r(n)] \approx 0 \tag{3.5}$$

$$y(n) = A[r(n)]e^{[j\phi[n]+\varphi[r(n)]]} \tag{3.6}$$

where $r(n)$ is the amplitude of the input signal, $\phi(n)$ is the phase of the input signal, V_{sat} is the saturated output voltage of the PA, p is the smoothness factor (a greater value of p indicating a higher PA linearity), and $y(n)$ is the output of Rapp model.

The power series model can be expressed as

$$y(n) = \sum_{k=1}^{K} a_k x(n)|x(n)|^{k-1} \tag{3.7}$$

where $x(n)$ is the input signal, $y(n)$ is the output signal, K is the nonlinear order and a_k is the coefficient of the model. The superiorities of the power series model lie in its simple structure, relatively easy method for coefficient identification and intuitive description of distortion for each order. The selected nonlinear order for the power series model is low, so the fitting accuracy is limited.

3.2 MEMORY BEHAVIORAL MODEL

The memoryless behavioral model is investigated under the assumption that the PA has no memory effects. However, the real PA comes with electrical memory effects and thermal memory effects, which are especially obvious in wideband systems. The presence of memory effects produces a great impact on the accuracy of the behavioral model. Therefore, the traditional memoryless behavioral model can no longer be used for wideband PA modeling. A memory behavioral model with similar characteristics of the wideband PA must be established [2].

3.2.1 Volterra Series Model and Memory Polynomial Model

Volterra series [3] is a solution to integral equations. Wiener applied it to the analysis of nonlinear circuits in 1942. Bussgang et al. discussed its application in nonlinear circuits in detail in 1974. For most nonlinear systems with mixed nonlinear characteristics, the Volterra series method is used to analyze the responses of the communication receiver, noises and RF interference, which is especially effective. As the general method to analyze practical nonlinear systems or circuits, the Volterra series addresses the drawbacks of the power series and considers the impact of memory effects.

In the Volterra series model, the relationship between the excitation $x(t)$ and the response $y(t)$ of a nonlinear system can be expressed by the differential equation as follows:

$$y(t) = \int_{-\infty}^{+\infty} h_1(\tau_1)x(t-\tau_1)\mathrm{d}\tau_1 + \int\int_{-\infty}^{+\infty} h_2(\tau_1, \tau_2)x(t-\tau_1)x(t-\tau_2)\mathrm{d}\tau_1\mathrm{d}\tau_2 + \int\int\int_{-\infty}^{+\infty} h_3(\tau_1, \tau_2, \tau_3)x(t-\tau_1)x(t-\tau_2)x(t-\tau_3)\mathrm{d}\tau_1\mathrm{d}\tau_2\mathrm{d}\tau_3 + \cdots \tag{3.8}$$

where $h_n(\tau_1, \tau_2, \cdots \tau_n)$ is the nth-order Volterra kernel, also known as the transfer function. The 0th-order kernel is the response of the system to the DC input, which can be ignored if the system has no DC input. The first-order kernel is the linear response of the system. Fourier transform of Equation (3.8) yields

$$Y(\mathrm{j}\omega) = H_1(\mathrm{j}\omega_1)\cdot X(\mathrm{j}\omega_1) + H_2(\mathrm{j}\omega_1, \mathrm{j}\omega_2)\cdot X(\mathrm{j}\omega_1)\cdot X(\mathrm{j}\omega_2) + \cdots \tag{3.9}$$

The above equation is a linear algebraic equation, where

$$H_n(\omega_1, \omega_2, \cdots, \omega_n) = \int\int \cdots \int\int_{-\infty}^{\infty} h_n(\tau_1, \tau_2, \cdots \tau_n) e^{-j(\omega_1\tau_1 + \omega_2\tau_2 + \cdots + \omega_n\tau_n)} d\tau_1 d\tau_2 \cdots d\tau_n \tag{3.10}$$

$H_n(\omega_1, \omega_2, \omega_3, \cdots, \omega_n)$ is the nth-order nonlinear transfer function in the frequency domain, also known as the nth-order Volterra kernel spectrum. The inverse transformation h_n of H_n is as follows:

$$h_n(\tau_1, \cdots, \tau_n) = \int_{-\infty}^{\infty} \cdots \int_{-\infty}^{\infty} H_n(f_1, \cdots f_n) e^{j2\pi(f_1\tau_1 + \cdots f_n\tau_n)} df_1 \cdots df_n \tag{3.11}$$

Here, $\omega = 2\pi f$ applies. The input–output (IO) relationship can also be written as:

$$y(t) = \sum_{n=1}^{\infty} y_n(t) = \sum_{n=1}^{\infty} \int_{-\infty}^{\infty} \cdots \int_{-\infty}^{\infty} H_n(f_1, \cdots, f_n) \prod_{i=1}^{n} X(f_i) e^{j2\pi f_i t} df_i \tag{3.12}$$

In the equation, $X(f_i)$ is the spectrum function of $x(t)$. Fourier transform on the above equation yields the nth-order output spectrum function:

$$Y(f) = \sum_{n=1}^{\infty} Y_n(f) = \sum_{n=1}^{\infty} \int_{-\infty}^{\infty} \cdots \int_{-\infty}^{\infty} H_n(f_1, \cdots, f_n) \delta (f - f_1 - \cdots - f_n) \prod_{i=1}^{n} X(f_i) df_i \tag{3.13}$$

where the Selta spectrum function is

$$\delta(f - f_1 - \cdots - f_n) = \int_{-\infty}^{\infty} e^{-j2\pi(f - f_1 - \cdots - f_n)} dt \tag{3.14}$$

Once the transfer function for each order is obtained, all nonlinear indicators can be obtained only by using the method for linear circuit analysis, thereby simplifying the research work on nonlinear distortion.

The Volterra series method separately describes nonlinear components of each order in the system and allows comparisons among different components. For systems with weak nonlinearity, the terms of series can be reduced. If the band-pass model is described, the DC term and even term can also be eliminated. However, the convergence rate of this method is slow, the formula is complicated in the strong nonlinearity context, and it is difficult to compute the high-order Volterra kernel from the measured data. These drawbacks limit the application of the Volterra series method.

The memory polynomial (MP) model [4] is the most commonly used simplified Volterra series model. It only retains the diagonal terms of the Volterra series and can be regarded as a trade-off between complexity and accuracy of Volterra series to some extent. The number of coefficients of the model is substantially reduced, enabling easy implementation and wide application. The MP model can be expressed as

$$y(n) = \sum_{k=1}^{K} \sum_{q=0}^{Q} a_{kq} x(n-q) |x(n-q)|^{k-1} \tag{3.15}$$

where $x(n)$ is the input signal, $y(n)$ is the output signal, K is the nonlinear order, Q is the memory depth and a_{kq} is the coefficient of the MP submodel.

For weakly nonlinear systems, Oualid Hammi proposed an Envelope Memory Polynomial (EMP) model. The mathematical expression of the model is as follows:

$$y(n) = \sum_{k=1}^{K} \sum_{q=0}^{Q} a_{kq} x(n) |x(n-q)|^{k-1} \tag{3.16}$$

where $x(n)$ is the input signal, $y(n)$ is the output signal, K is the nonlinear order, Q is the memory depth and a_{kq} is the coefficient of the MP submodel. Compared with the MP model, the EMP model uses fewer coefficients, suitable for weakly nonlinear systems.

Based on the MP model and the EMP model, Oualid Hammi proposed a hybrid memory polynomial (HME, short for Hybrid MP-EMP) model. The block diagram of the model is shown in Figure 3.1.

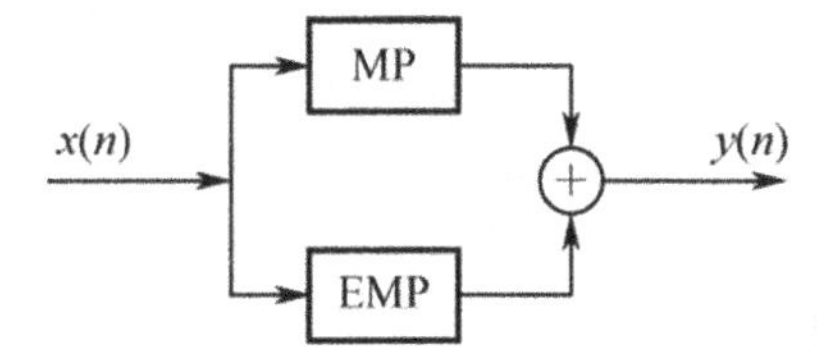

FIGURE 3.1 Block diagram of HME model.

The mathematical expression of HME model is as follows:

$$y(n) = \sum_{k=1}^{K_{\text{MP}}} \sum_{q=0}^{Q_{\text{MP}}} a_{kq} x(n-q)|x(n-q)|^{k-1} + \sum_{k=1}^{K_{\text{EMP}}} \sum_{q=0}^{Q_{\text{EMP}}} b_{kq} x(n)|x(n-q)|^{k-1} \tag{3.17}$$

where $x(n)$ is the input signal of the model; $y(n)$ is the output signal of the model; K_{MP} is the nonlinear order of the MP submodel; K_{EMP} is the nonlinear order of the EMP submodel; Q_{MP} is the memory depth of the MP submodel; Q_{EMP} is the memory depth of the EMP submodel; and a_{kq} and b_{kq} are the coefficients of the MP submodel and EMP submodel, respectively. HME model combines the advantages of the memory polynomial and the envelope memory polynomial. The model achieves higher accuracy without requiring a large number of coefficients, thus reducing the complexity.

The GMP model introduces the cross terms between the signal and its leading and lagging exponentiated envelope terms, based on the MP model. Therefore, the GMP model uses more coefficients but describes nonlinear systems more accurately. The mathematical expression of GMP model is as follows:

$$y(n) = \sum_{k=1}^{K_1} \sum_{q=0}^{Q_1} a_{kq} x(n-q)|x(n-q)|^{k-1} + \sum_{k=3}^{K_2} \sum_{q=0}^{Q_2} \sum_{l=1}^{L_1} b_{kql} x(n-q)|x(n-q-l)|^{k-1} + \sum_{k=3}^{K_3} \sum_{q=0}^{Q_3} \sum_{l=1}^{L_2} c_{kql} x(n-q)|x(n-q+l)|^{k-1} \tag{3.18}$$

where $x(n)$ and $y(n)$ are the input signal and output signal of the model, respectively. K_1, Q_1 and a_{kq} are the nonlinear order, memory depth and coefficient of the MP submodel, respectively. K_2, Q_2, L_1 and b_{kql} are the nonlinear order and memory depth of the cross term between the signal and its lagging envelope term, exponent of the lagging cross term and the model coefficient, respectively. K_3, Q_3, L_2 and c_{kql} are the nonlinear order and memory depth of the cross term between the signal and its leading envelope term, exponent of the leading cross term and the model coefficient, respectively. The introduction of signal cross-terms improves the accuracy of GMP model and produces better descriptions of the nonlinearity and memory effects of PA, but requires more model coefficients.

3.2.2 Hammerstein Model and Wiener Model

Hammerstein model [5] is designed to separate the nonlinear characteristics from the memory effects of PA. That is, the model considers the PA as a combination of a memoryless nonlinear subsystem cascaded with a memory linear subsystem. The block diagram of Hammerstein model is outlined in Figure 3.2. where $x(n)$ is the input signal of the model, $y(n)$ is the output signal of the model, $z(n)$ is the input of the linear subsystem, $F(\cdot)$ is the nonlinear function, and $H(\cdot)$ is the transfer function of the linear subsystem. The mathematical expression of Hammerstein model is as follows:

$$z(n) = \sum_{k=1}^{K} a_k x(n)|x(n)|^{k-1} \tag{3.19}$$

$$y(n) = \sum_{q=0}^{Q} h(q) z(n-q) \tag{3.20}$$

Substituting Equation (3.19) into Equation (3.20) yields

$$y(n) = \sum_{k=1}^{K} \sum_{q=0}^{Q} a_{kq} h(q) x(n-q)|x(n-q)|^{k-1} \tag{3.21}$$

where K is the nonlinear order of the model, Q is the memory depth of the model and a_{kq} is the coefficient of the MP submodel. Although the Hammerstein model combines the nonlinearity and memory effects of PA by means of cascading in a simple manner, the Hammerstein model cannot achieve high accuracy in most cases. Therefore, a parallel Hammerstein model is proposed, and its structure is outlined in Figure 3.3. The parallel

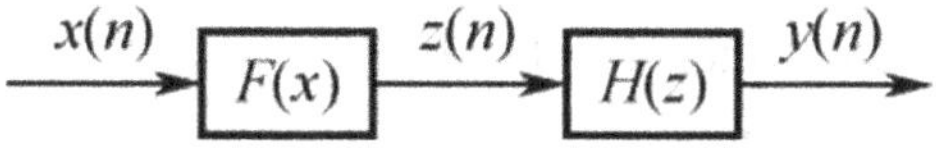

$$\xrightarrow{x(n)} \boxed{F(x)} \xrightarrow{z(n)} \boxed{H(z)} \xrightarrow{y(n)}$$

FIGURE 3.2 Block diagram of Hammerstein model.

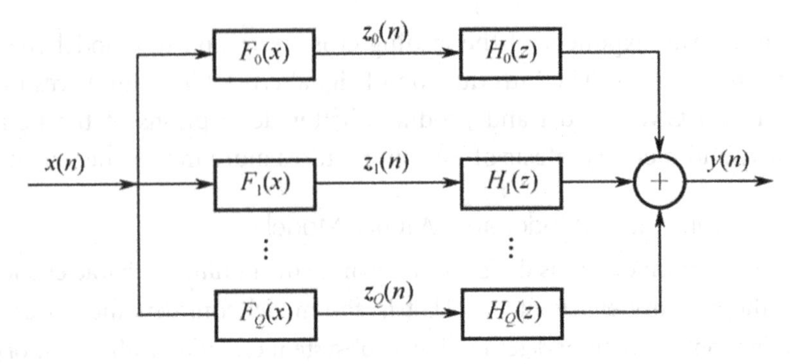

FIGURE 3.3 Block diagram of parallel Hammerstein model.

Hammerstein model is an extension based on the traditional Hammerstein model, enabling nonlinear components of different orders to pass through different linear filters.

The traditional Hammerstein model uses a linear FIR filter to compensate for memory effects of PA. Specifically, the linear FIR filter acts on the electrical memory effects caused by the nonconstant frequency response generated near the transmitter carrier, and does not act on the electrical memory effects caused by the impedance changes of the bias circuit and harmonics in the transmitter. To address this drawback, an improved Hammerstein model was proposed to improve the accuracy of PA modeling, based on the traditional Hammerstein model. Its mathematical expression is as follows:

$$x(n) = G(|u(n)|)u(n) \tag{3.22}$$

$$y(n) = \sum_{i=0}^{M_1-1} a_i x(n-i) + \sum_{i=0}^{M_2-1} b_i |x(n-i)| x(n-i) \tag{3.23}$$

where M_1 and M_2 are the number of FIR filter taps, respectively. In the improved Hammerstein model, a strong nonlinear static subsystem is cascaded with a weak nonlinear dynamic subsystem, and a parallel branch is added to the weak nonlinear dynamic subsystem. In addition, the model uses an FIR-based dynamic filter to eliminate spectral regrowth, which is caused by the transmitter driven by modulated signals. The structure of the improved Hammerstein model is outlined in Figure 3.4.

Wiener model [6], as opposed to Hammerstein model, equates a PA to a cascade of a memory linear subsystem with a memoryless nonlinear

FIGURE 3.4 Block diagram of improved Hammerstein model.

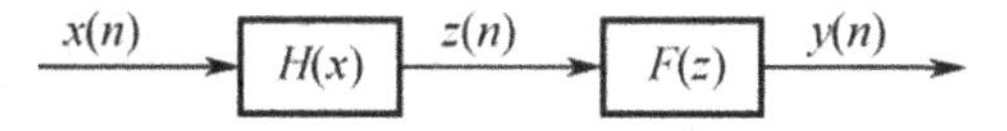

FIGURE 3.5 Block diagram of Wiener model.

subsystem. Its structure is outlined in Figure 3.5, where $x(n)$ is the input signal of the model, $y(n)$ is the output signal of the model and $z(n)$ is the input of the memoryless nonlinear system. The mathematical expression of the Wiener model is as follows:

$$
\begin{aligned}
y(n) &= F\left[\sum_{q=0}^{Q} b_q x(n-q)\right] \\
&= \sum_{k=1}^{K} a_k \left[\left|\sum_{q=0}^{Q} b_q x(n-q)\right|^{k-1} \sum_{q=0}^{Q} b_q x(n-q)\right]
\end{aligned}
\tag{3.24}
$$

where K is the nonlinear order of the model, Q is the memory depth of the model, a_k is the coefficient of the MP submodel, and b_q is the coefficient of different EMP models. Wiener model can combine the nonlinearity and memory effects of PA in a simple manner, but its accuracy is not high.

3.2.3 Neural Network Model

The artificial neural network (ANN) [7], also known as neural network (NN), is a mathematical model that simulates the biological neural network for information processing. It is based on the physiological research findings of the brain, aiming to simulate some mechanisms of the brain and achieve specific functions. At present, the ANN has been applied to device modeling, pattern control, data processing and other fields.

A neural network has the following three basic characteristics:

1. Parallel distributed processing: The neural network has a high-level parallel structure and parallel implementation capability, with the ability to find the optimal solution at high speed. It can give full play to the high-speed computing capability of the computer to find the optimal solution quickly.
2. Nonlinear processing: A large number of neurons have complex connections, and the connection weights between neurons can be changed. The neural network fits its input–output relationship by an approximate function with high nonlinearity. That is, the neural network as a whole can be considered as a complex nonlinear dynamical system.
3. Self-learning function: By learning the historical data, a specific neural network with the induction of all data is trained. The self-learning function is particularly important for prediction.

Inspired by biological neurons, people developed the artificial neuron model through simplification. The artificial neuron is the basic information processing unit of an ANN. A large number of artificial neurons constitute the basic topologies of neural networks through complex connections. Figure 3.6 shows a simplified diagram of a generic neuron, which is the basis for the design of an ANN.

$\boldsymbol{p}$ denotes the input vector of the neural network; that is, $p_1, p_2, p_3, \ldots, p_R$. It is a column vector, indicating R input values of neurons. The input of neurons can be expressed as

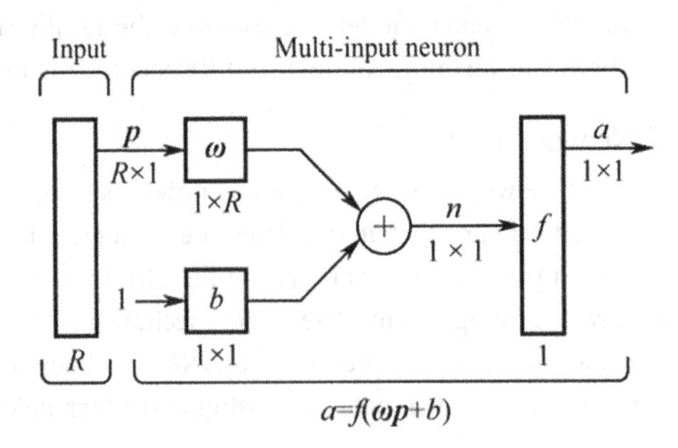

FIGURE 3.6 Simplified diagram of generic neuron.

$$\boldsymbol{p} = [p_1, p_2, \cdots, p_R]^{\mathrm{T}} \tag{3.25}$$

$\boldsymbol{\omega}$ denotes the weight matrix. The corresponding dimension is 1 row × R column(s). It must be ensured that the input vector and the weight matrix dimension can be multiplied. Record the weight matrix as

$$\boldsymbol{\omega} = [\omega_{1,1}, \omega_{1,2}, \cdots, \omega_{1,R}] \tag{3.26}$$

where $\omega_{i,j}$ denotes the weight from input j to neuron i, and the magnitude of the weight can change the signal strength.

Threshold b is a scalar. The weights are multiplied by the corresponding inputs, and then the products are accumulated plus the threshold value b to obtain the input to the hidden layer neurons:

$$n = \omega_{1,1}p_1 + \omega_{1,2}p_2 + \cdots + \omega_{1,R}p_R + b \tag{3.27}$$

This expression can also be written as a matrix:

$$n = \boldsymbol{\omega p} + b \tag{3.28}$$

The output of the hidden layer neurons is

$$a = f(n) = f(\boldsymbol{\omega p} + b) \tag{3.29}$$

where f is the hidden layer activation function and the output a is also a scalar. The activation function determines whether to activate the output according to the input and uses the function to change the input-output relationship, transforming the input into the output within the valid range.

Different ANNs use different transfer functions. The widely used activation functions are as follows:

1. Linear function

Figure 3.7 shows the function curve, and its expression is

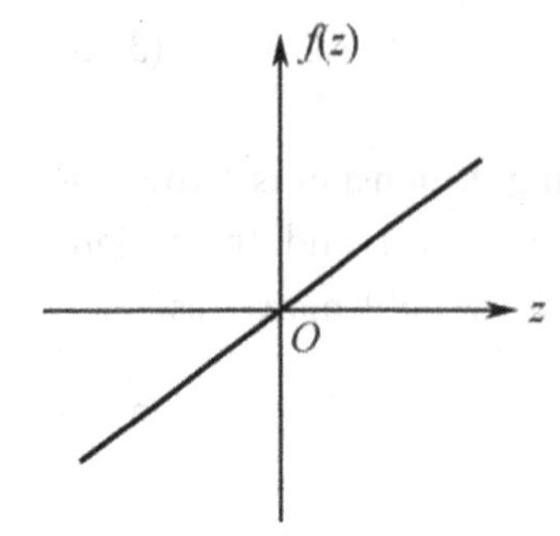

FIGURE 3.7 Linear function curve.

$$f(z) = z \tag{3.30}$$

2. Nonlinear function

 1. Sigmoid function. Figure 3.8 shows the function curve, and its expression is

$$f(z) = \frac{1}{1 + e^{-az}}, \quad -\infty < z < \infty \tag{3.31}$$

The sigmoid function has two asymptotes, which are $f(z) = 0$ and $f(z) = 1$. The sigmoid function is a smooth function, which is strictly monotone increasing and continuously differentiable. Adjusting the values of its parameters can enable the function to have characteristics similar to a threshold function. The sigmoid function is often used as the activation function for the BP network.

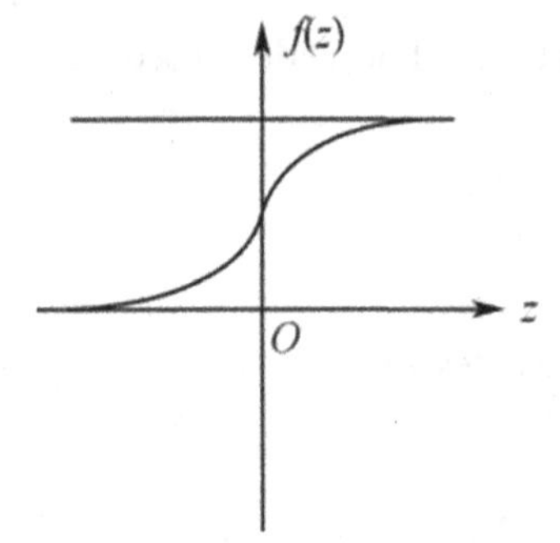

FIGURE 3.8 Sigmoid function curve.

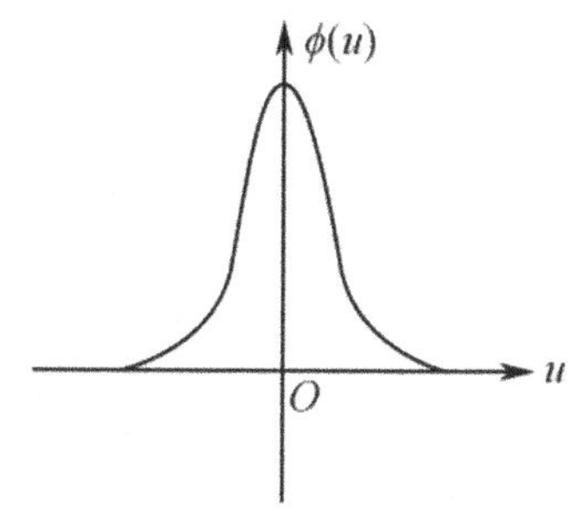

FIGURE 3.9 Gaussian function curve.

2. Gaussian function. Figure 3.9 shows the function curve, and its expression is

$$\phi(u) = \exp\left(-\frac{\|u\|^2}{2\sigma^2}\right) \tag{3.32}$$

where σ is the RBF width and $\|u\|$ is the Euclidean distance. Gaussian function is usually selected as the basic function of the RBF. Gaussian function is symmetric about $u = 0$.

3. Wavelet basis function. Figure 3.10 shows the function curve, and its expression is

$$\psi_{a,b}(x) = \frac{1}{\sqrt{a}}\psi\left(\frac{x-b}{a}\right) \tag{3.33}$$

where a is the scale parameter and b is the translation parameter. $\psi_{a,b}(x)$ decreases as a increases, and the center of the wavelet function is determined by b.

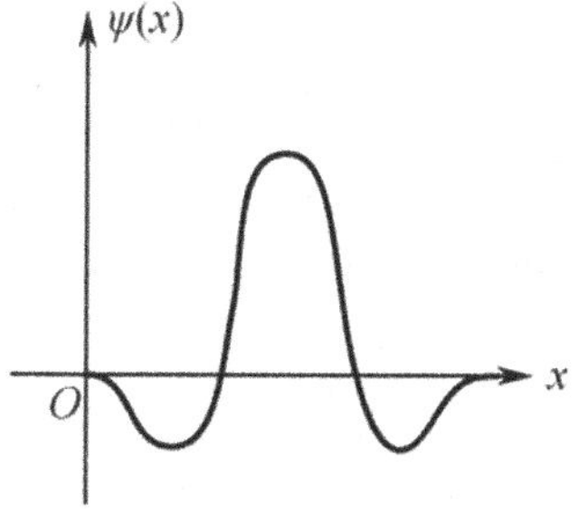

FIGURE 3.10 Wavelet basis function curve.

An ANN consists of massive neurons connected to each other. Different connection patterns among neurons constitute a variety of network models with various structures, which are categorized into layered networks and mutually connected networks.

1. Layered network

Layered network has a layered structure for all neurons in a neural network model. The neurons are divided into layers according to their functions, and the layers are sequentially connected, from the input layer, the hidden layer (also called "intermediate layer"), to the output layer.

The input layer connects the external input pattern, and the input unit transmits information to the connected hidden layer unit. The hidden layer is the internal processing unit layer of the neural network. The pattern transformation capabilities of the neural network, such as pattern classification, pattern refining, characteristic extraction, etc., are exhibited basically in the processing of hidden layer units. Regarding different pattern transformation functions, the hidden layer may come with multiple layers or no layer. The output layer generates the output pattern of the neural network.

Layered networks are subdivided into the following three types, regarding the connection mode.

1. Simple forward network (Figure 3.11): The input pattern enters the network through the input layer and is finally output in the output layer after a sequential pattern transformation through various intermediate layers. The network state is updated once this process is complete.

2. Forward network with feedback (Figure 3.12): A closed-loop in the form of feedback structure. The unit with feedback is also called a hidden unit, and its output is called internal output. The network itself is still a forward type.

3. Forward network with intra-layer connection (Figure 3.13): The interconnection between units in the same layer makes them restrict each other and limits the number of units that can be

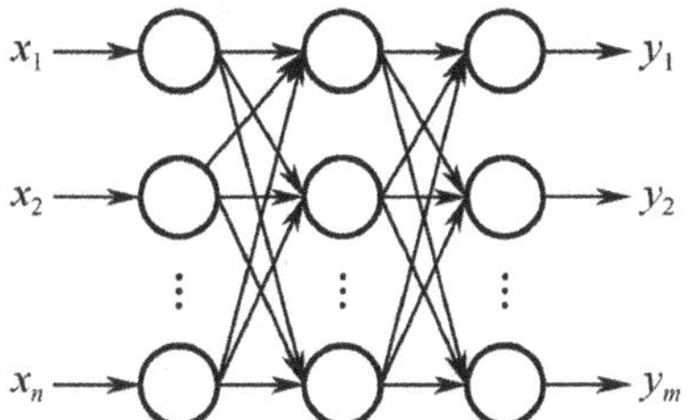

FIGURE 3.11 Simple forward network.

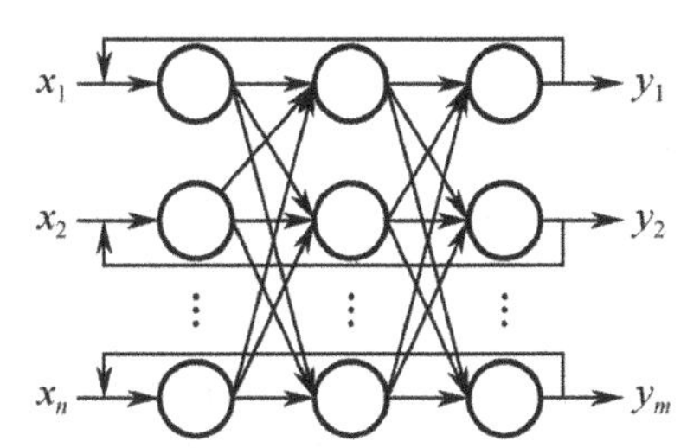

FIGURE 3.12 Forward network with feedback.

activated simultaneously in the same layer. However, from the outside, the network still appears to be a forward network. Some self-organizing competitive networks adopt this topology.

2. Mutually connected network

Any two units in a mutually connected network have a connection in between them. See Figure 3.14. The mutually connected networks are categorized into locally mutually connected networks and fully mutually connected networks. In a fully mutually connected network, the output of each neuron is connected with other neurons. In a locally mutually connected network, some neurons are not mutually connected.

Given a certain input pattern, the simple forward network can quickly generate a corresponding output pattern and keep it unchanged. However, in the mutually connected network, a given input pattern that starts from a certain initial state may experience a constantly changing process at the same time. The network may eventually generate a certain stable output pattern, or may enter a periodic oscillation or chaotic state.

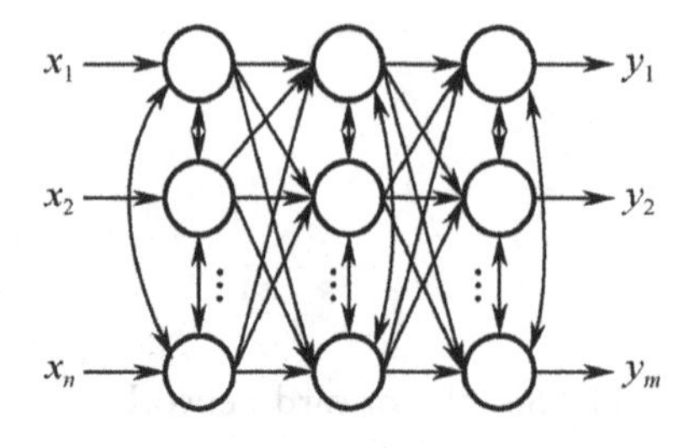

FIGURE 3.13 Forward network with intra-layer connection.

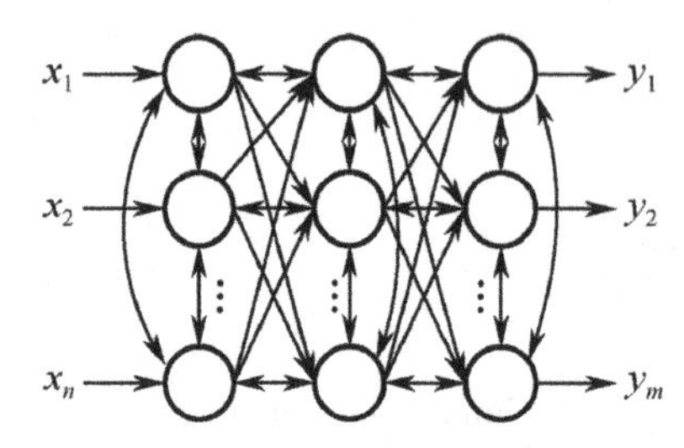

FIGURE 3.14 Mutually connected network.

Neural network learning, also called training, refers to a process in which the neural network adjusts the parameters of the network itself under the stimulation of the external environment, so that the neural network responds to the external environment in a new manner.

Generally, neural network learning is classified as supervised learning, unsupervised learning and reinforcement learning. Supervised learning is also known as teacher learning. Under supervised learning, the desired output is used as a model reference in the process of model training. Unsupervised learning, as the name suggests, is a type of learning done without the supervision of a teacher. Under unsupervised learning, there is no feedback from the environment as to what should be the desired output and whether it is correct or incorrect in the process of model training, and adjustments of network parameters can only be performed according to the input data. Reinforcement learning, also known as evaluation learning, is a learning method between supervised learning and unsupervised learning.

Learning algorithms refer to the definite learning rules. The selection of learning algorithms needs to take into account the structure of the neural network and the connection between the neural network and the external environment. At present, the commonly used learning algorithms are divided into the following four cases.

1. Hebb learning
   Hebb learning is the oldest and most famous learning algorithm. Almost all the learning rules of ANN can be attributed to the deformations of the Hebb learning algorithm. The idea of Hebb learning is that, if the neurons on both sides of a synapse (connection) are activated simultaneously, the strength of that synapse will be selectively enhanced; if the neurons on both sides of a synapse are activated asynchronously, that synapse will be selectively weakened.

2. δ learning rule
   δ learning rule is also known as error-correction learning rule. It has a wide applicable scope and can be used in the learning process of nonlinear neurons, setting no restriction on the number of learning samples. Furthermore, it even tolerates contradictions in the training samples, which is one of the fault-tolerance characteristics of neural networks. The learning rule adopts a gradient descent method, resulting in local minimum problems. The rule compares the actual ANN output with the desired output value and uses the error value, which is considered as a reference indicator for adjusting the network weight, to direct the training until the desired requirements are met.

3. Stochastic learning algorithm
   Stochastic learning algorithm addresses local minimum problems by introducing the instability factor. Generally, the instability factor changes stepwise from large to small. As long as the change is slow enough and the learning time is long enough, there is always a state that allows the neural network to jump out of the local minimum point but not out of the global minimum point, so that the neural network converges to the global minimum point. The well-known stochastic learning algorithms are simulated annealing algorithm and genetic algorithm.

4. Competitive learning algorithm
   Competitive learning refers to the competitive ability of the output neurons in the network to respond to external stimuli. In competitive learning, the weight of the winning neuron is modified. That is, if the input state of the winning neuron is 1, the corresponding

weight increases; if the state is 0, the weight decreases. In the learning process, the weight of the winning neuron gets more and more approximate to the corresponding input state, while the remaining neurons that fail the competition are suppressed. Kohomen's self-organization map (SOM) and adaptive resonance theory (ART) both adopt this algorithm.

An ANN model is basically divided into two stages: training and simulation. ANN training aims to extract the hidden rules and knowledge from the training data samples and store them in the network for use in the simulation phase. Essentially, ANN simulation is a process in which the network uses the input data to obtain the corresponding network output by means of complex numerical computations. ANN simulation results enable people to understand the performance of the current neural network in time and to determine whether further training of the neural network is required.

There are basically two methods for RF circuit modeling by the use of neural networks: one is neural network forward modeling, which means the forward modeling of circuits using neural networks; the other is neural network inverse modeling, which means the inverse modeling of circuits using neural networks.

1. Neural network forward modeling

Essentially, this method is to select an appropriate neural network model to approximate or replace the real circuit. That is, the neural network forward model is trained to exhibit the forward dynamic characteristics of the circuit. The process for forward modeling of a circuit using a neural network is shown in Figure 3.15.

As indicated by Figure 3.15, the neural network forward model and the RF circuit have the same input X. The network training objective is to reduce the error E between H (output of the forward model) and Y (output of the RF circuit). After the training is complete, the obtained forward model will have the same input/output characteristics as the real RF circuit, if the accuracy of the model is relatively high. Therefore, in system simulation applications, the neural network forward model is often used in place of the physical circuit model, which is helpful to simplify the analysis process.

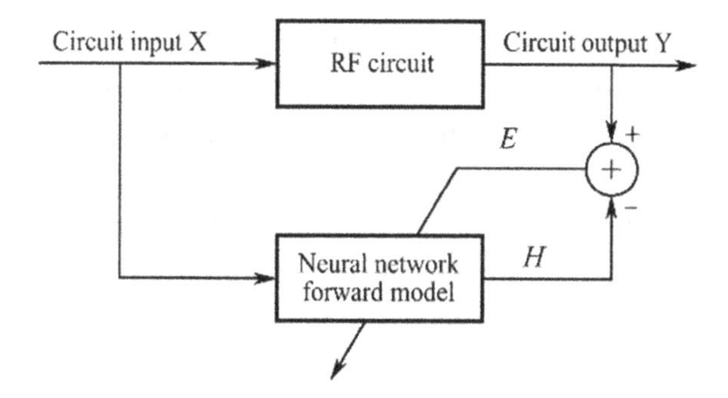

FIGURE 3.15 Neural network forward modeling process.

2. Neural network inverse modeling

There are two methods for inverse modeling of circuits by use of neural networks. The first is neural network function inverse modeling, which means a neural network is used to learn the inverse dynamic characteristics of the circuit. The process for neural network function inverse modeling is shown in Figure 3.16. It takes Y (output of the RF circuit) as the input of the neural network inverse model, compares H (output of the inverse model) with X (input of the RF circuit) and uses E (error between them) to train the neural network, thereby establishing the inverse model of the circuit. The function inverse model does not consider the properties of the input and output parameters of the circuit, but only exchanges the input and output of the real circuit, which can also be considered as an exchange of input and output of the forward model.

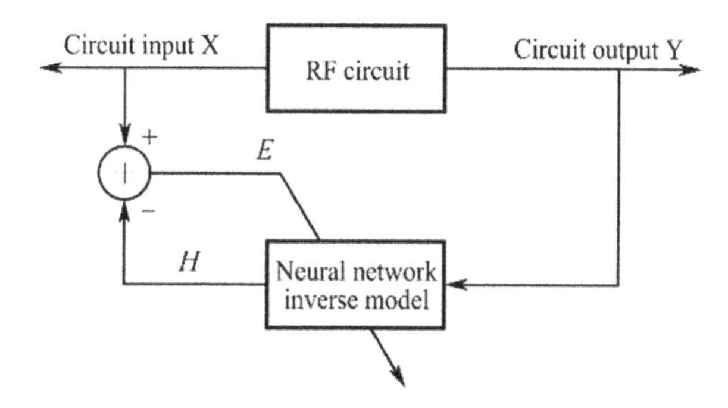

FIGURE 3.16 Neural network function inverse modeling process.

The second is neural network parameter inverse modeling. The previous function inverse model does not set requirements for the input/output parameters of the circuit, simply exchanging the input and output of the circuit. However, when a neural network is used to design a real RF circuit, the target electrical parameters that the circuit can achieve are usually known and the geometric parameters of the circuit are to be solved. Electrical parameters refer to scattering parameters, characteristic impedance, quality factor, etc. Geometric parameters refer to certain dimensions of the circuit, such as length and width. This process is the inverse of the CAD software simulation process. Therefore, the neural network inverse model that solves the geometric parameters of the circuit with the known target electrical parameters is called the parameter inverse model, also called the circuit synthesis. A relatively simple parameter inverse modeling method is direct inverse modeling. Its modeling process is similar to the function inverse modeling process in Figure 3.16. The difference lies in that, in the direct inverse modeling method, the input X and output Y of the RF circuit in Figure 3.16 are not parameters of arbitrary properties. Instead, X should be the geometric parameter and Y should be the electrical parameter of the circuit. In other words, the parameter inverse model exchanges the input and output of the forward model, in which the input is a geometric parameter and the output is an electrical parameter. There are also other types of parameter inverse modeling methods, such as grouping inverse modeling method, algorithm optimization inverse modeling method and novel inverse modeling method.

3.2.4 Input–Output Relationship of Nonlinear Power Amplifier

When the output signal of PA reaches saturation, the nonlinear characteristic between the input and output will be observed. The discretized output of the nonlinear PA behavioral model is expressed in Equation (3.34). A PA with memory effects is usually a nonlinear dynamic system [8], whose output depends both on the current input and on the previous inputs. Its dynamic characteristics are expressed in Equation (3.35). Since PA behavioral modeling realizes the fitting of characteristic curves based on the PA input and output data, the relationship between signal amplitude and phase has been used to establish the RF PA behavioral model, with Equation (3.35) expressing its mathematical model.

$$v_{\text{out}}(n) = f(|v_{\text{in}}(n)|)\exp\{j[g(|v_{\text{in}}(n)|) + \arg[v_{\text{in}}(n)]]\} \tag{3.34}$$

$$\begin{cases} |v_{\text{out}}(n)| = f(|v_{\text{in}}(n)|, |v_{\text{in}}(n-1)|, \cdots, |v_{\text{in}}(n-q)|) \\ \arg[v_{\text{out}}(n)] - \arg[v_{\text{in}}(n)] = g(|v_{\text{in}}(n)|, \cdots, |v_{\text{in}}(n-q)|) \end{cases} \tag{3.35}$$

where v_{in} and v_{out} are the input and output voltages of PA, respectively; $f(\cdot)$ and $g(\cdot)$ are the nonlinear distortion functions of AM/AM and AM/PM, respectively; q is the memory depth, which exhibits the characteristic of memory effects of PA.

3.2.5 Support Vector Machine Model

Support Vector Machine (SVM) [9] was proposed by Vapnik et al. according to the statistical learning theory. Based on the Vapnik Chervonenkis (VC) dimension and structural risk minimization (SRM), SVM is able to elegantly address the practical problems, such as small sample, nonlinearity or high dimension and local minimum in neural network algorithms. Moreover, SVM features a simple structure and a strong generalization ability. It is considered as an effective nonlinear regression modeling technique to specifically address the small sample problem. SVM has been widely used in various fields such as pattern recognition, regression estimation and function approximation.

Similar to neural networks, SVM also adopts a learning mechanism. However, unlike neural networks, SVM uses optimization technology and the mathematical method, and furthermore, SVM fundamentally addresses three major problems inherent in neural networks, which are over-fitting, structure determination and local extremum. The introduction of kernel function enables SVM to prevent "dimension curse" and to properly process high-dimensional data. Therefore, SVM has been highlighted by more researchers in the machine learning field.

The use of the kernel function is the key to the implementation of SVM. To implement SVM, the input vectors are mapped into a high-dimensional feature space through a certain preselected nonlinear mapping, and the optimal separating hyperplane is constructed in the space. Herein, "nonlinear mapping" refers to the kernel function.

By defining the kernel function, SVM skillfully replaces the inner product operation in the high-dimensional feature space with the kernel function of the original space, i.e., $k(x_i, x_j) = \Phi(x_i) \cdot \Phi(x_j)$, thereby preventing dimension curse.

According to the related functional theory, the symmetric functions meeting the mercer condition are considered kernel functions. At present, the kernel functions that are most applied and investigated in SVM are q-order polynomial kernel function, radial basis function and sigmoid function.

1. q-order polynomial kernel function: $K(x, x_i) = [(x, x_i) + 1]^q$
2. Radial basis function: $K(x, x_i) = \exp\left\{-\frac{|x - x_i|^2}{\sigma^2}\right\}$
3. Sigmoid function: $K(x, x_i) = \tanh[v(x, x_i) + a]$

SVM applications basically incorporate regression and classification, and regression is categorized into linear regression and nonlinear regression. The nonlinear characteristic of PA determines that PA modeling is a nonlinear regression problem.

SVM is suitable for addressing the problems such as small sample and nonlinear regression estimation. The basic idea of nonlinear SVM regression is to map data x into a high-dimensional feature space through a nonlinear mapping Φ, and perform linear regression in this feature space. In this way, the linear regression of the high-dimensional feature space corresponds to the nonlinear regression of the low-dimensional input space, which can be implemented by the kernel function $k(x_i, x_j) = \Phi(x_i) \cdot \Phi(x_j)$. In this way, complicated dot product operations in the high-dimensional feature space can be avoided, and the optimization problem $W(\alpha, \alpha^*)$ in linear regression becomes an optimization process in nonlinear regression with the introduction of kernel function. That is,

$$W(\alpha_i, \alpha^*) = -\frac{1}{2}\sum_{i,j=1}^{m}(\alpha_i - \alpha_i^*)(\alpha_j - \alpha_j^*)k(x_i \cdot x_j) + \sum_{i=1}^{m}(\alpha_i - \alpha_i^*)y_i - \sum_{i=1}^{m}(\alpha_i + \alpha_i^*)\varepsilon \tag{3.36}$$

$$\text{s.t.} \quad \sum_{i=1}^{m}(\alpha_i - \alpha_i^*) = 0, \quad \alpha_i, \alpha_i^* \in [0, C] \tag{3.37}$$

where,

$$w = \sum_{i=1}^{m}(\alpha_i - \alpha_i^*)\Phi(x_i) \tag{3.38}$$

The objective function $f(x)$ can be expressed as

$$f(x) = \sum_{i=1}^{m}(\alpha_i - \alpha_i^*)(\Phi(x_i)\cdot\Phi(x)) = \sum_{i=1}^{m}(\alpha_i - \alpha_i^*)k(x_i \cdot x) + b \tag{3.39}$$

Maximizing Equation (3.36) under the constraint (3.37) yields $\alpha_i - \alpha_i^*$ in linear regression.

After computing α_i^*, solve the value of b. According to the KKT condition, the product of the constraint and the dual variable is 0, x_i and y_i are the training data and ξ_i and ξ_i^* are relaxation variables. The following equation applies:

$$\begin{cases} \alpha_i(\xi_i + \varepsilon - y_i + w \cdot x_i + b) = 0 \\ \alpha_i^*(\xi_i^* + \varepsilon - y_i + w \cdot x_i + b) = 0 \end{cases} \tag{3.40}$$

Namely

$$\begin{cases} (C - \alpha_i)\xi_i = 0 \\ (C - \alpha_i^*)\xi_i^* = 0 \end{cases} \tag{3.41}$$

In the case of $\alpha_i^* \in (0, C)$, C is a constant and $C > 0$, $\xi_i^* = 0$, and ε is the error. Therefore

$$b = y_i - w \cdot x_i - \varepsilon,\ \alpha_i \in (0, C) \tag{3.42}$$

$$b = y_i - w \cdot x_i + \varepsilon,\ \alpha_i^* \in (0, C) \tag{3.43}$$

The value of b can be solved from Equations (3.42) and (3.43).

Research demonstrates that SVM is a proper solution to regression problems, so a concept of support vector machine regression (SVMR) is derived. SVMR is categorized into linear regression and nonlinear regression. In the case of linear regression, the linear regression function is used as follows:

$$f(x) = w \cdot x + b \tag{3.44}$$

The data samples $(x_1, y_1), (x_2, y_2), \cdots, (x_i, y_i), \cdots, (x_l, y_l),\ \ x_i, y_i \in R$ are fitted and regressed by use of this function. Assuming the training samples can be fitted by use of the linear function $f(x)$ with the accuracy ε, the problem of finding the minimum w can be expressed as a convex optimization problem. That is,

$$\begin{cases} \min\ \frac{1}{2}\|w\|^2 \\ \text{s.t.}\ \ y_i - w \cdot x_i - b \le \varepsilon \\ \qquad w \cdot x_i + b - y_i \le \varepsilon \end{cases} \tag{3.45}$$

Use relaxation variables ξ_i and ξ_i^* to process the samples that cannot be estimated by the function f with the accuracy ε. (3.45) can be written as:

$$\begin{cases} \min\ \frac{1}{2}\|w\|^2 + C\sum_{i=1}^{m} (\xi_i + \xi_i^*) \\ \text{s.t.}\ \ y_i - w \cdot x_i - b \le \varepsilon + \xi_i \\ \qquad w \cdot x_i + b - y_i \le \varepsilon + \xi_i^* \\ \qquad \xi_i, \xi_i^* \ge 0 \end{cases} \tag{3.46}$$

(3.46) is written as Lagrangian function form:

$$L = \frac{1}{2}\|w\|^2 + C\sum_{i=1}^{m}(\xi_i + \xi_i^*) - \sum_{i=1}^{m}\alpha_i(\xi_i + \varepsilon - y_i + w\cdot x_i + b)$$
$$- \sum_{i=1}^{m}\alpha_i^*(\xi_i^* + \varepsilon + y_i - w\cdot x_i - b) - \sum_{i=1}^{m}(\eta_i\xi_i + \eta_i^*\xi_i^*) \tag{3.47}$$

where η_i, η_i^*, α_i, $\alpha_i^* \geq 0$ and $C > 0$. According to the KKT condition, we have

$$\frac{\partial L}{\partial b} = \sum_{i=1}^{m}(\alpha_i - \alpha_i^*) = 0, \quad 0 \leq \alpha_i, \alpha_i^* \leq C, \quad i = 1,\cdots,m$$
$$\frac{\partial L}{\partial w} = w - \sum_{i=1}^{m}(\alpha_i - \alpha_i^*)x_i = 0 \tag{3.48}$$

which means that

$$w = \sum_{i=1}^{m}(\alpha_i - \alpha_i^*)x_i$$
$$\frac{\partial L}{\partial \xi_i^*} = C - \alpha_i^* - \eta_i^* \tag{3.49}$$

$$\text{Hence,}\ C = \alpha_i^* + \eta_i^* \tag{3.50}$$

The dual form of the optimization problem is obtained from Equations (3.47), (3.48) and (3.49),

$$W(\alpha_i, \alpha^*) = -\frac{1}{2}\sum_{i,j=1}^{m}(\alpha_i - \alpha_i^*)(\alpha_j - \alpha_j^*)(x_i\cdot x_j)$$
$$+ \sum_{i=1}^{m}(\alpha_i - \alpha_i^*)y_i - \sum_{i=1}^{m}(\alpha_i + \alpha_i^*)\varepsilon \tag{3.51}$$

Under the constraint of (3.48), the parameters α_i, α_i^* obtained by the maximized (3.51) are substituted into (3.49), and the following regression function is obtained from Equation (3.44):

$$f(x) = \sum_{i=1}^{m} (\alpha_i - \alpha_i^*)(x_i \cdot x) + b \tag{3.52}$$

where the sample data corresponding to a non-zero $\alpha_i - \alpha_i^*$ is the support vector.

3.2.6 X-Parameter Model

As a theoretical extension of S-parameters, X-parameters provide a method to characterize each frequency component relationship between the reflected wave and incident wave at both ports in a strongly non-linear condition, and can accurately describe the behavior characteristics of large-signal nonlinear microwave devices.

X-parameter measurements are enforced in the actual working state of the nonlinear device, including each harmonic component and the response between harmonics. Therefore, X-parameters can accurately reflect the real characteristics of the nonlinear device. By use of the tuner, X-parameters at any load impedance are obtained, so that X-parameters can cover the entire Smith circle diagram.

Under the excitation of small signals, X-parameters degenerate to S-parameters. A comparison of the X-parameter expression and the S-parameter expression indicates that the signal's conjugate component has been notably added to X-parameters. The conjugate component indicates the phase relationship between the reflected wave and incident wave, which is no longer a simple linear relationship. Therefore, X-parameters can be used to express the nonlinear system. The superiorities of X-parameters over S-parameters are described as follows:

1. Compared with S-parameters dedicated to small signals, X-parameters can describe not only the small-signal characteristics of components but also the characteristics under the large-signal working condition.

2. X-parameters can describe the relationships between all harmonics and intermodulation spectrums under the large-signal working condition, which are more powerful than S-parameters.

3. X-parameters enable accurate modeling of cascaded modules on fundamental, harmonic, and even intermodulation frequencies in the case of mismatch.

4. X-parameters accurately simulate the characteristics of real circuits with a fast simulation speed.
5. X-parameters truly protect the designer's intellectual property and support design sharing without disclosing the design topology.

X-parameter modeling [10] is a data-based modeling scheme, and therefore the extraction of X-parameters is a critical step in the X-parameter modeling process. There are basically two methods to extract X-parameters. One is to use the device, nonlinear vector network analyzer (NVNA), and the other is to use software, advanced design system (ADS). The device-based X-parameter extraction scheme requires a real test bench setup, including various additional accessories (PA, high-power directional coupler, high-power attenuator, etc.).

After connecting the additional accessories, perform the following steps to extract X-parameters:

1. Set the DC sweep range, frequency sweep range, large-signal power sweep range, number of harmonics, intermediate frequency bandwidth, etc.
2. Set the receiver's attenuator, calibration power, etc.
3. Enable the X-parameter measurement function and set the forward and reverse powers of small signals.
4. Complete the calibration following the NVNA calibration wizard.

Since the 2009 edition of ADS, a module for X-parameter simulation has been added. By use of the X-parameter simulation control contained in the module, X-parameters of the circuit and the system can be easily extracted. The extracted circuit structure is shown in Figure 3.17. The specific steps are as follows:

1. Create an X-parameter extraction template and save it in the template library, for the ease of X-parameter extraction in the future.
2. Import the transistor model of the DUT into ADS.

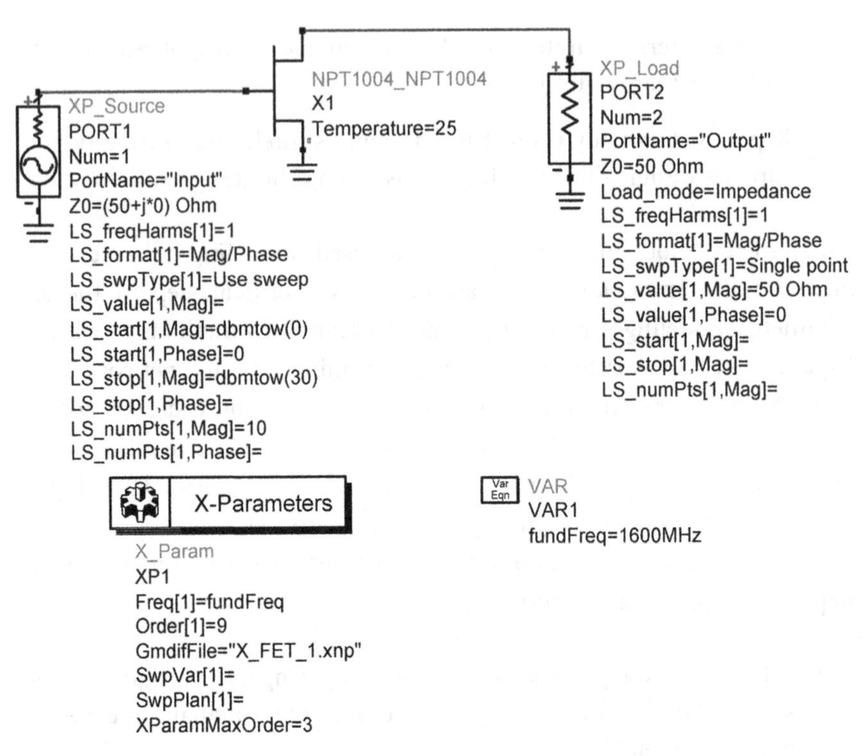

FIGURE 3.17 Extracted X-parameter circuit structure using ADS.

3. Create a new schematic diagram, insert the X-parameter extraction template, and add the DUT model.

4. Modify the relevant parameters of the X-parameter control based on the relevant parameters of the DUT.

5. Run the simulation to obtain the X-parameter file of the DUT.

Index and package the extracted X-parameter model, and then design the PA in ADS. This method can also be applied to PA modeling in the case of large signal nonlinearity, thereby achieving accurate PA modeling.

3.2.7 Dynamic X-Parameter Theory

As a reasonable extension of S-parameters in a nonlinear and large-signal operating environment, X-parameters describe the mapping relationship between the outgoing wave and the incident wave at each port of a nonlinear device. X-parameters are considered as a linear approximation of the scattering function under large-signal excitation conditions and come from the poly-harmonic distortion (PHD) framework. Dynamic X-parameters extend the application of X-parameters to memory effects. The dynamic X-parameter model is expressed as follows:

$$B(t) = \left(F_{\mathrm{CW}}(|A(t)|) + \int_0^{\infty} G(|A(t)|, |A(t-u)|, u)\,\mathrm{d}u \right) \cdot \exp^{j\phi(A(t))} \tag{3.53}$$

where $B(t)$ is the superposition of $F_{\mathrm{CW}}(\cdot)$ (indicating the static nonlinear part) and $G(\cdot)$ (indicating the dynamic nonlinear part). $F_{\mathrm{CW}}(\cdot)$ and $G(\cdot)$ are both the functions of $A(t)$ (instantaneous amplitude). Compared with the traditional PHD model, the dynamic X-parameter model introduces the dynamic part, thereby characterizing the nonlinearities of PA more accurately. In addition, $G(\cdot)$ is a three-dimensional function of the current input signal amplitude, historical input signal amplitude and time, which can characterize the dynamic part of PA.

3.3 THEORETICAL BASIS OF NONLINEAR CIRCUIT ANALYSIS METHOD

The nonlinear analysis method [11] is the core technique of RF/microwave CAD software. The simulation engine module of CAD software consists of the nonlinear analysis method and linear analysis method, which work together to perform the simulation of RF/microwave circuits and systems. At present, the most widely used analysis method for RF nonlinear circuits is the harmonic balance method (HBM), which divides an RF circuit into linear part and nonlinear part and forms the harmonic balance equation by applying Kirchhoff's law. The algorithms for solving a harmonic balance equation basically include the quasi-Newton method, ant colony algorithm and bee colony algorithm. The HBM and its three solving algorithms are described in the following sections.

3.3.1 Harmonic Balance Method

The HBM [12] is suitable for the nonlinear circuit with a single frequency input. It requires a rearrangement of components for the nonlinear circuit, dividing them into a linear network and a nonlinear network. Next, it

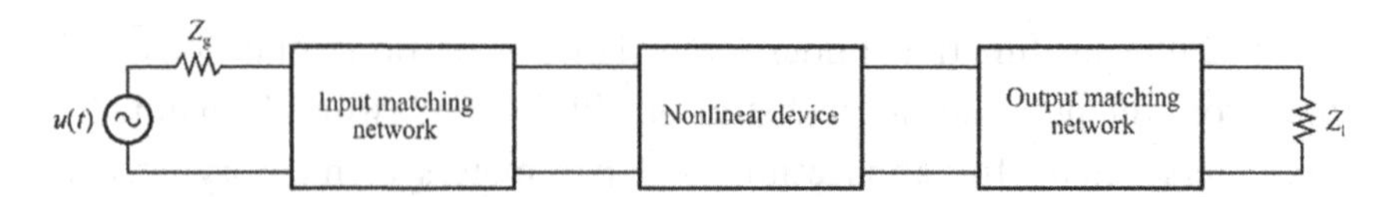

FIGURE 3.18 Equivalent circuit of general nonlinear system.

establishes the harmonic balance equation based on the equality of currents at ports, and then solves the equation by an appropriate method.

The microwave network containing nonlinear devices is a nonlinear microwave network. For the nonlinear microwave network under continuous-wave operation, pay attention to its working conditions in the steady state, in addition to knowing its initial transient working conditions. The HBM is an effective method to analyze the characteristics of microwave networks, especially when the network topology and the nonlinear characteristics of devices are known. The equivalent circuit of a general nonlinear system is shown in Figure 3.18, where Z_g and Z_l respectively indicate the source impedance and the load impedance, and u indicates the excitation source. The matching networks are used to optimize the overall performance of the circuit, where a bias is applied to the nonlinear solid device (such as a PA) to filter out unwanted harmonics.

The basic idea of HBM is to find a set of port voltage waveforms or harmonic voltage components so that its linear subnetwork equation and nonlinear subnetwork equation can obtain the same current. Essentially, the HBM aims to establish the harmonic balance equation and apply an appropriate method to solve it. Next, the establishment of a harmonic balance equation is described.

The circuit is divided into a nonlinear network and a linear network, as shown in Figure 3.19. The nonlinear network only contains nonlinear components, while the linear network contains linear components, source impedance, load impedance, DC bias and excitation in the circuit.

Take the MESFET large-signal equivalent model as an example. The model is firstly decomposed into a linear network and a nonlinear network. As shown in Figure 3.20, the dashed box part is the linear network and the solid box part is the nonlinear network.

In the figure, the source impedance and load impedance are added to make the equivalent model more accurate. That is, two voltage sources are added to the linear network so that the entire circuit is simplified into

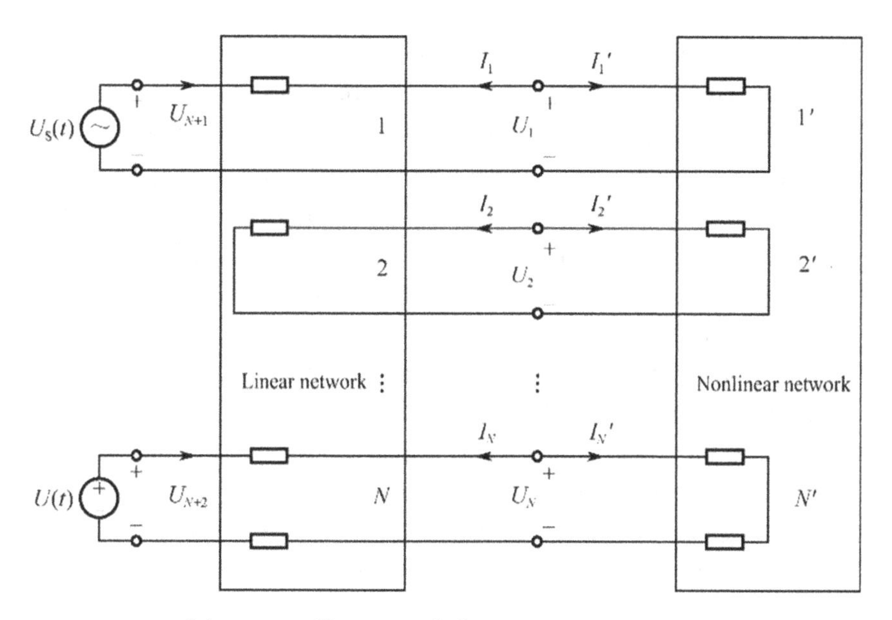

FIGURE 3.19 Schematic of harmonic balance.

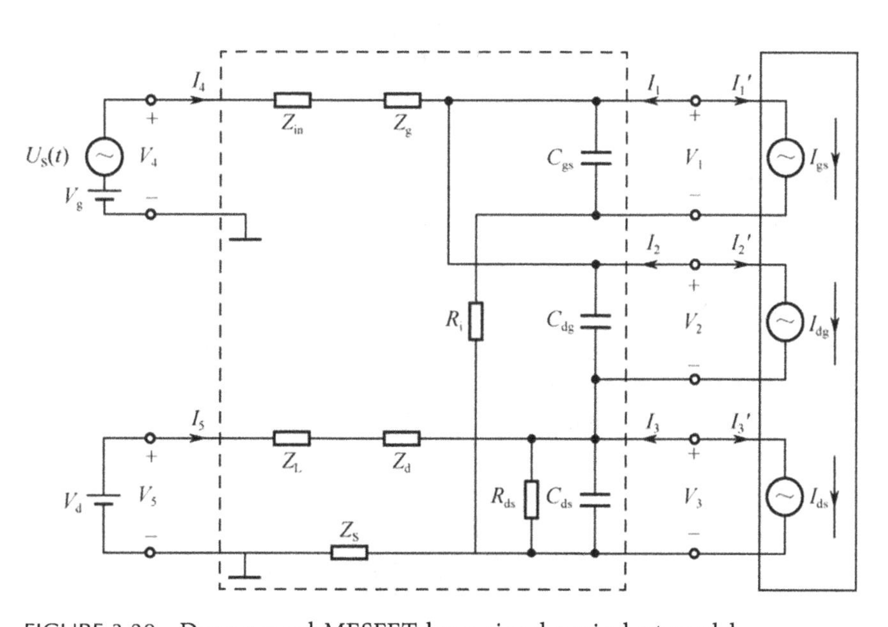

FIGURE 3.20 Decomposed MESFET large-signal equivalent model.

3 + 2 ports. In the figure, Z_{in} is the input matching impedance, Z_L is the output matching impedance, V_g and V_d are respectively the DC bias voltages of the gate and the drain, $U_s(t)$ is the input signal, $Z_g = R_g + j\omega L_g$ is the gate parasitic impedance, $Z_d = R_d + j\omega L_d$ is the drain parasitic impedance, $Z_S = R_S + j\omega L_S$ is the source parasitic impedance, U_1 is the voltage at node 1, and I_1 and I'_1 are the currents flowing into the linear network and nonlinear network, respectively. Similarly, U_2 and U_3 are respectively the voltages at nodes 2 and 3, I_2 and I'_2 are respectively the currents flowing into the linear network and nonlinear network at node 2, and I_3 and I'_3 are respectively the currents flowing into the linear network and nonlinear network at node 3.

In the linear network, vector $\boldsymbol{I}_{nk}$ denotes the kth-order harmonic current component on the nth port. In the nonlinear network where the current is determined by the port voltage and the properties of nonlinear components, vector $\boldsymbol{I'}_{nk}$ denotes the current.

The linear network currents at nodes 1, 2 and 3 are

$$\boldsymbol{I} = \begin{bmatrix} I_1 \\ I_2 \\ I_3 \end{bmatrix} = \begin{bmatrix} Y_{11} & Y_{12} & Y_{13} \\ Y_{21} & Y_{22} & Y_{23} \\ Y_{31} & Y_{32} & Y_{33} \end{bmatrix} \begin{bmatrix} V_1 \\ V_2 \\ V_3 \end{bmatrix} + \begin{bmatrix} Y_{14} & Y_{15} \\ Y_{24} & Y_{25} \\ Y_{34} & Y_{35} \end{bmatrix} \begin{bmatrix} V_4 \\ V_5 \end{bmatrix} \tag{3.54}$$

In this equation, the admittance parameter Y_{ij} can be obtained based on the small-signal equivalent parameters and the method for establishing the indefinite admittance matrix. That is, $Y_{ij} = \text{diag}[Y_{mn}(k\omega)]$, $i = 1, 2, \cdots, 5$, $j = 1, 2, \cdots, 5$ and $k = 0, 1, \cdots, 3$.

The time-domain expression of nonlinear current is converted to the frequency domain by a simplified method such as Fourier transform, and the vector in the frequency domain is obtained, which is

$$\boldsymbol{I'} = \begin{bmatrix} I'_1 \\ I'_2 \\ I'_3 \end{bmatrix} = \begin{bmatrix} I_{gs} \\ I_{dg} \\ I_{ds} \end{bmatrix} \tag{3.55}$$

The harmonic balance equation is obtained by applying Kirchhoff's current law (KCL), which is

$$F(V) = I + I' = 0 \tag{3.56}$$

In the equation, enable $F(V) = 0$ to apply by obtaining the reasonable V_1, V_2 and V_3, indicating that the set of V_1, V_2 and V_3 is the correct solution. By use of this set of solutions, the current and voltage values on any component can be obtained and the results contain multiple harmonics, which can favorably describe the nonlinear phenomena in the circuit.

(3.56) can be decomposed as

$$F(V) = I_S + Y_{N\times N}V + I' = 0 \tag{3.57}$$

In (3.57),

$$I_s = \begin{bmatrix} Y_{14} & Y_{15} \\ Y_{24} & Y_{25} \\ Y_{34} & Y_{35} \end{bmatrix} \begin{bmatrix} V_4 \\ V_5 \end{bmatrix}, \quad Y_{N\times N}V = \begin{bmatrix} Y_{11} & Y_{12} & Y_{13} \\ Y_{21} & Y_{22} & Y_{23} \\ Y_{31} & Y_{32} & Y_{33} \end{bmatrix} \begin{bmatrix} V_1 \\ V_2 \\ V_3 \end{bmatrix}$$

Solving the above nonlinear equations gives V.

3.3.2 Quasi-Newton Method

The quasi-Newton method [13] is one of the most effective methods to solve nonlinear optimization problems. It addresses the drawbacks of the Newton method (such as derivation and inversion), reduces the computation cost and simplifies the computation process, and also ensures superlinear convergence of the iterative process.

The quasi-Newton method simplifies the Jacobian Matrix as

$$H^{p+1} = H^p + \Delta H^p,\ p = 1, 2, \cdots \tag{3.58}$$

Adopting the BFGS correction formula, the most effective approach of the quasi-Newton method, we get

$$H^{p+1} = H^p - \frac{H^p s^p (s^p)^{\mathrm{T}} H^p}{(s^p)^{\mathrm{T}} H^p s^p} + \frac{y^p (y^p)^{\mathrm{T}}}{(y^p)^{\mathrm{T}} s^p} \tag{3.59}$$

Regarding a harmonic balance equation, in the previous equation, $s^p = V^{p+1} - V^p$, $y^p = F(V^{p+1}) - F(V^p)$.

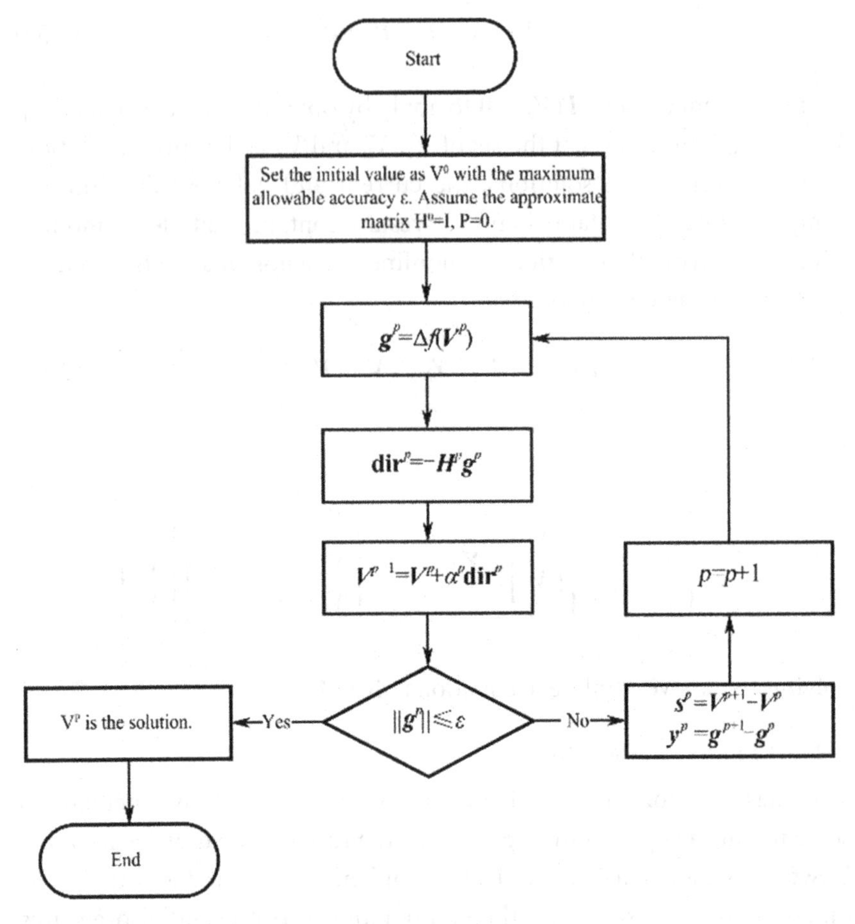

FIGURE 3.21 Flowchart of quasi-Newton method.

The iterative formula for solving the voltage is

$$\boldsymbol{V}^{p+1} = \boldsymbol{V}^p - \boldsymbol{H}^p\boldsymbol{F}(\boldsymbol{V}^p) \tag{3.60}$$

The flowchart of the quasi-Newton method to solve the harmonic balance equation is given in Figure 3.21.

The specific solution steps are as follows:

Step 1: Assume the initial value $\boldsymbol{V}^0$ with the maximum allowable accuracy ε. Assume the initial approximate matrix $\boldsymbol{H}^0 = \boldsymbol{I}$ with $p = 0$.

Step 2: Compute $\boldsymbol{g}^p = \nabla f(V^p)$. If $\|\boldsymbol{g}^p\| \leq \varepsilon$ applies, the algorithm ends and V^p is the output approximate solution that meets the accuracy requirement. Otherwise, go to Step 3.

Step 3: Compute the search direction $\mathbf{dir}^p$ and let $\mathbf{dir}^p = -\boldsymbol{H}^p\boldsymbol{g}^p$ apply.

Step 4: Utilize the step size factor α^p satisfying the backward step size rule and the line search method to obtain a new iteration point V^{p+1}; that is, $V^{p+1} = V^p + \alpha^p\mathbf{dir}^p$.

Step 5: Compute $\boldsymbol{g}^{p+1}$. Assume $\boldsymbol{s}^p = V^{p+1} - V^p$ and $\boldsymbol{y}^p = \boldsymbol{g}^{p+1} - \boldsymbol{g}^p$. Compute a new approximate matrix $\boldsymbol{H}^{p+1}$ by use of the BFGS correction formula. Assume $p = p + 1$. Go to Step 2.

3.3.3 Ant Colony Algorithm

The basic idea of the ant colony algorithm [14] is as follows. Suppose there are m ants. Each ant starts from a randomly selected position within a given interval. At each construction step, the solution corresponding to the ant when the pheromone concentration is the maximum is taken as the current optimal solution, a local update is performed, and the solutions corresponding to the remaining ants are globally updated; next, a pheromone update is performed for each ant. This procedure is repeatedly applied until the termination criterion is satisfied.

The specific steps are as follows:

Step 1: Computation of the initial pheromone concentration.

The pheromone concentration of each ant is computed according to (3.61):

$$\tau(k) = \mathrm{e}^{-F(\boldsymbol{x}^k)} \tag{3.61}$$

where F is a nonlinear function. As indicated from the above equation, the solution corresponding to the ant when the pheromone concentration is the maximum is selected as the current optimal solution, which is recorded as $\boldsymbol{x}^*$. The corresponding pheromone concentration is recorded as $\tau(\text{Best})$.

Step 2: Local update.

If the solution corresponding to the ant is the current optimal solution, the local update is performed by using (3.62):

$$\begin{cases} \boldsymbol{x} = \boldsymbol{x}^* + w \times \textbf{step} \\ \boldsymbol{x} = \boldsymbol{x}^* - w \times \textbf{step} \end{cases} \tag{3.62}$$

where w is the step size for local search, which is obtained by (3.63):

$$w = w_{\max} - (w_{\max} - w_{\min}) \times \frac{\mathrm{NC}}{\mathrm{NC}_{\max}} \tag{3.63}$$

where **step** is an n-dimensional column vector composed of random numbers in the interval [0, 0.1], and $w_{\max}$ and $w_{\min}$ are constants. Compared with x^*, the better solution is recorded as the solution corresponding to the ant.

Step 3: If the solution corresponding to the ant is not the current optimal solution, a global update is performed by using (3.64):

$$\begin{cases} \boldsymbol{x}^k = \boldsymbol{x}^k + \lambda \times (\boldsymbol{x}^* - \boldsymbol{x}^k), & p(k) < p_0 \\ \boldsymbol{x}^k = \boldsymbol{x}^k + \boldsymbol{\mu} \times \mathrm{len}, & p(k) \geq p_0 \end{cases} \tag{3.64}$$

where $\boldsymbol{\mu}$ is a random n-dimensional column vector in the variable interval $[\mu_i, l_i]$, λ is a random number of the interval [0, 1], len is the interval length for n variables equally divided into m sections within a given range, and $p(k)$ is selected according to the following global transition probability:

$$p(k) = \frac{e^{(\tau(\mathrm{Best}) - \tau(k))}}{e^{\tau(\mathrm{Best})}} \tag{3.65}$$

p_0 is the global state transition factor, which is computed according to Equation (3.84):

$$p_0 = \begin{cases} 0.9e^{\frac{\mathrm{NC}}{\mathrm{NC}_{\max}} \times 2 \times \log\left(\frac{1}{2}\right)}, & \mathrm{NC} < \frac{\mathrm{NC}_{\max}}{2} \\ 0.25e^{\frac{\mathrm{NC}}{\mathrm{NC}_{\max}} \times 2 \times \log(2)}, & \mathrm{NC} \geq \frac{\mathrm{NC}_{\max}}{2} \end{cases} \tag{3.66}$$

Step 4: Pheromone update.

The pheromone corresponding to each ant is updated. The expression is

$$\tau(k) = (1 - \rho)\tau(k) + \Delta\tau(k) \quad (3.67)$$

where

$$\Delta\tau(k) = \frac{a}{b + R(\boldsymbol{x}^k)} \quad (3.68)$$

3.3.4 Bee Colony Algorithm

The artificial bee colony (ABC) algorithm [15] handles the function optimization problem by simulating the honey collection mechanism of bees. An artificial bee colony basically consists of three parts: leaders, followers and scouters. The basic idea of the algorithm is to select half of the individuals with better fitness values from a randomly generated initial population as the leaders, perform searches at their nearby places, compute the fitness values before and after each search, and follow a one-to-one competition rule to select better individuals. Then, followers choose individuals with better fitness values to follow by means of a roulette scheme, and greedy searches are performed around the chosen individuals to generate the other half of individuals. The fitness values before and after each search are computed and individuals with better fitness values are retained. Leaders and the individuals generated by followers constitute a new population. To avoid the loss of population diversity, scouters perform variation searches. The algorithm approaches the global optimal solution through continuous iterations.

The harmonic balance equation can be regarded as the minimum-value problem of a nonlinear function, which is expressed as min$f(\boldsymbol{V})$. Critical steps of the ABC algorithm are described as follows.

Step 1: Population initialization.

Set the initial evolutionary generation $t = 0$ and the maximum number of cycles $m_{\max}$. Randomly generate NP individuals $\boldsymbol{V}$ in the feasible solution space according to (3.69), constituting the initial population.

$$\boldsymbol{V}_i^0 = \boldsymbol{V}_i^{\min} + \mathrm{rand}() \times (\boldsymbol{V}_i^{\max} - \boldsymbol{V}_i^{\min}) \; i = 1, 2, \cdots, \mathrm{NP} \quad (3.69)$$

where each solution $\boldsymbol{V}_i$ is a D-dimensional vector. $\boldsymbol{V}_i^0$ denotes the ith individual in the 0th-generation population. $\boldsymbol{V}_i^{\max}$ and $\boldsymbol{V}_i^{\min}$ are the upper and lower limits of V, respectively. rand() is a random number in the interval [−1, 1].

Step 2: Search algorithm for the leaders.

The leader population consists of half of the individuals with smaller fitness values in the initial population. New individuals X_i^j are selected according to the search formula (3.70).

$$X_i^j = V_i^j + \text{rand}(\,) \times (V_i^j - V_k^j) \tag{3.70}$$

where rand() is a random number in the interval [−1, 1] and k is a randomly selected value not equaling to i.

The ABC algorithm follows the idea of "survival of the fittest" to select better individuals. Specifically, the computed fitness values of newly generated individuals X_i^j and target individuals V_i^j are compared, and the individuals with smaller fitness values are retained according to (3.71).

$$V_i^{j+1} = \begin{cases} X_i^j, & f(X_i^j) < f(V_i^j) \\ V_i^j, & f(V_i^j) < f(X_i^j) \end{cases} \tag{3.71}$$

Step 3: Search algorithm for the followers.

Better individuals (V_k^{j+1}, $k \in [1, \cdots, \text{NP}/2]$) are selected from the new leader population as the followers by means of a roulette scheme, according to the probability formula (3.72). After the following process is complete, (3.69) and (3.70) are applied to search for better individuals selected as new individuals (V_k^{j+1}, $k \in [\text{NP}/2 + 1, \cdots, \text{NP}]$), constituting the follower population.

$$P_i = \frac{\text{fit}_i}{\sum_{i=1}^{\text{NP}/2} \text{fit}_i} \tag{3.72}$$

Step 4: Search algorithm for the scouters.

To avoid the loss of population diversity, the ABC algorithm introduces an important parameter "limit". If a certain solution remains unchanged for "limit" consecutive generations, the solution will be eliminated. That is, the corresponding individuals will be converted into scouters. In this situation, (3.68) is applied to search for new individuals, which are compared with the original individuals according to (3.70). Better individuals with smaller fitness values are retained.

The ABC algorithm performs the above steps to make the populations evolve continuously until the algorithm ends when the number of iterations reaches the predefined maximum cyclic coefficient $m_{\max}$ or the optimal solution of the population reaches the predefined error accuracy.

REFERENCES

1. 侯道琪, 杨正. 无记忆功放的预失真数学模型[J]. 舰船电子对抗, 2015, 38(04):56–61.
   HOU D, YANG Z. Pre-Distortion Mathematical Model of Memoryless Power Amplifier[J]. Shipboard Electronic Countermeasure, 2015, 38(04):56–61.
2. 陈庆霆, 王成华, 朱德伟, 等. 改进 BP 神经网络的功放有记忆行为模型[J]. 微波学报, 2012, 28(02):90–93.
   CHEN Q, WANG C, ZHU D, et al. Behavioral Model of Power Amplifiers Based on Improved BP Neural Network Considering Memory Effects[J]. Journal of Microwaves, 2012, 28(02):90–93.
3. 郑凯帆, 梁猛. 逆 Volterra 级数的 O-OFDM 系统均衡方法[J]. 信息通信, 2018, 185(05):23–24.
   ZHENG K, LIANG M. Equalization Method of Multi-Span O-OFDM System Based on Inverse Volterra Series[J]. Information & Communications, 2018(05):23–24.
4. 惠明, 张萌, 张新刚, 等. 基于时间功率混合分段模型的磁共振功放非线性建模[J]. 微波学报, 2018, 34(05):35–40.
   HUI M, ZHANG M, ZHANG X, et al. Time and Power Hybrid Piecewise Model Based Nonlinear Modeling for Magnetic Resonance Power Amplifier[J]. Journal of Microwaves, 2018, 34(05):35–40.
5. 杨新宇, 张臻, 谭清远, 等. 基于 Hammerstein 结构的电子节气门动态非线性建模[J]. 北京航空航天大学学报, 2018, 44(12):2605–2612.
   YANG X, ZHANG Z, TAN Q, et al. Dynamic Nonlinear System Modeling of Electronic Throttle Body Based on Hammerstein Structure[J]. Journal of Beijing University of Aeronautics and Astronautics, 2018, 44(12):2605–2612.
6. 张明泽, 刘骥, 陈昕, 等. 基于 Wiener 模型的变压器油纸绝缘老化剩余寿命评估方法[J]. 电工技术学报, 2018, 33(21):5098–5108.
   ZHANG M, LIU J, CHEN X, et al. Residual Life Assessment Method of Transformer Oil-Paper Insulation Aging Based on Wiener Model[J]. Transaction of China Electrotechnical Society, 2018, 33(21):5098–5108.
7. 吴忠强, 尚梦瑶, 申丹丹, 等. 基于神经网络和 MS-AUKF 算法的蓄电池荷电状态估计 [J/OL]. 中国电机工程学报, 2019, 39(21):6336–6344.

WU Z, SHANG M, SHEN D, et al. SOC Estimation of Battery by MS-AUKF Algorithm and BPNN[J/OL]. Proceedings of the Chinese Society for Electrical Engineering, 2019, 39(21):6336–6344.

8. 胡正高, 赵国荣, 李飞, 等. 基于自适应未知输入观测器的非线性动态系统故障诊断[J]. 控制与决策, 2016, 31(05):901–906.
HU Z, ZHAO G, LI F, et al. Fault Diagnosis for Nonlinear Dynamical System Based on Adaptive Unknown Input Observer[J]. Control and Decision 2016, 31(05):901–906.
9. 郭业才, 王超. 基于支持向量机可控功率响应的 MUSIC-DOA 估计方法[J/OL]. 南京理工大学学报, 2019, 43(02):237–243.
GUO Y, WANG C. Research on MUSIC-DOA Estimation Method Based on SVM Steered Response Power[J/OL]. Journal of Nanjing University of Science and Technology, 2019, (02):237–243.
10. 南敬昌, 樊爽, 高明明. 基于负载牵引和记忆效应的X参数的功放建模[J]. 计算机应用, 2018, 38(10):2982–2989.
NAN J, PAN S, GAO M. Power Amplifier Modeling of X-Parameter Based on Load-Pulling and Memory Effect[J]. Journal of Computer Applications, 2018, 38(10):2982–2989.
11. 高翔, 冯正进. 电液伺服系统研究中的非线性分析方法[J]. 上海交通大学学报, 2002, (03):306–310.
GAO X, FENG Z. Nonlinear Analysis Method for Research of Electro-Hydraulic Servo System[J]. Journal of Shanghai Jiaotong University, 2002(03):306–310.
12. 罗骁, 张新燕, 张珺, 等. 基于谐波平衡法的尾流激励的叶片振动降阶模型方法[J]. 应用数学和力学, 2018, 39(08):892–899.
LUO X, ZHANG X, ZHANG J, et al. A Reduced-Order Model Method for Blade Vibration Due to Upstream Wake Based on the Harmonic Balance Method[J]. Applied Mathematics and Mechanics, 2018, 39(08):892–899.
13. 孙娜, 刘继文, 肖东亮. 基于 BFGS 拟牛顿法的压缩感知SL0重构算法[J]. 电子与信息学报, 2018, 40(10):2408–2414.
SUN N, LIU J, XIAO D. SL0 Reconstruction Algorithm for Compressive Sensing Based on BFGS Quasi Newton Method[J]. Journal of Electronics and Information Technology, 2018, 40(10):2408–2414.
14. 叶小莺, 万梅, 唐蓉, 等. 基于图聚类与蚁群算法的社交网络聚类算法[J/OL]. 计算机应用研究, 37:1–7.
YE X, WAN M, TANG R, et al. Clustering Algorithm of Social Network Based on Graph Clustering and Ant Colony Optimization Algorithm[J/OL]. Application Research of Computers: 1–7.
15. 简献忠, 吴明伟, 肖儿良, 等. 蜂群算法在太阳电池组件参数辨识中的应用[J]. 太阳能学报, 2019, 40(03):741–747.
JIAN X, WU M, XIAO E, et al. Application of Bee Colony Algorithm in Parameter Identification of Solar Cell Module[J]. Acta Energiae Solaris Sinica, 2019, 40(03):741–747.

CHAPTER 4

Overview of Power Amplifier Predistortion

4.1 PRINCIPLE AND CLASSIFICATION OF PREDISTORTION TECHNOLOGY

4.1.1 Principle of Predistortion Technology

Commonly used linearization technologies include power back-off, feedforward, negative feedback and predistortion. Out of these technologies, digital predistortion has been widely used owing to its applications in wideband communications, favorable stability, superior linearization performance and strong adaptive capability.

Predistortion [1] was initially used in analog systems. With the development of digital signal processing technology, digital predistortion emerged. It is currently the most widely used linearization method for RF PA. Its basic principle originates from the inverse function in mathematics. That is, a nonlinear unit complementary to the PA characteristic curve is placed at the front end of a nonlinear PA, which basically achieves the PA linearization.

Table 4.1 describes the advantages and disadvantages of various linearization technologies.

The basic principle of predistortion linearization is to utilize an auxiliary distortion unit to generate the auxiliary distortion signal containing

TABLE 4.1 Advantages and Disadvantages of Various Linearization Technologies

Linearization Technology	Advantage	Disadvantage
Power back-off	Simple implementation	Very poor efficiency
Feedforward	Very wide applicable bandwidth, favorable linearity, fast speed, basically not affected by memory effects	Complex structure, high cost, low efficiency
Negative feedback	High accuracy, low price	Limited applicable bandwidth, poor stability, narrow application range
EE&C	High efficiency, moderately wide bandwidth	Difficult time delay calibration
LINC	High efficiency, difficult accurate matching, suitable for narrowband communications	Quite sensitive to imbalance between signal amplitude and phase, resulting in adjacent-channel interference
Predistortion	Absence of the stability problem, moderately wide applicable bandwidth, moderately high accuracy	Slow adjustment

an intermodulation interference component, thereby suppressing the nonlinear distortion of PA. The schematic diagram of predistortion technology is outlined in Figure 4.1.

The predistorter is a nonlinear unit that is placed at the front end of a nonlinear PA. The predistorter's characteristic function $F(|x_i|)$ is set

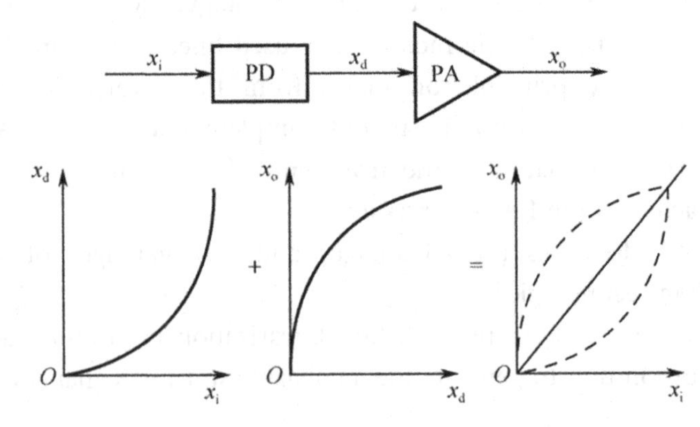

FIGURE 4.1 Schematic diagram of predistortion technology.

according to the characteristic function $H(|x_d|)$ of PA, so that the predistorter's characteristic curve and the PA's characteristic curve are complementary to each other. As a result, the whole system where the two parts are cascaded exhibits the linear characteristics, which can be expressed as

$$F(|x_i|) \cdot H(|x_d|) = G \tag{4.1}$$

where the system gain is expressed by constant G. After the input signal x_i passes through the cascaded system, it is linearly amplified as the output signal $x_o = Gx_i$.

In this way, predistortion technology implements PA linearization. However, the predistorter is not effective across the whole frequency band range. The predistorter can correct the input signals below the saturation level of PA. If the input signal of PA exceeds the saturation level, the predistorter cannot completely correct its nonlinearity.

4.1.2 Classification of Predistortion Technology

Regarding the signals processed and the relative placement of predistorter and modulator, common predistortion technologies are categorized into RF predistortion [2] and baseband predistortion [3]. Baseband predistortion has been widely used owing to its low operating frequency, strong adaptability and easy adoption of the rapidly developing digital signal processing techniques. However, memory effects of PA must be considered when the predistorter is applied, which makes the predistortion algorithm more difficult.

The typical structure of analog RF predistortion is shown in Figure 4.2. The upper branch exhibits linear characteristics while the lower branch exhibits compression characteristics. The predistortion signal can be obtained by subtracting the lower branch signal from the upper branch signal. All signal processing is accomplished at the RF end. The advantages of this method lie in its simple circuit structure, high power efficiency and low cost. However, the inverse characteristics of PA constructed by analog devices are relatively complex and difficult to control and adjust adaptively, resulting in a limited linearization effect. Figure 4.3 shows a typical structure of digital baseband predistortion. The signal output by the PA is converted into a baseband signal through

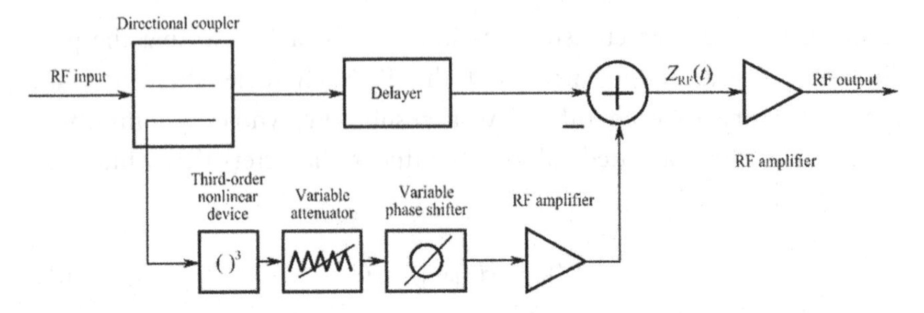

FIGURE 4.2 Typical structure of analog RF predistortion.

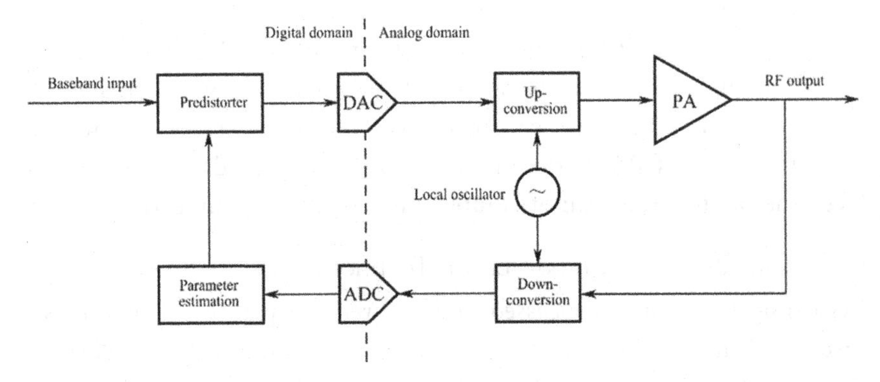

FIGURE 4.3 Typical structure of digital baseband predistortion.

down-conversion and analog-to-digital conversion, and then the predistortion parameters are estimated. Next, the baseband input signal is predistorted by using the estimated predistortion parameters. The signal after digital predistortion experiences digital-to-analog conversion and up-conversion, and then is transmitted to the RF PA. The equipment used for the digital baseband predistortion system is complex, but digital baseband predistortion processes baseband signals in the digital domain, so it has high flexibility and adaptive capability.

4.2 MAINSTREAM TECHNIQUES OF DIGITAL PREDISTORTION

4.2.1 LUT and Polynomial Predistortion

Digital baseband predistortion is implemented by means of either the LUT method or the polynomial method. The former uses one or more

LUTs to store the nonlinear compensation parameters of PA, thereby indicating the nonlinear relationship of the predistorter. The latter uses polynomials to simulate the characteristics of PA and enables the predistorter to have the inverse characteristics of PA also expressed by polynomials, thereby counteracting the PA nonlinearity.

The LUT method [4] is a type of predistortion technology based on data analysis. It accurately characterizes discontinuous models in the analog domain and is easy to implement in the baseband. However, it is quite difficult for the LUT predistorter to characterize the memory effects of PA. Therefore, the LUT method is currently only the mainstream scheme of memoryless digital predistortion. The basic principle of the LUT method is shown in Figure 4.4. During normal operation, the corresponding compensation amount is obtained from the LUT according to a certain parameter of the input signal used as the index address. The compensation amount is computed beforehand and stored in the table. In addition, the LUT is updated in real-time by the use of an adaptive algorithm to implement predistortion compensation for the input baseband signal, thereby achieving linearization.

Polynomial predistortion technology [5], by means of computations, uses a certain functional form to express the nonlinearity for each order of PA and uses the same functional form to express the predistorter, which aims to compensate the nonlinearity distortion for each order of PA. In this way, the overall linearization is achieved. The amplitude nonlinearity and phase nonlinearity of PA can be expressed by

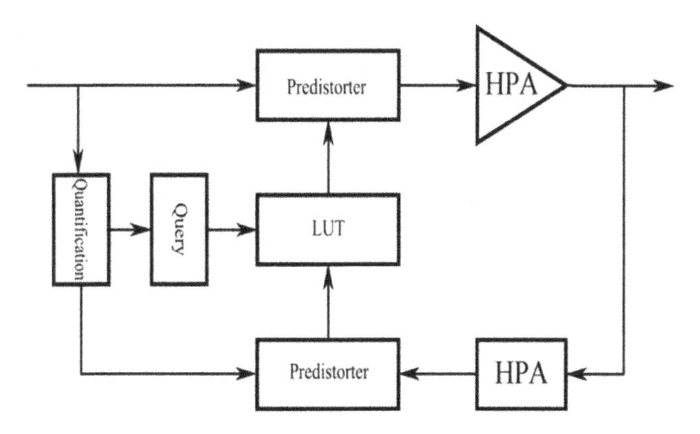

FIGURE 4.4 Basic principle of the LUT method.

polynomials respectively, or their common nonlinearity can be directly expressed by a polynomial with complex coefficients.

As the signal bandwidth and peak-to-average ratio increase, the nonlinear characteristics and memory effects of PA have become more significant. If the LUT method is used, the number of LUT entries required will substantially increase, occupying massive storage space. Moreover, the probability of peak value of a high peak-to-average power ratio (PAPR) signal is very small, and the LUT entry corresponding to the peak value is updated less often, resulting in the entry values unable to converge in a long period of time. Consequently, an unfavorable predistortion effect is caused, and furthermore, noise may be introduced to make distortion more significant. In contrast, polynomial predistortion technology can accurately simulate the inverse memory effects of PA by means of the constructed polynomial model, thereby obtaining a favorable linearization effect with a small amount of storage space. Therefore, polynomial predistortion has been more widely used in wideband communications.

4.2.2 Adaptive Learning Structure

In adaptive PA predistortion technology [6], the predistorter needs to enable the ability of adaptive adjustment, and its parameter estimation commonly utilizes a direct learning structure [7] and an indirect learning structure [8].

The block diagram of a direct learning structure is shown in Figure 4.5. The input signal $x(n)$ is converted into the desired response $d(n)$ of the system by gain/amplification. $d(n)$ is compared with the output signal $y(n)$ of PA to obtain the error $e(n)$, which is used as

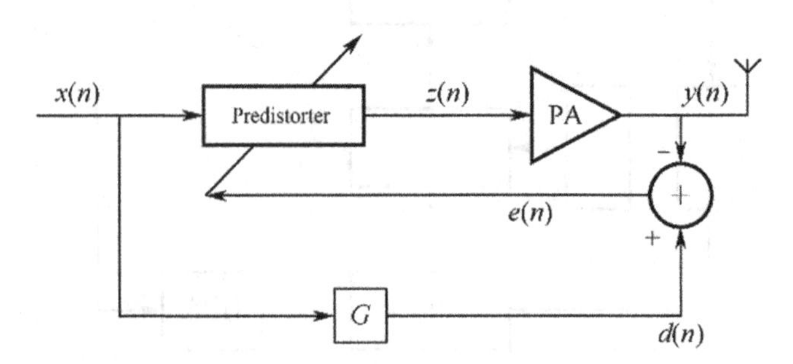

FIGURE 4.5 Block diagram of direct learning structure.

the coefficient for the adaptive adjustment predistorter. When $e(n) = 0$, the PA output is $y(n) = Gx(n)$. The direct learning structure requires the identification of PA models and cannot use or directly use the efficient least square method.

For linearization of the entire system, the preinverse can be cascaded at the front end of the PA system H, or the postinverse can be cascaded at its back end, which are called predistortion system and postdistortion system, respectively. Since the postdistortion works at the RF end behind the PA, it is not feasible in terms of complexity and flexibility of system design. However, in the Volterra structure system, the preinverse H_{pre}^{-1} and postinverse H_{post}^{-1} of system H are equivalent. In Figure 4.6(b), the postinverse H_{post}^{-1} linearizes the entire system S. If the preinverse H_{pre}^{-1} in Figure 4.6(a) is directly replaced by the postinverse H_{post}^{-1}, the entire system Q will also be linearized. Therefore, the coefficients of the preinverse H_{pre}^{-1} nonlinear function can be solved indirectly in the design of predistortion. The principle of indirect learning structure is to solve the coefficients of the postinverse H_{post}^{-1} nonlinear function of PA first, and then use it to directly replace the preinverse H_{pre}^{-1}, thereby achieving the linearization of system Q.

The block diagram of the indirect learning structure is shown in Figure 4.7, where the PA is followed by a training estimator. The training estimator and the predistorter adopt the same nonlinear behavioral model. The PA output signal $y(n)$ is input into the training estimator after a G-fold attenuation. Under ideal conditions, $y(n)/G = x(n)$ is desired, which requires that the error $e(n) = z(n) - \hat{z}(n)$ between the

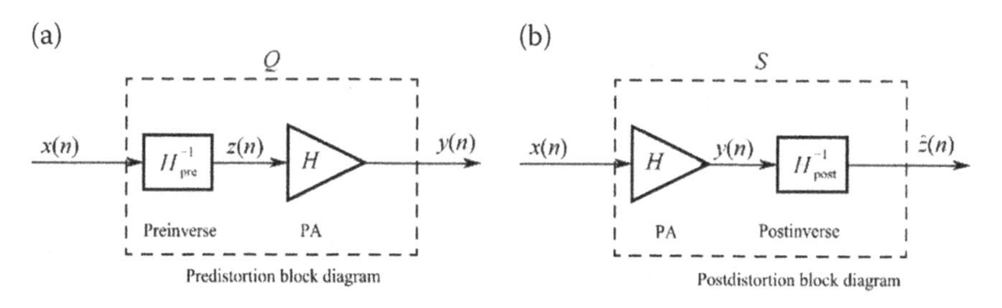

FIGURE 4.6 Block diagram of predistortion and postdistortion.

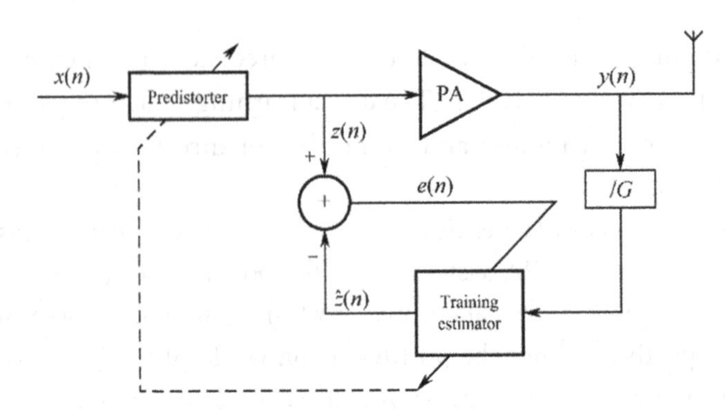

FIGURE 4.7 Block diagram of indirect learning structure.

predistorted signal $z(n)$ and the output signal $\hat{z}(n)$ of the training estimator to be close to 0. When the algorithm converges, the coefficients of the training estimator are duplicated and passed to the predistorter, thereby achieving the PA linearization. Compared with the direct learning structure, the indirect learning structure does not require the identification of PA models and can directly use the advanced adaptive algorithm. Therefore, the indirect learning structure has been widely used.

REFERENCES

1. 吴林煌, 苏凯雄, 王琳, 等. 正交匹配追踪和BIC准则的自适应双频段预失真模型优化算法[J]. 电子学报, 2018, 46(09):2149–2156.
   WU L, SU K, WANG L, et al. Adaptive Dual-band Predistortion Model Optimization Algorithm Based on Orthogonal Matching Pursuit and Bayesian Information Criterion[J]. Acta Electronica Sinica, 2018, 46(09):2149–2156.
2. 邓海林, 张德伟, 白维达,等. 一种行波管用毫米波射频预失真线性化器[J]. 电子学报, 2017, 45(07):1784–1791.
   DENG H, ZHANG D, BAI W, et al. A Millimeter-Wave RF Predistortion Linearizer for Travelling Wave Tube Amplifier[J]. Acta Electronica Sinica, 2017, 45(07):1784–1791.
3. 胡欣, 王刚, 王自成, 等. 射频预失真器与基带预失真算法结合对行波管功率放大器线性化改善的影响[J]. 通信学报, 2012, 33(07):158–163.
   HU X, WANG G, WANG Z, et al. Effect on the Linearity Improvement of TWTA by Combining RF Predistortion Linearizer and Digital Predistortion Algorithm[J]. Journal on Communications, 2012, 33(07): 158–163.

4. 郑帅. 基于新型查找表的数字预失真技术的设计与实现[D]. 武汉; 华中科技大学, 2016.
ZHENG S. Design and Realization of Digital Predistortion Based on Novel LUT[D]. Wuhan: Huazhong University of Science and Technology, 2016.
5. 马晓波, 杨韶, 孙慧萍. 基于多项式数字基带自适应预失真技术的研究[J]. 山西大同大学学报(自然科学版), 2014, 30(04):22–24,27.
MA X, YANG S, SUN H. Study on the Polynomial Digital Baseband Adaptive Predistortion[J]. Journal of Shanxi Datong University, 2014, 30(04):22–24,27.
6. 詹鹏, 秦开宇, 蔡顺燕.新的射频功放预失真线性化方法[J]. 电子科技大学学报, 2011, 40(05):676–681.
ZHAN P, QIN K, CAI S. New Predistortion Method for RF Power Amplifier Linearization[J]. Journal of University of Electronic Science and Technology of China, 2011, 40(05):676–681.
7. 张月, 黄永辉. 一种基于直接学习结构的数字预失真方法[J]. 电子设计工程, 2018, 26(11):91–94+99.
ZHANG Y, HUANG Y. A Digital Predistortion Method Using the Direct Learning Architecture[J]. Electronic Design Engineering, 2018, 26(11):91–94+99.
8. 沈忠良, 张子平. 基于间接学习结构的改进功放非线性失真补偿算法[J]. 通信技术, 2016, 49(10):1320–1325.
SHEN Z, ZHANG Z. Modified Distortion Compensation Algorithm for Nonlinear Power Amplifier Based on Indirect-Learning Architecture[J]. Communications Technology, 2016, 49(10):1320–1325.

CHAPTER 5

Volterra Series Modeling for Power Amplifier

BEHAVIORAL MODELS FOR narrowband systems are based on amplitude/amplitude and amplitude/phase memoryless polynomial models, which are not accurate enough for wideband communication systems, especially for systems having modulated signal inputs with complex envelopes. The Volterra series model (Volterra model) can accurately represent the RF/microwave PAs in the context of wideband and long-term memory effects, but this requires sufficient nonlinear kernels. At present, works of literature have reported various behavioral models with memory effects[1–3], such as memory polynomial model, parallel Hammerstein model and parallel Wiener model. These models can favorably characterize PAs with memory effects, although all of them are expanded or simplified forms of the Volterra series behavioral model. Therefore, the analysis of the Volterra series model is of significance to the analysis of other behavioral models with memory effects.

Some simplified Volterra models basically include memory polynomial (MP) model, orthogonal matching pursuit (OMP)[4], envelope memory polynomial (EMP) model[5], general memory polynomial (GMP) model[6], pruned Volterra model[7] and dynamic deviation reduction (DDR) Volterra model[8]. These models remove redundant items in the Volterra model and retain modeling-related items, thus reducing the complexity of identification. There are also some two-box models, such as the Wiener model, the Hammerstein model [9], the expanded Hammerstein model [10] and twin nonlinear two-box (TNTB) model[11]. There is still a three-box model[12] as a combination of the Wiener model and the Hammerstein model.

5.1 ANALYSIS AND BUILDUP OF EXPANDED VOLTERRA MODEL FOR NONLINEAR POWER AMPLIFIER WITH MEMORY EFFECTS

The general Volterra series model is limited to weak nonlinear amplifiers due to its high computational complexity. To reduce the computational complexity of and the number of kernel coefficients for the Volterra series, two schemes are proposed, which are the Volterra series based on Chebyshev orthogonal polynomial function and Laguerre orthogonal polynomial function, respectively. This section starts from the analysis of the mathematical expressions of the general Volterra series model with different input signals (continuous or discrete signals, carrier and complex envelope signals), further theoretically explains how to change the structure of the general Volterra series model by means of special functions (Laguerre and Chebyshev orthogonal functions) and obtains the mathematical expression and simplified model structure based on Laguerre and Chebyshev orthogonal functions, respectively. In addition, this section describes the simulation of the Volterra–Laguerre model, discusses the number of parameters to be extracted for computation and compares the Volterra–Laguerre model with the Volterra model. Mathematical analysis and simulation result demonstrate that Volterra–Chebyshev and Volterra–Laguerre behavioral models have simplified structures and fewer coefficients, compared with the ordinary Volterra series model.

5.1.1. Volterra–Chebyshev Model Derivation and Analysis

The general Volterra series is often used for nonlinear systems of models. The Volterra series expression in continuous form is

$$y(t) = \sum_{n=0}^{\infty} y_n [x(t)] \tag{5.1}$$

where,

$$y_n[x(t)] = \int_{-\infty}^{\infty} h_n(\tau_1, \tau_2, \cdots, \tau_n) \cdot x(t-\tau_1)x(t-\tau_2)\cdots x(t-\tau_n)d\tau_1 d\tau_2 \cdots d\tau_n \quad (5.2)$$

where $x(t)$ is the system input, $y(t)$ is the system response, $y_n[x(t)]$ is the nth-order component of the system response and the multidimensional function $h_n(\tau_1, \tau_2, \cdots, \tau_n)$ is called the nth-order kernel or the nth-order nonlinear impulse response.

The Volterra series in discrete form can be expressed as

$$y(l) = \sum_{n=0}^{\infty} y_n[x(l)] \quad (5.3)$$

where,

$$y_n[x(l)] = \sum_{m_1=-\infty}^{\infty} \cdots \sum_{m_n=-\infty}^{\infty} h_n(m_1, m_2, \cdots, m_n)x(l-m_1)x(l-m_2)\cdots x(l-m_n) \quad (5.4)$$

$x(l)$ and $y(l)$ are the input and output signals in the discrete case, respectively.

In practical applications, the Volterra series usually uses finite terms and finite-length memory and can also achieve sufficient accuracy. The odd-order discrete Volterra series model is expressed as

$$\begin{aligned} y(l) &= \sum_{m=0}^{M-1} h_1(m)x(l-m) + \sum_{m_1=0}^{M-1}\sum_{m_2=0}^{M-1}\sum_{m_3=0}^{M-1} h_3(m_1, m_2, m_3)x(l-m_1) \\ &\quad x(l-m_2)x(l-m_3)+\cdots \\ &= \sum_{k=0}^{K}\sum_{m_1=0}^{M-1}\cdots\sum_{m_{2k+1}=0}^{M-1} h_{2k+1}(m_1, \cdots, m_{2k+1}) \prod_{j=1}^{2k+1} x(l-m_j) \end{aligned} \quad (5.5)$$

The impact of even-order kernels in the band-limited modulation system is negligible. Herein, K and M denote the truncation order and memory depth of the nonlinearity system, respectively.

A nonlinear system based on Volterra series model usually uses the complex envelope signal (baseband signal) as the input signal. Assuming

$x(t) = Re[\tilde{x}(t) \cdot e^{j\omega_0 t}]$ and $y(t) = Re[\tilde{y}(t) \cdot e^{j\omega_0 t}]$ are, respectively, the input and output signals of PA, where ω_0 denotes the carrier angular frequency and $\tilde{x}(t)$ and $\tilde{y}(t)$, respectively, denote the complex envelopes of the input and output signals. Then, the discrete finite-memory odd-order complex baseband Volterra model is expressed as follows:

$$\begin{aligned}
\tilde{y}(l) &= \sum_{m=0}^{M-1} h_1(m)\tilde{x}(l-m) + \sum_{m_1=0}^{M-1}\sum_{m_2=0}^{M-1}\sum_{m_3=0}^{M-1} h_3(m_1, m_2, m_3)\tilde{x}(l-m_1) \\
&\quad \tilde{x}(l-m_2)\tilde{x}^*(l-m_3)+\cdots \\
&= \sum_{k=0}^{K}\sum_{m_1=0}^{M-1}\cdots\sum_{m_{k+1}=m_k}^{M-1}\sum_{m_{k+2}=0}^{M-1}\cdots\sum_{m_{2k+1}=m_{2k}}^{M-1} h_{2k+1}(m_1,\cdots,m_{2k+1}) \\
&\quad \cdot \prod_{j=1}^{k+1}\tilde{x}(l-m_j)\prod_{j=k+2}^{2k+1}\tilde{x}^*(l-m_j)
\end{aligned} \tag{5.6}$$

To reduce the number of kernel coefficients for the Volterra series, a frequency-domain Volterra kernel approximation based on the Chebyshev orthogonal polynomial can be adopted.

To derive the Volterra–Chebyshev model, the multidimensional time-domain signal of each component in the complex envelope $\tilde{y}(t)$ of the amplifier's output signal is transformed to the frequency domain[13]:

$$\begin{aligned}
\tilde{Y}_{2k+1}(\omega_1,\cdots,\omega_{2k+1}) &= \int_0^{\infty}\cdots\int_0^{\infty}\tilde{y}_{2k+1}(t_1,\cdots,t_{2k+1}) \\
&\quad \cdot exp\left(-j\sum_{i=1}^{2k+1}\omega_i t_i\right)\mathrm{d}t_1\cdots\mathrm{d}t_{2k+1}
\end{aligned} \tag{5.7}$$

where the multidimensional time-domain signal $\tilde{y}_{2k+1}(t_1,\cdots,t_{2k+1})$ is

$$\begin{aligned}
\tilde{y}_{2k+1}(t_1,\cdots,t_{2k+1}) &= \int_0^{\infty}\cdots\int_0^{\infty}\tilde{h}_{2k+1}(\tau_1,\cdots,\tau_{2k+1}) \\
&\quad \cdot\prod_{i=1}^{k+1}\tilde{x}(t-\tau_i)\prod_{i=k+2}^{2k+1}\tilde{x}^*(t-\tau_i)\mathrm{d}\tau_1\cdots\mathrm{d}\tau_{2k+1}
\end{aligned} \tag{5.8}$$

Fourier transform of the combination of (5.7) and (5.8) yields

$$\widetilde{Y}_{2k+1}(\omega_1, \cdots, \omega_{2k+1}) = \widetilde{H}_{2k+1}(\omega_1, \cdots, \omega_{2k+1}) \cdot \prod_{i=1}^{k+1} \widetilde{X}(\omega_i) \prod_{i=k+2}^{2k+1} \widetilde{X}^*(-\omega_i) \tag{5.9}$$

where $\widetilde{H}_{2k+1}$ is the multidimensional Fourier transform of the baseband kernel $\widetilde{h}_{2k+1}$ and $\widetilde{X}$ is the Fourier transform of the time-domain baseband signal $\widetilde{x}$. The frequency-domain kernel $\widetilde{H}_{2k+1}(\omega_1, \cdots, \omega_{2k+1})$ can be approximated as a multidimensional polynomial series in the $\pm B$ input signal bandwidth, which is

$$\hat{H}_{2k+1}(\omega_1, \cdots, \omega_{2k+1}) = \sum_{m_1=0}^{M_{2k+1}} \cdots \sum_{m_{2k+1}=0}^{M_{2k+1}} c_{2k+1}(m_1, \cdots, m_{2k+1}) \cdot T_{m_1}(\omega_1) \cdots T_{m_{2k+1}}(\omega_{2k+1}) \tag{5.10}$$

$\hat{H}_{2k+1}$ has a complete set of real orthogonal polynomials T_i with the quantity specified by $M_{2k+1} + 1$, where $0 \le i \le M_{2k+1}$. Converting the approximated signal $\hat{Y}$ from the frequency domain to the time domain yields

$$\hat{y}_{2k+1}(t_1, \cdots, t_{2k+1}) = F^{-1}\{\hat{Y}_{2k+1}(\omega_1, \cdots \omega_{2k+1})\} \tag{5.11}$$

where the operator F^{-1} denotes the inverse Fourier transform. The frequency-domain signal $\hat{Y}$ in (5.11) can be obtained from (5.9), and the frequency-domain kernel $\widetilde{H}$ is replaced by $\hat{H}$ in (5.10). In combination with (5.9) and (5.10), (5.11) derives the multidimensional time-domain signal; with the interval of $2k + 1$-dimensional time-domain functions in (5.8) being ignored, the $2k + 1$-order approximate baseband output signal is obtained:

$$\hat{y}_{2k+1}(t) = \sum_{m_1=0}^{M_{2k+1}} \cdots \sum_{m_{2k+1}=0}^{M_{2k+1}} c_{2k+1}(m_1, \cdots m_{2k+1}) \cdot w_{m_1}(t) \cdots w_{m_{k+1}}(t) u_{m_{k+2}}(t) \cdots u_{m_{2k+1}}(t) \tag{5.12}$$

Superimposing all the terms of (5.12) yields the complete output signal

$$\hat{y}(t) = \sum_{k=1}^{K} \hat{y}_{2k+1}(t) \tag{5.13}$$

In the case of $0 \le i \le M_{2k+1}$, $w_i(t)$ and $u_i(t)$ in (5.12) are as follows:

$$w_i(t) = F^{-1}\{T_i(\omega)\widetilde{X}(\omega)\} \tag{5.14}$$

$$u_i(t) = F^{-1}\{T_i(\omega)\widetilde{X}^*(-\omega)\} \tag{5.15}$$

According to the recurrence relationship $T_0(\omega) = 1$, $T_1(\omega) = \omega/B$ and $T_{n+1}(\omega) = (2\omega/B)T_n(\omega) - T_{n-1}(\omega)$ of Chebyshev polynomials, the complete form of Chebyshev polynomials is expressed as

$$T_i(\omega) = \sum_{m=0}^{[i/2]} \sum_{k=0}^{m} (-1)^k \binom{i}{2m}\binom{m}{k} \frac{\omega^{i-2k}}{B^{i-2k}} \tag{5.16}$$

According to (5.16), the time-domain signals of (5.14) and (5.15) can be expressed as

$$w_i(t) = \sum_{m=0}^{[i/2]} \sum_{k=0}^{m} (-1)^k \binom{i}{2m}\binom{m}{k} \frac{\widetilde{x}^{(i-2k)}(t)}{(jB)^{i-2k}} \tag{5.17}$$

$$u_i(t) = \sum_{m=0}^{[i/2]} \sum_{k=0}^{m} (-1)^k \binom{i}{2m}\binom{m}{k} \frac{\widetilde{x}^{*(i-2k)}(t)}{(jB)^{i-2k}} \tag{5.18}$$

where $\widetilde{x}$ is the input signal and $\widetilde{x}^*$ is the conjugate of the input signal. Parenthesized in the upper corner of the time-domain signal $(i - 2k)$ is the derivative order. The signal $u_i(t)$ can be simply transformed by the signal $w_i(t)$ into the following:

$$u_i(t) = g_i(w_i(t)) = (-1)^i w_i^*(t) \tag{5.19}$$

After the Chebyshev polynomial transformation for the frequency-domain kernel is complete, the nonlinear model structure of PA is eventually simplified, with the number of orthogonal polynomials reasonably selected and the number of coefficients reduced. In this way, the computational complexity is reduced as expected.

5.1.2 Volterra–Laguerre Model Analysis and Derivation

In the identification of linear systems, a Laguerre orthogonal basis function can be used to reduce the number of parameters required for model construction. In the model, the basis function for the finite impulse response filter uses the Laguerre complex orthogonal function $\{\varphi_p(m)\}$. p is the order of the orthogonal function $\{\varphi_p(m)\}$ and m is the discrete point. A z transformation applied to $\{\varphi_p(m)\}$ yields the discrete Laguerre function $L_p(z, \lambda)$ in the z domain[14]

$$L_p(z, \lambda) = \frac{\sqrt{1 - |\lambda|^2}}{1 - z^{-1}\lambda}\left\{\frac{-\lambda^{\mathrm{H}} + z^{-1}}{1 - z^{-1}\lambda}\right\}^p, \quad p \geq 0 \tag{5.20}$$

where λ is the pole of the Laguerre function and $|\lambda| < 1$. $(\cdot)^{\mathrm{H}}$ denotes the conjugate transpose. The linear behavioral model based on the Laguerre function is expressed as

$$Y(z) = \sum_{p=0}^{L-1} b_p L_p(z, \lambda) \cdot X(z) \tag{5.21}$$

where b_p is the pth-order regression coefficient and $L_p(z, \lambda)$ is the pth-order discrete Laguerre function, which is given by (5.20). L is the number of Laguerre orthogonal functions. The transfer function of the system is

$$H(z) = \sum_{p=0}^{L-1} b_p L_p(z, \lambda)$$

The linear Laguerre model is as follows:

$$Y(z) = \sum_{p=0}^{L-1} b_p q_p(z) \tag{5.22}$$

where $q_p(z)$ is defined as

$$q_0(z) = L_0(z, \lambda)x(z) \tag{5.23}$$

$$q_p(z) = B(z, \lambda)q_{p-1}(z), \quad p = 1, \cdots, L-1 \tag{5.24}$$

where

$$L_0(z, \lambda) = \frac{\sqrt{1 - |\lambda|^2}}{1 - z^{-1}\lambda}, \quad B(z, \lambda) = \frac{-\lambda^H + z^{-1}}{1 - z^{-1}\lambda} \tag{5.25}$$

The Laguerre linear behavioral model is constructed from discrete filters. The first part of the equation represents a first-order low-pass filter $L_0(z, \lambda)$, followed by L-1 all-pass $B(z, \lambda)$. See Figure 5.1.

Since the Laguerre-based model has favorable low-frequency characteristics, it can be used for PAs with long-term memory effects. To utilize the Laguerre orthogonal polynomial function, assume that the Volterra kernels $h_{2k+1}(m_1, \cdots m_{2k+1})$ in (5.6) have a decaying memory. That is, they are absolutely additive in the system memory $[0, M]$ and

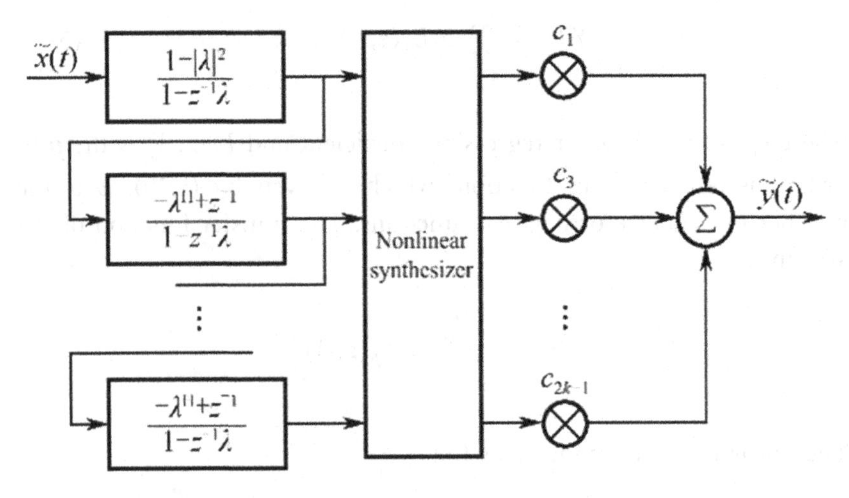

FIGURE 5.1 Volterra–Laguerre PA model structure.

they can be approximated by the complete basis $\{\varphi_p(m)\}$ of the Laguerre function defined in $[0, L]$:

$$h_1(m) = \sum_{p=0}^{L-1} c_1(p)\varphi_p(m) \tag{5.26}$$

$$h_3(m_1, m_2, m_3) = \sum_{p_1=0}^{L-1}\sum_{p_2=p_1}^{L-1}\sum_{p_3=0}^{L-1} c_3(p_1, p_2, p_3)\cdot\varphi_{p_1}(m_1)\varphi_{p_2}(m_2)\varphi^*_{p_3}(m_3) \tag{5.27}$$

In combination of (5.26) and (5.27), the Volterra–Laguerre model can be expressed as

$$\tilde{y}(l) = \sum_{p=0}^{L-1} c_1(p)q_p(l) + \sum_{p_1=0}^{L-1}\sum_{p_2=p_1}^{L-1}\sum_{p_3=0}^{L-1} c_3(p_1, p_2, p_3)q_{p_1}(l)q_{p_2}(l)q^*_{p_3}(l) + \cdots \tag{5.28}$$

where,

$$q_p(l) = \sum_{m=0}^{M-1} \varphi_p(m)\tilde{x}(l-m) \tag{5.29}$$

$c_{2k+1}(p_1, p_2, \cdots p_{2k+1})$ is the kernel expansion factor.

A z transformation of (5.29) yields the function expressed in the z domain. Obviously, the accuracy of the model mainly depends on the number of basis functions L. The nonlinear Laguerre model can be easily implemented by superimposing all the weighting terms together with the use of a nonlinear synthesizer. See Figure 5.1.

Once the model structure is determined, parameters λ and L can be specified based on the actual situation. At the same time, coefficients $c_{2k+1}(p_1, p_2, \cdots, p_{2k+1})$ can be extracted from measured input and output data.

5.1.3 Model Simulation Experiment

A simulation experiment is performed for the Volterra–Laguerre model and general Volterra model, for the purpose of performance validation. The input and output data used in the simulation are obtained from the

PA beta designed by Freescale semiconductor transistor MRF21030. The input and output data of PA are extracted from ADS. The detailed design parameters of PA are listed in Literature [15]. The amplifier operates in a nonlinear state; that is, the output power reaches a peak (power of the input signal being 28 dBm herein). The input signal adopts a WCDMA baseband signal with a bandwidth of approximately 5 MHz. It comes with a certain PAPR and a nonconstant envelope and can exhibit the memory effects of PA. The Laguerre orthogonal function adopts $\lambda = 0.25$ and $L = 3$ (number of orthogonal functions). The Volterra model adopts the nonlinear order $K = 2$ and memory depths $M = 3$ and $M = 4$. The simulation results are indicated in Figure 5.2 and Table 5.1.

The mathematical expression of the Volterra series model for RF PAs has been obtained from the previous analysis. The two forms of the Volterra series expansion model, namely Volterra–Laguerre model and the Volterra–Chebyshev model, are also discussed. In the Volterra–Laguerre model, the Volterra time-domain kernel is obtained by a direct expansion of the complete set of the Laguerre function.

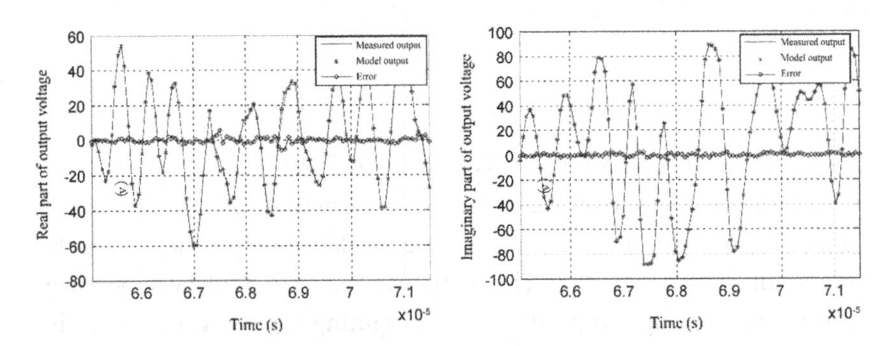

FIGURE 5.2 Waveform comparison of measured voltage and model output voltage for Volterra–Laguerre model with $K = 2$ and $L = 3$.

TABLE 5.1 Comparison of Volterra Model and Volterra-Laguerre Model

Volterra Model ($K = 2$)				Volterra–Laguerre Model ($K = 2$)	
$M = 3$		$M = 4$		$L = 3$	
RMS Error	Number of Coefficients	RMS Error	Number of Coefficients	RMS Error	Number of Coefficients
1.9842	81	1.845	244	1.6350	81

In the Volterra–Chebyshev model, the Volterra frequency-domain kernel is realized by the complete set of Chebyshev real orthogonal polynomials.

The above derivation, analysis and simulation results demonstrate a conclusion that the improved Volterra series model has a simplified structure and extracts much fewer coefficients, compared with the general Volterra series model. For example, in the case of a fifth-order Volterra series, the Volterra–Laguerre model with $L = 3$ needs to extract only 81 parameters. However, the typical Volterra model requires 244 to 605 parameters for achieving the same accuracy. The computation indicates that the Volterra–Chebyshev model with M_1, M_3, $M_5 = 2$ needs to extract 273 parameters or less. The two models approximate the Volterra series kernel with two orthogonal polynomial functions in the frequency domain and time domain respectively, by means of different derivation methods or orthogonal polynomials. Consequently, the obtained model performances are slightly different from each other. Regarding the principle and performance, the Volterra–Laguerre model demonstrates significant advantages in its simpler principle, easier implementation and fewer coefficients required.

Theoretical analysis and derivation of the two Volterra series expansion behavioral models for PA and the simulation of the Volterra–Laguerre model show that the Volterra–Laguerre model can efficiently and accurately regenerate the nonlinear distortion of PA, including memory effects. This model can be used to process the wideband complex modulation signals. The Volterra–Chebyshev model is an approximation to the Volterra kernel in the frequency domain based on multivariate orthogonal polynomials. This model can generate a time-domain model. If the frequency-domain order of the kernel is low, the number of extracted parameters will be greatly reduced. Both models deliver a simplified structure and extract fewer coefficients, thus reducing the complexity, improving the processing speed of system simulation and producing more benefits for system simulation.

5.2 PGSC MODELING AND DIGITAL PREDISTORTION OF WIDEBAND POWER AMPLIFIER

This section describes a new PA modeling and digital predistortion method, which is Parallel GMP-SCT-CIMT (PGSC), to accommodate the strong memory effects of wideband PAs. This method uses three basis functions to construct the PA behavioral model and digital predistorter. The basis functions are GMP, specific cross term (SCT) and cross items between memory times (CIMT). A testing platform is set up to validate the model accuracy and linearization effect of the new method. The PGSC model is used for PA modeling and digital predistortion (DPD), and the test results are compared with the Parallel MP-EMP-CIMT (PMEC) model and GMP model. The LDMOS Doherty PA with a 16QAM input signal is used in the test. The test results demonstrate that the PGSC method outperforms the PMEC method and GMP method, regarding the modeling accuracy and linearization effect. Compared with the PMEC method, the PGSC method improves the normalized mean square error (NMSE) by 2.1 dB in modeling and reduces the third-order ACPR of the output signal by 4.94/2.03 dB in DPD. Compared with the GMP method, the PGSC method obtains higher model accuracy and better linearization effect by using only 73% coefficients.

5.2.1 Novel PGSC Behavioral Model Analysis

Discrete-time finite-memory complex baseband Volterra series can be used to describe the nonlinearity of PA in a wireless communication system. Literature [16] gives its mathematical expression:

$$y(n) = \sum_{\substack{k=1 \\ k-odd}}^{K} \sum_{m_k=0}^{M_k} h_k(m_k)x(n-m_1) \prod_{m=1}^{(k-1)/2} x(n-m_{2m})x^*(n-m_{2m+1}) \tag{5.30}$$

where $x(n)$ and $y(n)$ are the system input and output, respectively; K is the nonlinear order; M_k is the memory depth; $h_k(q_k)$ is the kth-order Volterra kernel; consists of the integer-valued delays; and m_k for all $k = 1, 3, \ldots, K$. A large number of coefficients make the Volterra series model difficult to be applied. Therefore, a simplified Volterra series model is desirable in practical DPD.

Firstly, consider the case that only the signals all at the same sampling timepoint are retained. Assuming $m_1 = m_2 = \ldots = m_k = m$, (5.30) is simplified as a two-dimensional MP model:

$$y_{2D}(n) = \sum_{\substack{k=1 \\ k-odd}}^{K} \sum_{m=0}^{M_k} h_{km}x(n-m)|x(n-m)|^{k-1} \tag{5.31}$$

Then, relax the restriction condition. Consider another case that just one-time delay differs from the others for the input signal x. The special time delay is m and the other time delays are $m - l$ ($l \geq 1$). In this situation, (5.31) is simplified as the first 3-D array:

$$y_{3D,1}(n) = \sum_{\substack{k=1 \\ k-odd}}^{K} \sum_{m=0}^{M_k} \sum_{l=1}^{L} h_{kml}^{1} x(n - m)|x(n - m + l)|^{k-1} \tag{5.32}$$

If the other time delays for the input signal x are all $m + l$ ($l \geq 1$), then (5.31) is simplified as the second 3-D array:

$$y_{3D,2}(n) = \sum_{\substack{k=1 \\ k-odd}}^{K} \sum_{m=0}^{M_k} \sum_{l=1}^{L} h_{kml}^{2} x(n - m)|x(n - m - l)|^{k-1} \tag{5.33}$$

(5.32) together with (5.33) and (5.34) lead to the GMP model:

$$\begin{aligned} y_{GMP}(n) = & \sum_{\substack{k=1 \\ k-odd}}^{K_a} \sum_{m=0}^{M_a} a_{km} x(n - m)|x(n - m)|^{k-1} \\ & + \sum_{\substack{k=3 \\ k-odd}}^{K_b} \sum_{m=0}^{M_b} \sum_{l=1}^{L_b} b_{kml} x(n - m)|x(n - m + l)|^{k-1} \\ & + \sum_{\substack{k-3 \\ k=odd}}^{K_c} \sum_{m=0}^{M_c} \sum_{l=1}^{L_c} c_{kml} x(n - m)|x(n - m - l)|^{k-1} \end{aligned} \tag{5.34}$$

where K_a, K_b and K_c are the nonlinear orders of GMP model; M_a, M_b and M_c are the memory depths of GMP model; L_b and L_c are the cross terms indexes.

If the input signal corresponding to the special time delay m is x^* in (5.30) and the other time delays are all $m - l$ ($l \geq 1$), then (5.30) is simplified as the third 3-D array:

$$y_{3D,3}(n) = \sum_{\substack{k=3 \\ k-odd}}^{K} \sum_{m=0}^{M_k} \sum_{l=1}^{L} h_{kml}^{3} x^*(n - m)x^2(n - m + l)|x(n - m + l)|^{k-3} \tag{5.35}$$

If the other time delays for the input signal x^* are all $m + l$ ($l \geq 1$), then (5.31) is simplified as the fourth 3-D array:

$$y_{3D,4}(n) = \sum_{\substack{k=3 \\ k-odd}}^{K} \sum_{m=0}^{M_k} \sum_{l=1}^{L} h_{kml}^{4} x^*(n-m) x^2(n-m-l) |x(n-m-l)|^{k-3} \tag{5.36}$$

If this procedure continues, more coefficients will be generated, resulting in a complicated model structure. For flexible use in practice, the derivation procedure is stopped after the above four 3-D arrays are obtained. For a wideband system, the influence of CIMT cannot be ignored, and its mathematical expression is

$$y_{CIMT}(n) = \sum_{p=1}^{M} \sum_{\substack{m=1 \\ m \neq p}}^{M} \sum_{\substack{r=1 \\ r-odd}}^{N} c_{pmr} x(n-m) |x(n-m)|^{r-1} \tag{5.37}$$

where M and N are the cross memory depth and the cross-term order of the CIMT model, respectively; c_{pmr} is the coefficient of the CIMT model. The augmented cross-term order of the CIMT model leads to a fast increase in the number of model coefficients. Since the high-order nonlinear terms between signals at memory times produce little influence on the system, only the third-order intermodulation between signals at memory times is considered, which produces great influence on the system. In this situation, (5.37) can be simplified as (assuming $r = 3$):

$$y_{CIMT}^{*}(n) = \sum_{p=1}^{M} \sum_{\substack{m=1 \\ m \neq p}}^{M} c_{pm} x(n-p) |x(n-m)|^2 \tag{5.38}$$

Equation (5.31) together with (5.32), (5.33), (5.35), (5.36) and (5.38) lead to a new PGSC model. To reduce the complexity of the new model, assume $K = 3$ in (5.35) and (5.36). The mathematical expression of the PGSC model is

$$
\begin{aligned}
y_{\text{PGSC}}(n) = & \sum_{\substack{k=1 \\ k-odd}}^{K_a} \sum_{m=0}^{M_a} a_{km} x(n-m) \left| x(n-m) \right|^{k-1} \\
& + \sum_{\substack{k=3 \\ k-odd}}^{K_b} \sum_{m=0}^{M_b} \sum_{l=1}^{L_b} b_{kml} x(n-m) \left| x(n-m+l) \right|^{k-1} \\
& + \sum_{\substack{k=3 \\ k-odd}}^{K_c} \sum_{m=0}^{M_c} \sum_{l=1}^{L_c} c_{kml} x(n-m) \left| x(n-m-l) \right|^{k-1} \\
& + \sum_{m=0}^{M_d} \sum_{l=1}^{L_d} d_{ml} x^*(n-m) x^2(n-m+l) \\
& + \sum_{m=0}^{M_e} \sum_{l=1}^{L_e} e_{ml} x^*(n-m) x^2(n-m-l) \\
& + \sum_{p=1}^{M_f} \sum_{\substack{m=1 \\ m \neq p}}^{M_f} f_{pm} x(n-p) \left| x(n-m) \right|^2
\end{aligned}
\tag{5.39}
$$

where K_a, K_b and K_c are the nonlinear orders of (5.31) to (5.33), respectively; M_a, M_b, M_c, M_d and M_e are the memory depths of (5.31) to (5.33) and (5.35) to (5.36), respectively; M_f is the cross memory depth of the CIMT submodel; L_b, L_c, L_d and L_e are the respective cross-term indexes; a_{km}, b_{kml}, c_{kml}, d_{ml}, e_{ml} and f_{pm} are the coefficients of PGSC model. Compared with the GMP model, the PGSC model adds five new variables, but the expansion terms greatly improve the model accuracy. In addition, the model complexity can be reduced by choosing appropriate model coefficients.

5.2.2 PGSC Model Identification

First assume

$$
\boldsymbol{Y} = \boldsymbol{X} \cdot \boldsymbol{A} \tag{5.40}
$$

where $\boldsymbol{Y}$ is the output vector of the PGSC model, $\boldsymbol{X}$ is a matrix constructed from six polynomial basis functions and $\boldsymbol{A}$ is a vector consisting of coefficients of six submodels. $\boldsymbol{X}$ is defined as:

$$
\boldsymbol{X} = [\boldsymbol{X}_{\text{2D}}, \boldsymbol{X}_{\text{3D},1}, \boldsymbol{X}_{\text{3D},2}, \boldsymbol{X}_{\text{3D},3}, \boldsymbol{X}_{\text{3D},4}, \boldsymbol{X}_{\text{CIMT}}] \tag{5.41}
$$

where X_{2D}, $X_{3D,1}$, $X_{3D,2}$, $X_{3D,3}$, $X_{3D,4}$, X_{CIMT} are sub-matrices built from six basis functions, respectively. When fitted with the least mean square (LMS) error method, A becomes:

$$A = (X^H \cdot X)^{-1} \cdot X^H \cdot Y \tag{5.42}$$

5.2.3 Test Result

PA modeling and DPD are performed using the MP model to compare the performances among the three models (PGSC, PMEC and GMP). The vector signal generator (N5182A), spectrum analyzer (PXA N9030A), high-power LDMOS Doherty PA, attenuator and computer are used to set up a testing platform, as shown in Figure 5.3. The center frequency of the PA is 3.45 GHz, and a 16QAM signal with a bandwidth of 15 MHz is used as the input signal. The computer is used to download the signal to the N5182A to generate the RF signal driving the PA, and the PA output is attenuated and then transmitted to the PXA N9030A. VSA89600 software is used to collect input and output data of the PA, which are transmitted to MATLAB® for PA modeling and DPD. A total of 10000 sets of input and output data are collected. The former 5000 sets are used for PGSC model identification of the PA using the LMS error method, and the latter 5000 sets are used to validate the model accuracy.

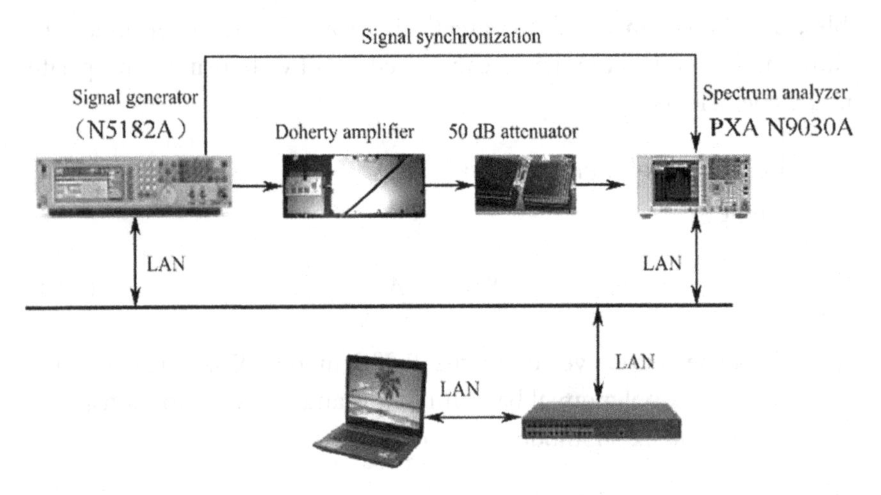

FIGURE 5.3 Testing platform.

In addition, PA modeling and DPD are also performed using the PMEC and GMP models for a comparison of the performances among the three models (PGSC, PMEC and GMP).

The accuracy of the model is evaluated by using NMSE in the time domain and exploit power spectral density (PSD) in the frequency domain. To reduce the complexity of models, the cross-term indexes, L_b, L_c, L_d, L_e and the memory depths M_d, M_e of special cross-terms in the M_d model are all set to 1 and the cross memory depth M_f in the CIMT submodel is set to 3. Selection of the values of K_a, K_b, K_c, M_a, M_b, M_c directly affects the accuracy of the model and sweeping these six parameters simultaneously results in a large number of unreasonable combinations which increase the computational load. Therefore, the parameters K_a and M_a of the MP submodel are estimated at the beginning. The MP model is built by using the collected data with the nonlinear order swept from 1 to 15 and the memory depth swept from 0 to 5. Figure 5.4 shows the NMSE values of the MP model with different values of K_a and M_a. Figure 5.4 indicates that both the accuracy of the model and the number of coefficients increase as the nonlinear order and memory depth increase. When $(K_a, M_a) = (15,5)$, it yields the best NMSE (−34.06) but increases the number of coefficients to 48. Considering model accuracy and complexity, we choose $(K_a, M_a) = (9,3)$ as the parameter for the MP submodel. Table 5.2 lists the respective NMSE values of the models PMEC, GMP and PGSC with different values of K_b, K_c, M_b and M_c. Table 5.2 indicates that PGSC model is superior to PMEC

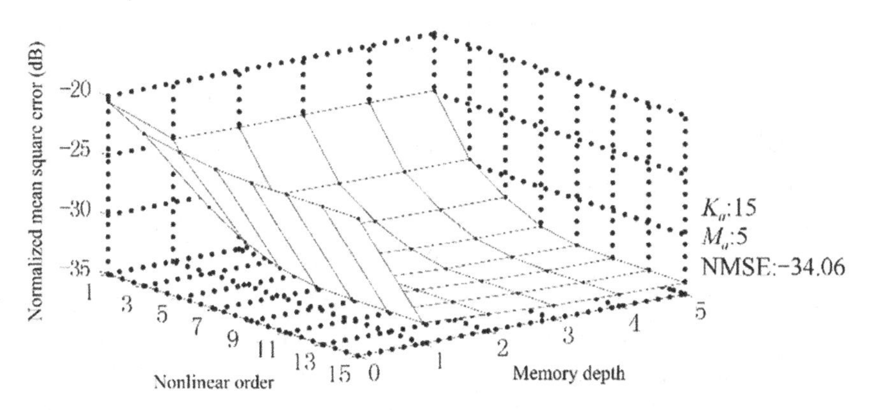

FIGURE 5.4 NMSE values of the MP model with different parameters.

TABLE 5.2 NMSE Comparison of Three Models

K_b/K_c	M_b/M_c	PMEC Model		GMP Model		PGSC Model	
		Number of Coefficients	**NMSE**	**Number of Coefficients**	**NMSE**	**Number of Coefficients**	**NMSE**
3	0	/	/	22	–34.50	32	–35.02
3	1	27	–33.26	24	–35.21	34	–35.62
3	2	28	–33.64	26	–35.82	36	–35.98
3	3	29	–33.93	28	–35.96	38	–36.01
5	0	/	/	24	–34.52	34	–36.11
5	1	28	–33.51	28	–35.26	**38**	**–37.02**
5	2	30	–33.90	32	–35.92	42	–37.22
5	3	32	–34.02	36	–36.03	46	–37.51
7	0	/	/	26	–34.55	36	–36.40
7	1	29	–33.96	32	–35.34	42	–37.25
7	2	32	–34.05	38	–35.95	48	–37.49
7	3	35	–34.38	44	–36.09	54	–37.58
9	0	/	/	28	–34.61	38	–36.54
9	1	30	–34.01	36	–35.42	46	–37.42
9	2	34	–34.65	44	–36.01	54	–36.71
9	3	**38**	**–34.92**	**52**	**–36.12**	62	–37.97

model with its NMSE 2.1 dB than that of the latter when 38 coefficients are applied. The NMSE of the GMP model is –36.12 when the number of coefficients hits its maximum (52), inferior to that (–37.02) of the PGSC model which requires only 73% coefficients.

For a more intuitive understanding of the model performance, Figure 5.5 compares the best NMSEs of the three models using different numbers of coefficients. The comparison indicates that the accuracy of the PGSC model in the time domain is far higher that of the other two models. The bold data in Table 5.2 are used as the respective model parameters to obtain the EPSD, as shown in Figure 5.6. Curve ① denotes the PA output spectrum, and curves ②, ③ and ④ denote the error power spectral density curves of models PMEC, GMP and PGSC, respectively. The comparison reveals that the performance of the PGSC model in the frequency domain is also better than that of the PMEC and GMP models.

The comparisons of the three models in terms of NMSE, EPSD and number of coefficients show that the PGSC model outperforms the other

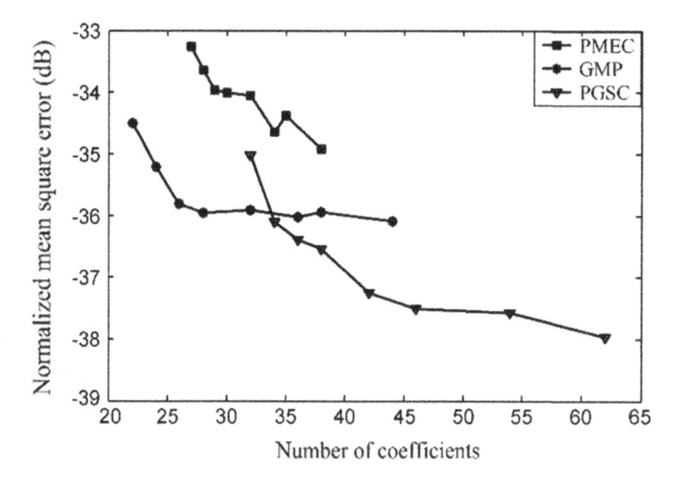

FIGURE 5.5 Comparison of best NMSEs of three models with different number of coefficients.

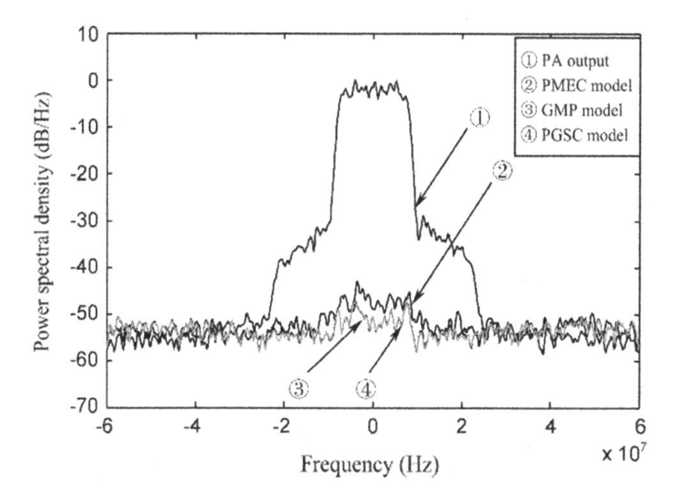

FIGURE 5.6 Error power spectral density curves.

two as it achieves the maximum accuracy in the time and frequency domains using the least number of coefficients.

The DPD performances of the three models (PGSC, PMEC and GMP) in the DPD context are also investigated. The bold model parameters in Table 5.2 are used to configure the predistorters. Figure 5.7 shows the power spectral densities before and after DPD in the case that the 16QAM signal is used. Curve ① denotes the original signal. Curve ②

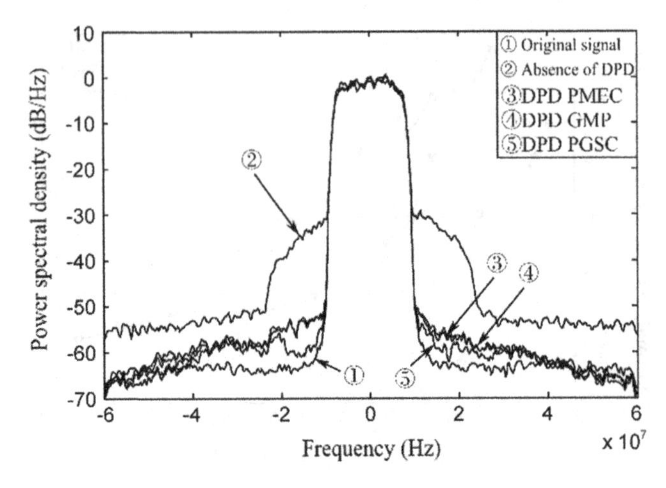

FIGURE 5.7 Power spectral density curves before and after DPD.

denotes the power spectral density of PA at an output level without DPD. Curves ③, ④ and ⑤ denote the power spectral densities at the output of PA with DPD for the PMEC, GMP and PGSC models, respectively. Table 5.3 lists the number of required coefficients and ACPR values when DPD is performed using the three models. For the same number of coefficients used in DPD, an improvement of 4.94/2.03 dB in the ACPR value is observed when the PGSC model is used, compared with the PMEC model. Compared with the GMP model, the PGSC model has an ACPR improvement by 4.05/0.93 dB when DPD is performed, with 27% less coefficients. The test results show that the PGSC model is much more superior to the PMEC and GMP models.

Figure 5.8 shows the AM/AM curves before and after DPD is performed for the PA using the PGSC model as a predistorter. Curve ① is the AM/AM curve of the PA, curve ② is the AM/AM curve of the

TABLE 5.3 Comparison of the Number of Coefficients and ACPRs in DPD

DPD Mode	Upper Band ACPR (dBc)	Lower Band ACPR (dBc)	Number of Coefficients
Absence of DPD	−33.42	−35.27	/
DPD PMEC	−53.26	−55.17	38
DPD GMP	−54.15	−56.27	52
DPD PGSC	−58.20	−57.20	38

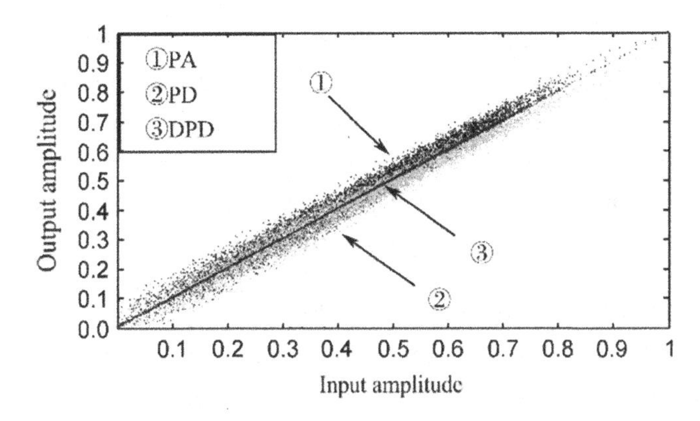

FIGURE 5.8 AM/AM curves before and after DPD.

predistorter and curve ③ is the AM/AM curve after DPD. The AM/AM curve after DPD is significantly improved and basically appears as a straight line, which meets the system requirements.

The test results on the wideband RF PA modeling and DPD for the new PGSC model indicate that DPD significantly improves the linearity of PA while improving the accuracy of the model. Because the Doherty amplifier's behavioral model is different from that of the traditional AB-type PA, we can build a more appropriate behavioral model based on the features of the Doherty amplifier to achieve higher performance. More work can be done such as working out the FPGA implementation scheme for the DPD system and constructing the predistorter model by means of the PGSC model. In addition, the adaptive predistortion algorithm needs to be optimized to improve its accuracy and convergence rate, which helps to reduce the cost on circuit implementation in the future.

5.3 LMEC RESEARCH AND PREDISTORTION APPLICATION

This section proposes a model with high accuracy and low complexity, which is LUT-MP-EMP-CIMT (LMEC), to accommodate to the strong nonlinearity of RF PAs in modern wireless communication systems. This model consists of the LUT, MP, EMP and CIMT aligned in parallel, which more accurately describes the strong memory effects of PA. Compared with the Parallel Twin Nonlinear Two-box (PTNTB) model, the LMEC model additionally uses the EMP and CIMT submodels. The test results demonstrate that the LMEC model produces better model accuracy and linearization effect. Compared with the MP model, the LMEC model improves the model accuracy by 2.9 dB and reduces the ACPR by about 5 dB. Compared with the PTNTB model, the LMEC model improves the model accuracy by 1.1 dB and reduces the ACPR by about 3 dB. Compared with the GMP model, the LMEC model delivers a similar performance with 48% less coefficients.

5.3.1 LMEC Behavioral Model Description

The PTNTB model is a two-box model, which consists of a memoryless nonlinear function (implemented by a LUT or polynomial function) and a low-order MP aligned in parallel. The outputs of the two submodels are combined as the output of the PTNTB model. Its structure is outlined in Figure 5.9. The two submodels of PTNTB model are both nonlinear functions, without an assumption to separate nonlinear behaviors from linear behaviors. Strong static nonlinearities are processed first, followed by the processing of moderate dynamic nonlinearities. In the PTNTB model, the MP submodel is expressed as

$$y_{\mathrm{MP}}(n) = \sum_{j=1}^{M_1} \sum_{i=1}^{N_1} a_{ji} x(n-j) |x(n-j)|^{i-1} \tag{5.43}$$

where x and y_{MP} are the input signal and output signal of the MP submodel, respectively; M_1 and N_1 are the memory depth and nonlinear order of the MP submodel, respectively; a_{ji} is the coefficient of the MP submodel. Since the LUT characterizes static nonlinearities in the PTNTB model, the MP submodel does not include memoryless terms; j starts from 1.

The LMEC model is an extension of the PTNTB model. Figure 5.10 outlines the structure of LMEC model. Basically, the superiority of LMEC model over the PTNTB model lies in the addition of EMP sub-model and CIMT submodel[17]. That is, the influence of historical

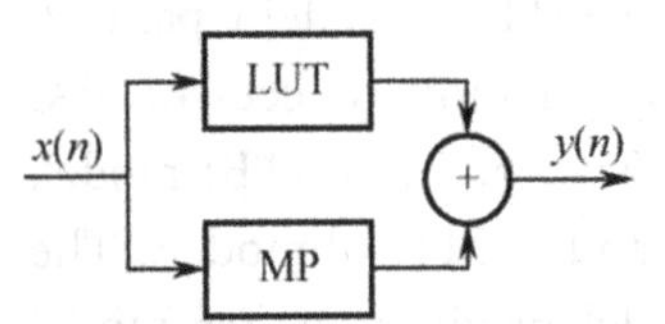

FIGURE 5.9 PTNTB model structure.

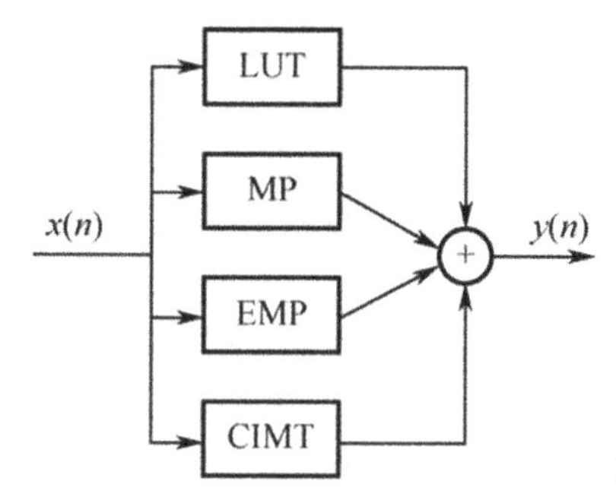

FIGURE 5.10 LMEC model structure.

envelopes on the current input signal and the influence of historical cross terms on the PA system are added respectively, thereby enhancing the model accuracy. In addition, the LMEC model increases the total number of coefficients in a moderate manner. The increase of (memory depth and nonlinear order) parameters is controlled by a reasonable selection of sizes for EMP and CIMT submodels, so that the coefficients do not increase significantly. The mathematical expression of the EMP submodel is

$$y_{\mathrm{EMP}}(n) = \sum_{k=1}^{M_2} \sum_{l=2}^{N_2} b_{kl} x(n) |x(n-k)|^{l-1} \tag{5.44}$$

where x and y_{EMP} are the input signal and output signal of the EMP submodel, respectively; M_2 and N_2 are the memory depth and nonlinear order of the EMP submodel, respectively; b_{kl} is the coefficient of the EMP submodel. Since the LUT characterizes static nonlinearities in the LMEC model, the EMP submodel does not include memoryless terms; k starts from 1 and l starts from 2.

For a wideband system, the influence of memory cross terms on the model cannot be ignored. Its expression is

$$y_{\mathrm{CIMT}}(n) = \sum_{p=1}^{M_3} \sum_{\substack{q=1 \\ p \neq q}}^{M_3} \sum_{r=1}^{N_3} c_{pqr} x(n-p) |x(n-q)|^{r-1} \tag{5.45}$$

where x and y_{CIMT} are the input signal and output signal of the CIMT submodel, respectively; M_3 and N_3 are the memory depth and nonlinear order of the CIMT submodel, respectively; c_{pqr} is the coefficient of the EMP submodel.

For the CIMT submodel, only the third-order intermodulation between signals at memory times is considered due to its significant influence, since a higher nonlinear order selected requires more coefficients to be identified and the high-order nonlinear terms between signals at memory times produce minor influence on the system. Therefore, the CIMT submodel is rewritten as

$$y_{\mathrm{CIMT}}(n) = \sum_{p=1}^{M_3} \sum_{\substack{q=1 \\ p \neq q}}^{M_3} c_{pq} x(n-p)|x(n-q)|^2 \tag{5.46}$$

The LMEC model enables a flexible characterization of memory effects by using three functions based on the MP model. If the MP-based model of each branch chooses the same nonlinear order, the model will become oversized, increasing the complexity of model computation. To address this drawback, the LMEC model introduces the output of a nonlinear LUT and the output based on the MP model in parallel. In addition, the LUT characterizes the high-order static nonlinear behavior of PA, and the MP-based submodels utilize low-order nonlinearity with their sizes respectively controlled, thereby generating the sum of all the numbers of coefficients.

The relationship between output waveform and input waveform of the GMP model is as follows:

$$\begin{aligned} y_{\mathrm{GMP}}(n) = & \sum_{m=0}^{M_a} \sum_{k=1}^{N_a} a_{mk} x(n-m)|x(n-m)|^{k-1} \\ & + \sum_{m=0}^{M_b} \sum_{k=2}^{N_b} \sum_{l=1}^{l_b} b_{mkl} x(n-m)|x(n-m-l)|^{k-1} \\ & + \sum_{m=0}^{M_c} \sum_{k=2}^{N_c} \sum_{l=1}^{l_c} c_{mkl} x(n-m)|x(n-m+l)|^{k-1} \end{aligned} \tag{5.47}$$

where x and y_{GMP} are the input and output signals of the GMP model, respectively; M_a, N_a and a_{mk} are the memory depth, nonlinear order and coefficient of the aligned signal envelope terms, respectively; M_b, N_b, l_b

and b_{mkl} are the memory depth, nonlinear order, cross term index and coefficient of the lagging cross terms, respectively; M_c, N_c, l_c and c_{mkl} are the memory depth, nonlinear order, cross terms index and coefficient of the leading cross terms, respectively.

It can be considered that the LMEC model uses a special number of delayed cross terms to determine the sizes of EMP submodel and CIMT submodel. These cross terms introduced are special cases of cross terms in the GMP model. In addition, an effective selection of special cross terms reduces the model complexity and requires less coefficients for the LMEC model to achieve the same performance as the GMP model.

5.3.2 Model Identification

LMEC model identification is divided into two steps. Firstly, the input and output waveforms are used to identify the first static nonlinear function of the DUT, by means of polynomial fitting or smoothing algorithm. Then, the measurement data are embedded to generate the input and output of the remaining three models while the submodel coefficients are identified.

The simulation identification of MP, EMP and CIMT models is a traditional linear identification problem:

$$\boldsymbol{Y} = \boldsymbol{\phi} \cdot \boldsymbol{A} \tag{5.48}$$

where $\boldsymbol{Y}$ is the output vector of three dynamic nonlinear polynomial submodels (MP, EMP and CIMT); $\boldsymbol{\phi}$ is a matrix composed of the basis functions of the three polynomials and their input signals; $\boldsymbol{A}$ is a vector containing the coefficients of the three submodels.

Matrix $\boldsymbol{\phi}$ is defined as

$$\boldsymbol{\phi} = [\phi_{\mathrm{MP}} \quad \phi_{\mathrm{EMP}} \quad \phi_{\mathrm{CIMT}}] \tag{5.49}$$

where $\boldsymbol{\phi}_{\mathrm{MP}}$, $\boldsymbol{\phi}_{\mathrm{EMP}}$ and $\boldsymbol{\phi}_{\mathrm{CIMT}}$ are the submatrices composed of the MP, EMP and CIMT basis functions, respectively.

By use of least square (LS) fitting, the coefficients of the three submodels are

$$\boldsymbol{A} = (\boldsymbol{\phi}^{\mathrm{H}} \cdot \boldsymbol{\phi})^{-1} \cdot \boldsymbol{\phi}^{\mathrm{H}} \cdot \boldsymbol{Y} \tag{5.50}$$

Determination of sizes for the three MP-based submodels should weigh the overall performance of the enhanced model and select appropriate model coefficients. Therefore, selection of the LMEC model size is divided into three steps. Firstly, determine the MP model size, and the LMEC model is considered as a PTNTB model in this step; secondly, determine the EMP model size; finally, determine the memory depth of CIMT model. A general sweep method is applied to determine the sizes of the three submodels, and an NMSE indicator is used to measure the accuracy of each submodel. The selected submodel sizes should require a minor number of coefficients and lead to small NMSE values.

$$\mathrm{NMSE_{dB}} = 10\log_{10}\left[\frac{\sum_{n=1}^{N}|y_{\mathrm{means}}(n) - y_{\mathrm{est}}(n)|^2}{\sum_{n=1}^{N}|y_{\mathrm{means}}(n)|^2}\right] \tag{5.51}$$

where y_{means} is the real PA output waveform, y_{est} is the output waveform of the model and N is the number of samples of the output waveforms.

One thousand (1000) pairs of data points are selected to solve the model coefficients, and 2000 pairs of data points are selected to test the performance of the behavioral model. The test signal is a 16QAM signal with a chip rate of 15 Mcps and a signal bandwidth of 15 MHz. An LDMOS Doherty PA is driven with a gain of 50 dB and the center frequency of 1.96 GHz. The input and output signals of PA are collected for identification of behavioral models. The finally selected memory depth and nonlinear order of MP submodel are 5 and 5, respectively; the memory depth and nonlinear order of EMP submodel are 4 and 3, respectively; the memory depth of CIMT submodel is 3. Therefore, the LMEC model has a total of 51 coefficients and the NMSW value of the model is -37.1 dB.

5.3.3 Model Performance Evaluation

This subsection compares the proposed LMEC model with the MP, PTNTB and GMP models to validate the accuracy of the proposed LMEC model.

A sweep method is applied to determine the size of each model. Figure 5.11 shows the NMSE values of the MP, PTNTB, GMP and LMEC models with varying total coefficients. As indicated in the figure, the PTNTB model delivers higher accuracy and requires far less model coefficients, compared with the MP model. More importantly,

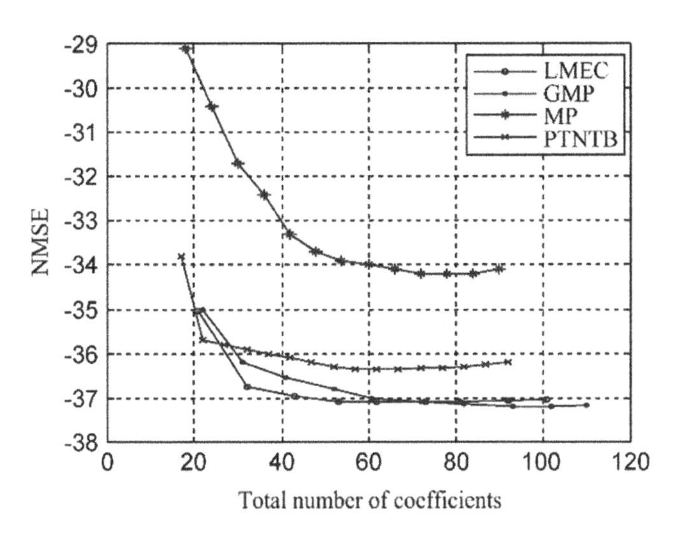

FIGURE 5.11 Comparison of NMSE performance among models.

the NMSE value of the PTNTB model remains less than or equal to −36.4 dB with the increase of the nonlinear order, while the NMSE value of the proposed LMEC model reaches −37.1 dB. A comparison of the NMSE performance between LMEC model and GMP model indicates that they deliver similar model accuracy but the LMEC model requires less coefficients. In a word, the proposed LMEC model introduces the EMP and CIMT submodels based on the PTNTB model, further enhancing the model accuracy compared with the MP and PTNTB models and reducing the model complexity compared with the GMP model.

Figure 5.12 shows the spectrum comparison between the LMEC model and the PTNTB model. Curve ① is the real output power spectrum of PA, curve ② is the output power spectrum computed by the LMEC model and curve ③ is the output power spectrum computed by the PTNTB model. As indicated by the simulation result, the output power spectrum computed by the LMEC model gets more approximate to the real power spectrum of PA and the LMEC model delivers a better performance, compared with the PTNTB model.

Table 5.4 gives a comparison of LMEC, MP, PTNTB and GMP models regarding the model size, total number of coefficients and NMSE performance. Compared with the MP model, the LMEC model improves the

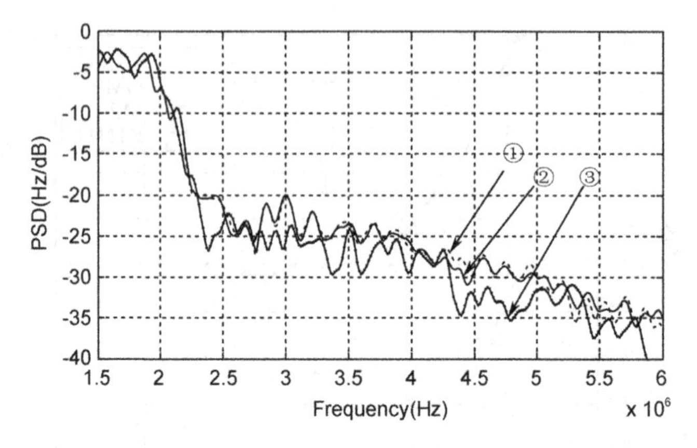

FIGURE 5.12 Spectral comparison between LMEC model and PTNTB model.

TABLE 5.4 Performance Comparison of MP, PTNTB, GMP and LMEC Models

Model	Model Size	NMSE Performance(dB)	Total Number of Coefficients
MP	$(M, N)(5,12)$	−34.2	72
PTNTB	$(M, N)(5,5)$	−36.0	42
GMP	$(M_a, N_a)(5,12)$	−37.4	102
	$(M_b, N_b, l_b)(4,4,1)$		
	$(M_c, N_c, l_c)(4,4,1)$		
LMEC	$(M_1, N_1)(5,5)$	−37.1	53
	$(M_2, N_2)(5,3)$		
	$M_3 = 3$		

NMSE performance by 2.9 dB with 26.38% less coefficients. Compared with the GMP model, the LMEC model achieves similar accuracy (only 0.3 dB lower) with 48.04% less coefficients. Compared with the PTNTB model, the LMEC model has an increase of NMSE value by 1.1 dB but requires more coefficients. Anyway, the total number of coefficients remains in a reasonable range, and the LMEC model achieves higher accuracy at a cost of acceptable increased complexity.

5.3.4 Predistortion Application

This subsection describes the predistortion application of LMEC, MP, PTNTB and GMP models. Figure 5.13 shows the spectral comparison of

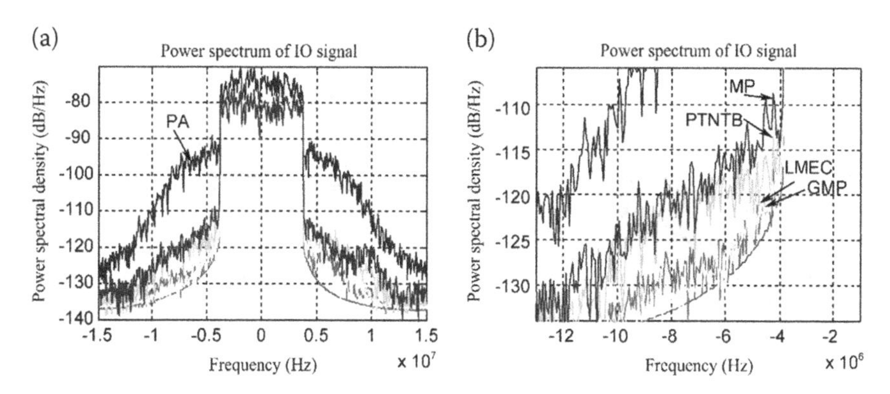

FIGURE 5.13 Spectral comparison of models after predistortion.

different predistortion functions (LMEC, MP, GMP and PTNTB). As indicated in the figure, the four predistortion functions all effectively reduce the spectrum regrowth, compared with the output spectrum of the signal that only passes through a PA. The PTNTB predistortion model has approximately a 5-dB improvement in the ACPR compared with the MP model. The proposed LMEC model further improves the ACPR by approximately 3 dB on the basis of the PTNTB model. The spectral curves of GMP and LMEC predistortion models almost coincide with each other at frequencies near the carrier frequency. At frequencies far away from the carrier frequency, the GMP model delivers a slightly better ACPR performance although its model complexity is higher, compared with the LMEC model.

Figure 5.14 compares the AM/AM and AM/PM characteristics between PTNTB predistortion and LMEC predistortion. Both models improve the linearization after predistortion. In contrast, the AM/AM and AM/PM curves of LMEC predistortion appear more like straight lines, which means a better linearization effect of predistortion.

This subsection proposes an LMEC model with high accuracy and low complexity. The model consists of the LUT (characterizing static nonlinearities) and MP, EMP, CIMT submodels (characterizing dynamic nonlinearities) aligned in parallel. The test results demonstrate that the LMEC model performs very well in PA modeling and digital predistortion application. Compared with the MP model, the LMEC model delivers a better performance with 26.38% less coefficients. Compared with the GMP model, the LMEC model achieves similar accuracy with

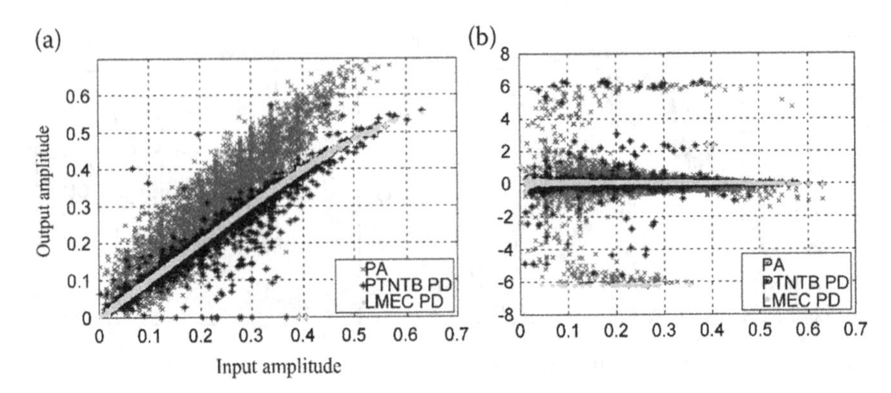

FIGURE 5.14 AM/AM and AM/PM characteristics of LMEC and PTNTB predistortion.

48.04% less coefficients. Compared with the PTNTB model, the LMEC model obtains higher accuracy using more coefficients. In addition, the PTNTB model cannot obtain the same accuracy of LMEC model simply by increasing the coefficients.

5.4 IMPROVED DYNAMIC MEMORY POLYNOMIAL MODEL OF POWER AMPLIFIER AND PREDISTORTION APPLICATION

This subsection proposes an improved multi-slice combined behavioral model - improved generalized-dynamic memory polynomial (G-DMP) model, accommodating to the strong nonlinearity and strong memory effects of RF PA in wireless communication systems. Based on the dynamic memory polynomial model, the G-DMP model introduces the cross-terms of the input signal envelope at memory times to the current input signal and the cross-terms of the current input signal envelope to the input signal at memory times, thereby supporting more flexible modeling for the strong memory effects of PA. In addition, the improved Recursive Least Squares-Least Means Square (RLS_LMS) joint algorithm is used to simulate the adaptive predistortion system for the G-DMP model. As indicated by the results, the G-DMP model requires 14.29% less coefficients and has an ACPR improvement by 4 dB when its accuracy is 0.5 dB higher compared with the DMP model. Therefore, the G-DMP model can deliver higher modeling accuracy and a better linearization effect.

5.4.1 Improved Multi-Slice Combined Behavioral Model of Power Amplifier

The multi-slice combined behavioral model has a relatively novel PA model structure. This modeling method is able to reduce the model coefficients and complexity while constructing a PA behavioral model with strong nonlinearity and strong memory effects in a modern wireless communication system. In addition, this modeling method considers the interaction between input signals at different memory times when the nonlinear characteristics of strong memory effects are modeled, so that memory effects can be flexibly characterized and model accuracy can be improved.

The DMP model introduces the dynamic nonlinear order to ensure model accuracy with fewer model coefficients. However, the DMP model does not consider the influence of coupling between input signals at memory times on model accuracy; the DMP model only simulates part of memory effects and therefore is not suitable for a PA model with strong memory effects in a wideband system. Moreover, a predistorter using the DMP model cannot appropriately compensate for the nonlinear distortion of a PA with strong memory effects.

The G-DMP model is an extension of the DMP model. Figure 5.15 outlines the structure of G-DMP model. Based on the traditional DMP model, the G-DMP model introduces the cross-terms of the input signal envelope at memory times to the current input signal and the cross-terms of the current input signal envelope to the input signal at memory times. That is, the G-DMP model adopts a special form of cross-terms of memory times to enhance the model accuracy, while the total number of

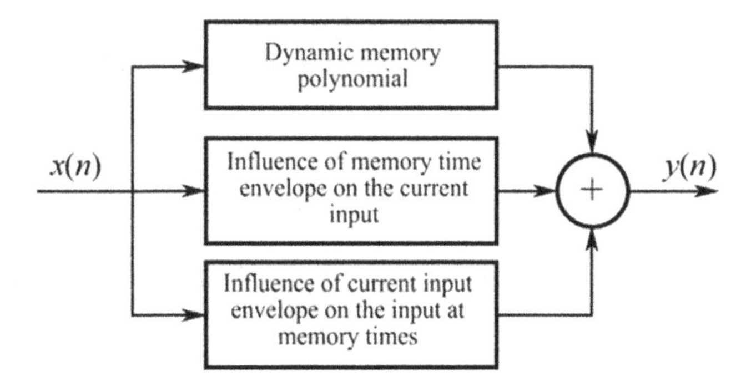

FIGURE 5.15 G-DMP model structure.

coefficients is appropriately controlled by a selection of the memory depth and nonlinear order of cross terms. The G-DMP model is more suitable for PAs with strong memory effects.

The G-DMP expression is as follows:

$$\begin{aligned} y(n) = & \sum_{m=0}^{M} \sum_{k=1}^{K_m} a_{qk} x(n-m) |x(n-m)|^{k-1} \\ & + \sum_{j=1}^{M_1} \sum_{i=2}^{N_1} b_{ji} x(n) |x(n-j)|^{i-1} \\ & + \sum_{s=1}^{M_2} \sum_{t=2}^{N_2} c_{st} x(n-s) |x(n)|^{t-1} \end{aligned} \tag{5.52}$$

where x and y are the input and output signals of the G-DMP model, respectively; M_1, N_1 and b_{ji} are the memory depth, nonlinear order and coefficient of the product term of the envelope power series of the memory input and the current input, respectively; M_2, N_2 and c_{st} are the memory depth, nonlinear order and coefficient of the product term of the envelope power series of the current input and the memory input, respectively.

It can be seen from the above that the G-DMP model is able to characterize memory effects more flexibly through a parallel connection of three polynomial functions based on the MP model. The three branch submodels adopt low-order nonlinearity and control their respective sizes to constitute a reasonable number of total coefficients, thereby reducing the complexity of model computation.

5.4.2 Power Amplifier Model Evaluation and Validation

The process of behavioral model identification is a process of identifying the optimal parameters of a behavioral model by use of the input and output data of the real PA on the basis of a selected "optimal" model structure. The model accuracy is determined according to some identification criteria so that the modeling results approximate the real PA results.

Coefficients of the multislice combined model meet the linear weighting relationship, and the ordinary least square (LS) method is commonly used to identify the model coefficients [18]. The NMSE indicator is used to measure the accuracy of the constructed behavioral model.

The test signal is a 16QAM signal with a chip rate of 15 Mcps and a signal bandwidth of 15 MHz. The input and output data of PA are collected for identification of the behavioral model. One thousand (1000) sets of data samples are selected for identification of model coefficients, and 2000 sets of data samples for validation of model performance.

The G-DMP and DMP models are compared in the aspects of model parameters, total coefficients and NMSE to validate the model accuracy. As indicated by Table 5.5, the G-DMP model requires 14.29% less coefficients when its accuracy is 0.5 dB higher, further improving the accuracy and complexity compared with the DMP model.

Figure 5.16 shows a comparison of the output power spectrum computed by G-DMP and DMP models. As indicated by the figure, both models appropriately simulate the spectrum characteristics of the real

TABLE 5.5 Performance Comparison of G-DMP and DMP Models

Model	Model Parameter	NMSE (dB)	Total Number of Coefficients
DMP	$M=5,\ K_m=6$	−35.2769	21
G-DMP	$M=4,\ K_m=5,\ M_1=2$ $N_1=2,\ M_2=1,\ N_2=2$	−35.7743	18

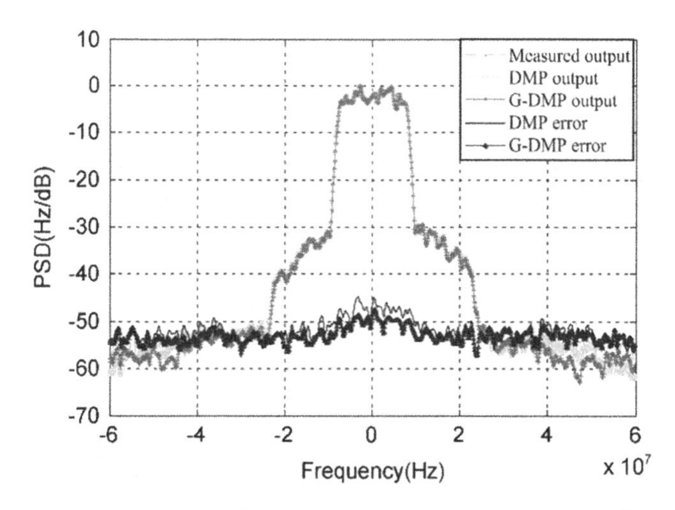

FIGURE 5.16 Comparison of output power spectrum computed by G-DMP and DMP models.

PA. Compared with the DMP model, the G-DMP model delivers a smaller error, higher accuracy and better performance.

5.4.3 Predistortion Application

The proposed G-DMP model is used as a predistorter model for adaptive predistortion of an indirect learning structure. Herein, the adaptive algorithm adopts an improved RLS_LMS joint algorithm (GRLS_LMS algorithm). The block diagram of GRLS_LMS joint adaptive predistortion system is outlined in Figure 5.17.

The GRLS_LMS algorithm integrates the Recursive Least Square (RLS) algorithm and the Least Mean Square (LMS) algorithm, combining the advantages of RLS and LMS. The GRLS_LMS algorithm solves the contradiction between convergence rate and steady-state error, providing a fast-tracking capability and a small parameter estimation error. A self-perturbation term $\boldsymbol{e}(n)$ is added to the inverse covariance matrix, preventing the algorithm from being no longer updated when the Kalman gain vector tends to zero after the algorithm enters steady-state convergence [19]. In addition, the GRLS_LMS algorithm is simpler than the RLS algorithm.

The GRLS_LMS algorithm improves the RLS algorithm (G-RLS algorithm), which is expressed as

$$\boldsymbol{e}(n) = \boldsymbol{d}(n) - \boldsymbol{X}^{\mathrm{T}}(n)\boldsymbol{W}(n-1) \tag{5.53}$$

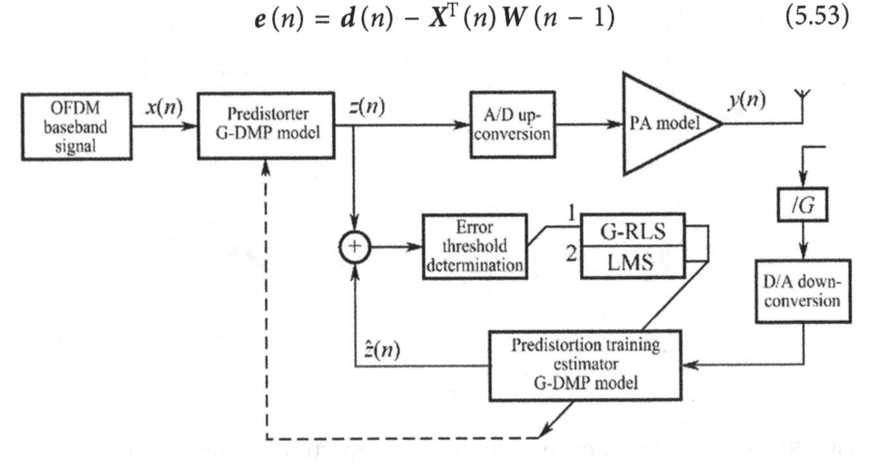

FIGURE 5.17 Block diagram of GRLS_LMS joint adaptive predistortion system.

$$K(n) = \frac{P(n-1)X(n)}{\lambda(n) + X^{\mathrm{T}}(n)P(n-1)X(n)} \tag{5.54}$$

$$W(n) = W(n-1) + K(n)e(n) \tag{5.55}$$

$$\lambda(n) = \lambda_{\min} + (1 - \lambda_{\min}) \times \log \mathrm{sig}(n/M) \tag{5.56}$$

$$P(n) = \frac{1}{\lambda(n)}\{P(n-1) - K(n)X^{\mathrm{T}}(n)P(n-1) + \mathrm{round}[\gamma e(n)]\} \tag{5.57}$$

where γ is the sensitive factor assumed as $\gamma = 1$; $\lambda_{\min}$ is the minimum value of forgetting factor; n is the number of iterations; M is an integer obtained by the test.

The proposed GRLS_LMS algorithm selects a specific algorithm based on the error threshold in the adaptive module. In the initial stage, the error is large and therefore the switch is connected to Terminal 1 enabling the G-RLS algorithm to obtain a fast convergence rate. In this way, the initial values of coefficients for the training estimator of predistortion can be rapidly estimated. When the G-RLS algorithm converges steadily and the error signal is less than the error threshold, the switch is automatically connected to Terminal 2 enabling the LMS algorithm. The initial values of coefficients for the training estimator are duplicated and passed to the polynomial predistortion model and also to the LMS algorithm module as the initial values of weight coefficients for the LMS algorithm. Further iterations by use of the LMS algorithm can maintain a small adjustment step size contributing to a low steady-state offset noise [20]. When the error exceeds the error threshold, the switch is reconnected to Terminal 1 and the above procedure is repeated. The definition of error signal threshold and its value affect the entire system performance. The amplitude of the error signal is used to define the algorithm switchover threshold $\mathrm{ET} = |e(n)|$, so that the ET threshold is only associated with the error signal $e(n)$ independent of other factors. Consequently, the algorithm switchover will not be too frequent, thereby ensuring the stability of the entire algorithm.

The G-DMP and DMP models are applied to predistortion system simulation, respectively. The simulation results are compared with the input signal and the PA output without predistortion. Figure 5.18 shows the spectrum comparison before and after predistortion. Parameters of each model are specified as per Table 5.5. It is noted that the predistorted power spectrum is more approximate to the spectrum characteristics of the input signal, significantly better than the PA output without predistortion. Both predistorter models produce favorable predistortion effects. Specifically, an improvement of approximate 4 dB in the ACPR value is observed when the G-DMP model is used, compared with the DMP model; the spectrum suppression capability after predistortion of the G-DMP model is significantly better than that of the DMP model.

Figure 5.19 shows the comparison of AM/AM characteristic curves produced by the DMP and G-DMP models used as the predistorter models. The EVM value of the DMP predistorter model is 3.2867%, while that of the G-DMP predistorter model is only 2.4200%. Compared with the DMP model, the G-DMP model produces better linearization effect, which basically appears as a straight line with little divergence. In addition, the G-DMP predistorter model delivers better performance in

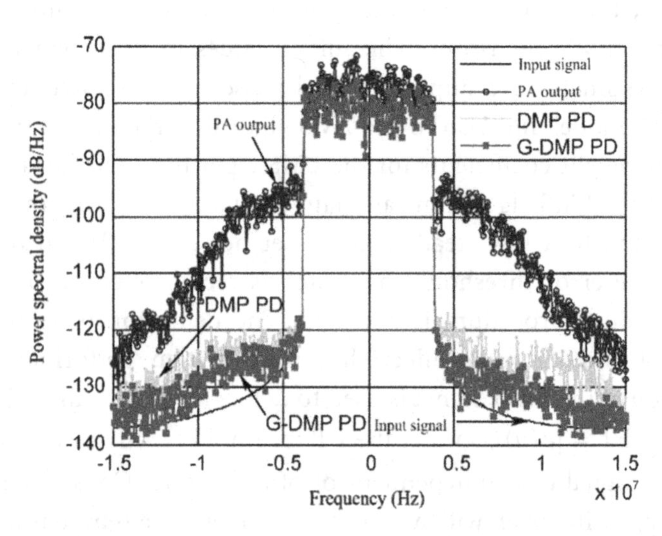

FIGURE 5.18 Spectrum comparison before and after predistortion for each model.

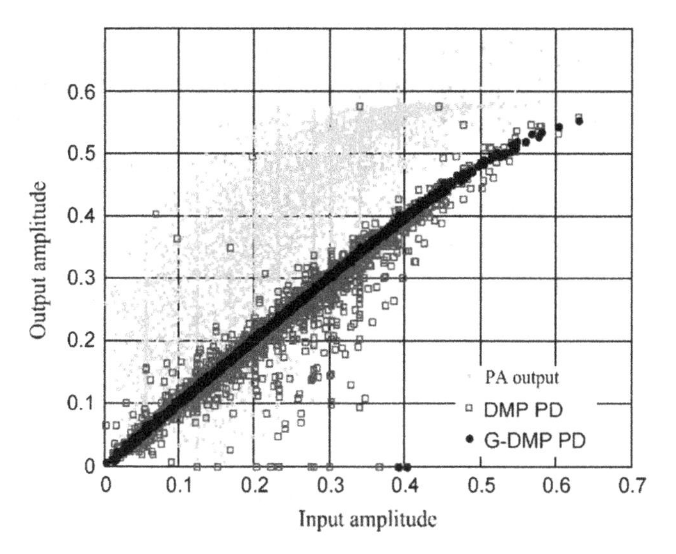

FIGURE 5.19 AM/AM characteristic curves after predistortion for each model.

producing linearization effect, inhibiting out-of-band spectral regrowth and reducing the in-band error.

This subsection proposes an improved multi-slice combined behavioral model, which is an improved dynamic memory polynomial model. The model fully considers the interaction between the input signals at memory times and the current input signal, and is suitable for modeling PAs with strong nonlinearity and strong memory effects in modern wideband wireless communications. Simulation results demonstrate that the improved model features higher accuracy and less complexity. Using the improved model as a predistorter model indicates that the G-DMP model provides the best spectrum suppression capability and the whole system obtains a favorable linearization effect when the GRLS_LMS joint predistortion algorithm is applied to adaptively update the predistorter coefficients.

5.5 RESEARCH ON SPLIT AUGMENTED HAMMERSTEIN MODEL

This section proposes a split augmented Hammerstein (SAH) model with an additional distortion path to accommodate the nonlinearity and memory effects of an RF PA. The proposed model simulates static nonlinearities and memory effects of a PA by utilizing polynomials in a memoryless subsystem and finite impulse response (FIR) filters in a memory effect subsystem, respectively. The parameters of FIR filters are determined by the RLS algorithm. The PA circuit is designed by Freescale semiconductor transistor MRF7S21170, and the input and output data are derived from ADS for model validation. The results demonstrate that the proposed model delivers significantly higher accuracy compared with the augmented Hammerstein model, and can simulate memory effects of a real PA.

5.5.1 Model Analysis

1. Augmented Hammerstein (AH) Model

Figure 5.20 outlines the structure of the AH model. The AH model includes a memoryless subsystem and a memory effect subsystem. The memoryless subsystem uses the nth-order memoryless polynomial to characterize AM/AM and AM/PM static nonlinearities of PA, and the memory effect subsystem uses FIR filters in parallel connection to characterize memory effects. The relationship between the output $x(n)$ of the memoryless polynomial and the output $y_{\mathrm{AH}}(n)$ of the AH model is expressed as

$$y_{\mathrm{AH}}(n) = \sum_{k=1}^{K_1} \sum_{l=1}^{L_1} a_{1kl} x(n - l + 1)|x(n - l + 1)|^{k-1} \tag{5.58}$$

where a_{1kl} represents the parameter of the memory effect subsystem; K_1 and L_1 represent the number of FIR filters and the memory depth of the AH model, respectively.

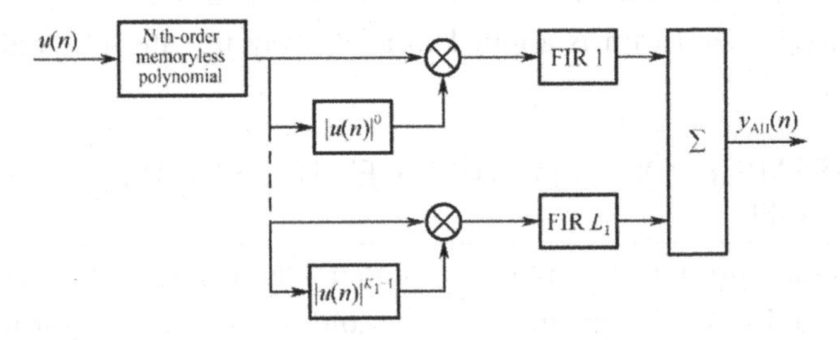

FIGURE 5.20 AH model structure.

2. SAH Model

Figure 5.21 outlines the structure of the SAH model. The model contains an additional distortion path. The error $d(n)$ between the input signal and the SAH model's output is expressed as

$$d(n) = u(n) - y_{\mathrm{AH}}(n) \tag{5.59}$$

The output signal of the SAH model is

$$\begin{aligned} y_{\mathrm{SAH}}(n) = & \sum_{k=1}^{K_2} \sum_{l=1}^{L_2} a_{2kl} x(n-l+1) |x(n-l+1)|^{k-1} \\ & + \sum_{k=1}^{K_3} \sum_{l=1}^{L_3} a_{3kl} d(n-l+1) |d(n-l+1)|^{k-1} \end{aligned} \tag{5.60}$$

where a_{2kl} and a_{3kl} represent the parameters of the memory effect subsystem in the main path and the additional distortion path, respectively; K_2 and L_2 represent the number of FIR filters and memory depth in the main path, respectively; K_3 and L_3 represent the number of FIR filters and the memory

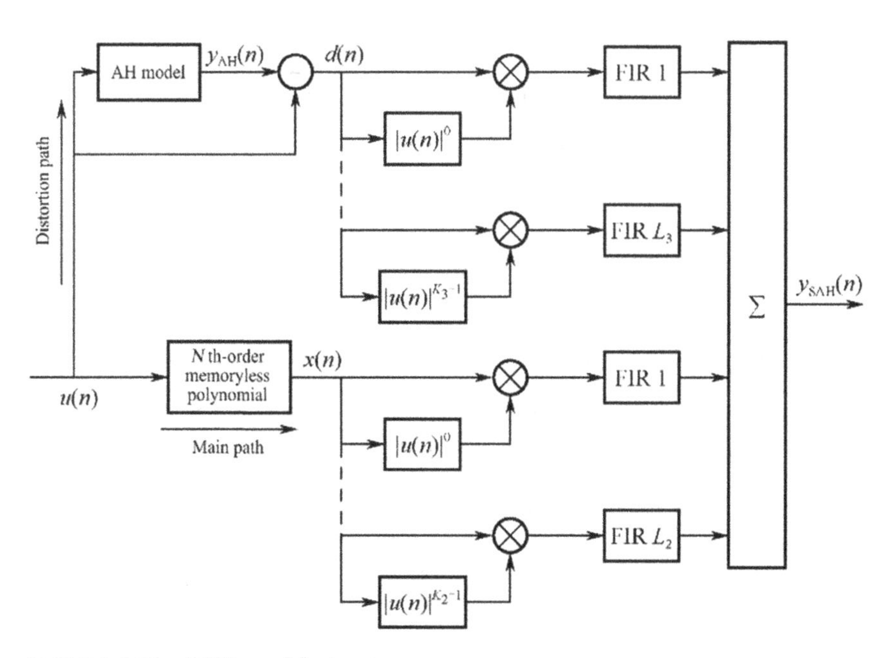

FIGURE 5.21 SAH model structure.

depth in the additional distortion path. Owing to the additional distortion path, the SAH model significantly improves the accuracy, compared with the traditional Hammerstein model and the improved AH model.

5.5.2 Power Amplifier Design and Parameter Extraction

For PA modeling by use of the proposed model, the input and output data of PA need to be extracted from ADS. The N-channel enhanced LDMOS field-effect transistor (FET) MRF7S21170H designed by Freescale specially for WCDMA base stations is selected for the PA design. Its operating frequency band is 2110–2170 MHz, with the static operating points $V_{DS} = 28$ V, $I_{DS} = 1370$ mA and $V_{GS} = 2.88$ V. The matching circuit design adopts the load-pull technique. The complete circuit schematic diagram is shown in Figure 5.22.

5.5.3 Model Simulation Experiment

The input and output data of PA are based on the data obtained from the simulation in Figure 5.22. One thousand (1,000) pairs of data points are

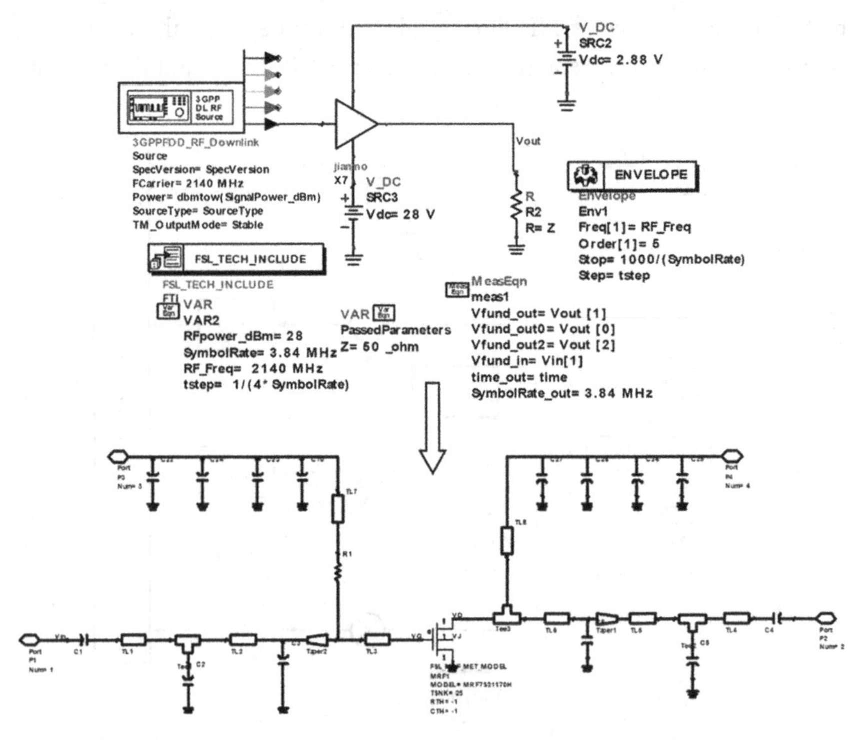

FIGURE 5.22 PA circuit schematic diagram based on MRF7S21170H.

selected to solve the model coefficients, and 2,000 pairs of data points are selected to test the performance of the behavioral model. The 13th-order memoryless polynomial is used to characterize the static nonlinearities of PA. The AH model includes 2 FIR filters, each containing 6 taps (i.e., $K_1 = 2,\ \ L_1 = 6$). The SAH model includes 4 FIR filters, each containing 3 taps (i.e., $K_2 = K_3 = 2,\ \ L_2 = L_3 = 3$). The parameters of FIR filters in the model are determined by the RLS algorithm.

Figure 5.23(a) shows the output power spectrums of the amplifier and the proposed SAH model. Figure 5.23(b) shows a local zoom-in part.

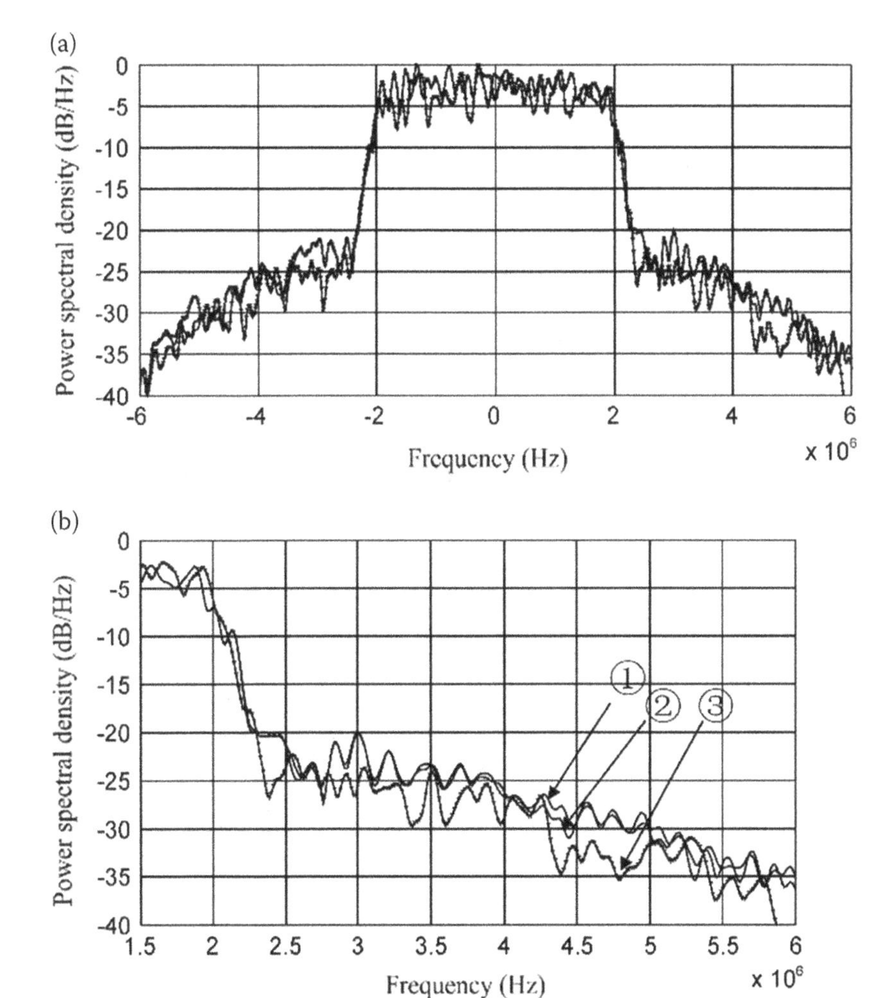

FIGURE 5.23 Output spectrum comparison.

Curve ① is the real output power spectrum of PA, curve ② is the output power spectrum computed by the SAH model, and curve ③ is the output power spectrum computed by the AH model. As indicated by the simulation result, the output power spectrum computed by the SAH model gets more approximate to the real power spectrum of PA and the SAH model delivers a better performance, compared with the AH model.

Figures 5.24 and 5.25 show the comparisons between the real output signal amplitude of PA and the computed output signal amplitude of AH and SAH models, respectively. As indicated by Figures 5.24 and 5.25, the

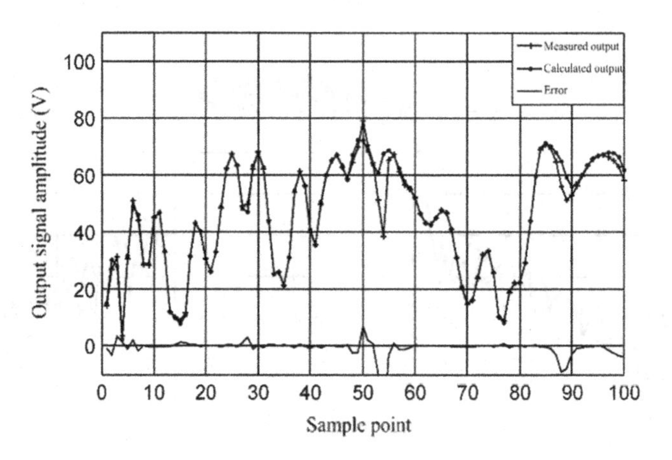

FIGURE 5.24 Amplitude comparison of output signal between PA and AH model.

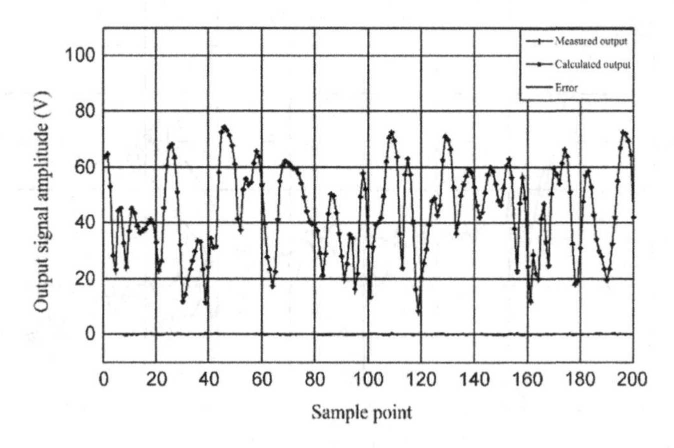

FIGURE 5.25 Amplitude comparison of output signal between PA and SAH model.

SAH model can better simulate the characteristics of real output signal of PA with a smaller error and higher model accuracy, compared with the AH model.

Figure 5.26 shows the comparison of AM/AM characteristics between the real PA output and the output computed by the SAH model. As indicated by the simulation results, the AM/AM characteristics of the SAH model basically coincide with the real AM/AM characteristics of PA. Therefore, the SAH model can represent memory effects of PA and is suitable for behavioral modeling of PA with memory effects.

The behavioral model of PA simulates the characteristics of a real amplifier, playing an important role in the simulation research of communication systems and PA linearization systems. An accurate simulation of PA characteristics and a further deep understanding of the model's applicability are critical to system simulation and of referential importance for the actual design. Based on the traditional Hammerstein model, this subsection proposes the AH and SAH models and describes the model simulation validation. The results demonstrate that the SAH

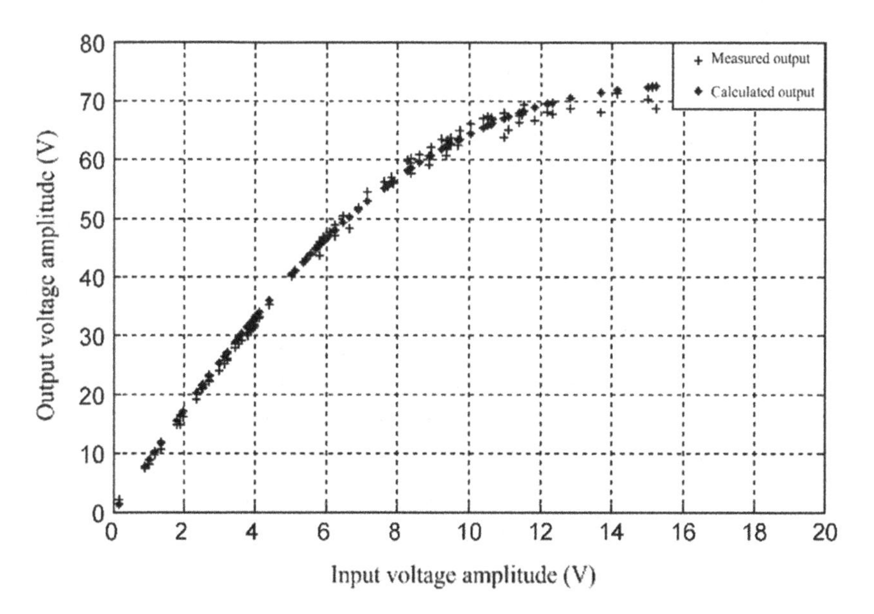

FIGURE 5.26 Comparison of AM/AM characteristics between real PA and SAH model.

model delivers a better performance and more accurately characterizes the nonlinearities and memory effects of an RF PA. The proposed SAH model can be applied to a PA linearization system to enable efficient spectrum utilization, shorten the development cycle and improve the communication quality.

5.6 NOVEL HAMMERSTEIN DYNAMIC NONLINEAR POWER AMPLIFIER MODEL AND PREDISTORTION APPLICATION

This section proposes an improved Hammerstein dynamic nonlinear model to accommodate to memory effects of wideband PA and high-efficiency PA. To improve modeling accuracy, the improved Hammerstein dynamic nonlinear model is established by using two paths: the main path and the additional path. On the main path, FIR filters are used to establish a weak memory effect subsystem and a strong memory effect subsystem. The two subsystems are modeled respectively, and then an LUT is used to establish a memoryless nonlinear model. On the additional path, the memory linear system and the memoryless nonlinear system are still modeled respectively. A simulation test of the NPT1004 PA indicates that the improved Hammerstein model not only compensates the in-band distortion caused by the short term memory (STM), but also inhibits the out-of-band spectral regrowth caused by the long term memory (LTM).

5.6.1 Improved Hammerstein Model

After the main path is constructed for the improved Hammerstein model, an additional path may be added to improve the model accuracy. On the additional path, the output signal of the model is subtracted from the input signal, for the purpose of a nonlinear model. After that, let the computed difference pass through a weak nonlinear model and a strong nonlinear model, respectively, to construct another part of the output. Figure 5.27 shows the topology of an improved Hammerstein model, where NL is a nonlinear subsystem and LTM is a cascaded linear causal subsystem.

The improved Hammerstein model is shown in Figure 5.28. The nonlinear part is implemented by an LUT with high stability, while the memory effect part is implemented by FIR filters.

The mathematical expression of improved Hammerstein model is as follows:

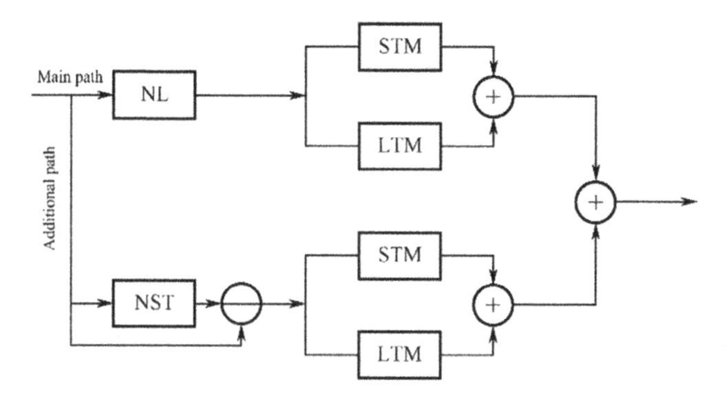

FIGURE 5.27 Topology of improved Hammerstein model.

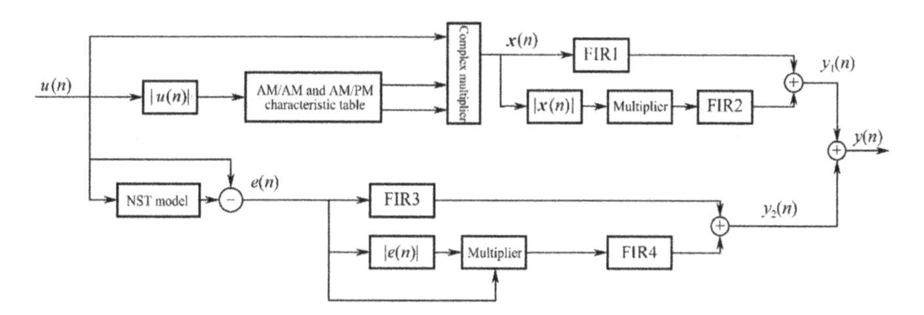

FIGURE 5.28 Improved Hammerstein model.

$$x(n) = [G_{\text{i,LUT}} + jG_{\text{o,LUT}}]u(n) = G_{\text{LUT}}u(n) \tag{5.61}$$

$$y_1(n) = \sum_{p=0}^{M_1-1} a_p x(n-p) + \sum_{k=0}^{M_2-1} b_k |x(n-k)| x(n-k) \tag{5.62}$$

$$e(n) = u(n) - y_{\text{NST}}(n) \tag{5.63}$$

$$y_2(n) = \sum_{s=0}^{M_3-1} a_s e(n-s) + \sum_{t=0}^{M_4-1} b_t |e(n-t)| e(n-t) \tag{5.64}$$

$$y(n) = y_1(n) + y_2(n) = \sum_{p=0}^{M_1-1} a_p x(n-p) + \sum_{k=0}^{M_2-1} b_k |x(n-k)| x(n-k) + \sum_{s=0}^{M_3-1} a_s e(n-s) + \sum_{t=0}^{M_4-1} b_t |e(n-t)| e(n-t) \tag{5.65}$$

where G_{LUT} denotes the complex gain of the LUT; M_1 and M_2 denote the memory depths of the weak and strong memory effect subsystems in the main path, respectively; a_p and b_k denote the system parameters of weak and strong memory effects in the main path, respectively; M_3 and M_4 denote the memory depths of the weak and strong memory effect subsystems in the additional path, respectively; a_s and b_t denote the system parameters of weak and strong memory effects in the additional path, respectively. The nonlinear module of the improved Hammerstein model is implemented by means of the LUT, because it is based on data analysis and able to favorably characterize discontinuous models in the analog domain. In addition, the LUT features good stability and easy implementation in the baseband section. The memory effect module of PA is implemented by a weak memory effect subsystem cascaded with a strong memory effect subsystem, thereby accurately characterizing the memory effects of PA and improving the model accuracy. Compared with the traditional Hammerstein model, the improved Hammerstein model with an additional path significantly improves the model accuracy.

5.6.2 Model Simulation and Validation

The NPT1004 (GaN high-electron-mobility device made by Nitronex) is selected to design the PA. Its operating frequency band is 0–4 GHz, with the static operating points V_{DS} = 28 V, I_{DS} = 350 mA, V_{GS} = −1.5 V. The matching circuit design adopts the load-pull technique by use of electromagnetic simulation software ADS2009. The complete circuit schematic diagram is shown in Figure 5.29.

A WCDMA signal with three carriers is used as the test signal, and the input and output data of PA are extracted by means of envelope simulation. A total of 10,000 sets of data are extracted, of which 5,000 sets of data are used to identify model parameters and the other 5,000 sets of data are used to validate the model accuracy. An improved Hammerstein

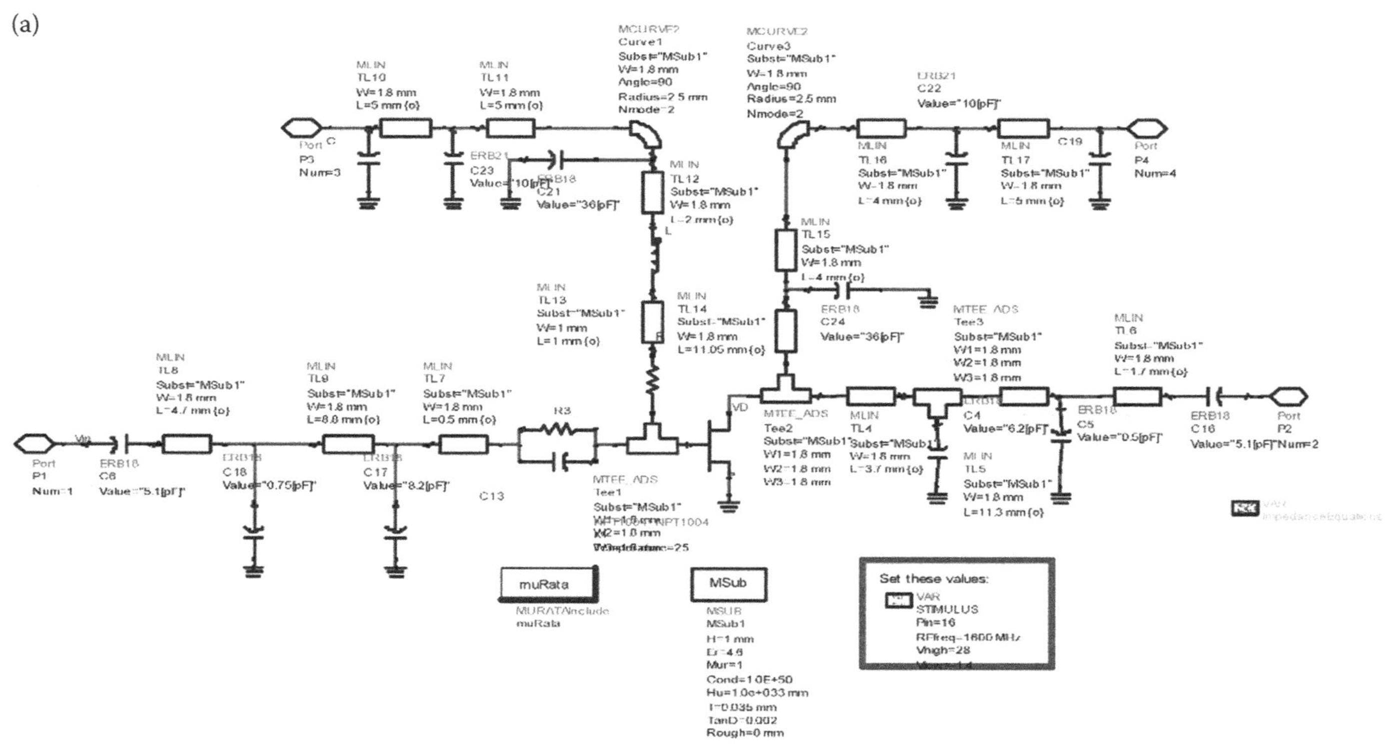

FIGURE 5.29 PA circuit schematic diagram based on NPT1004.

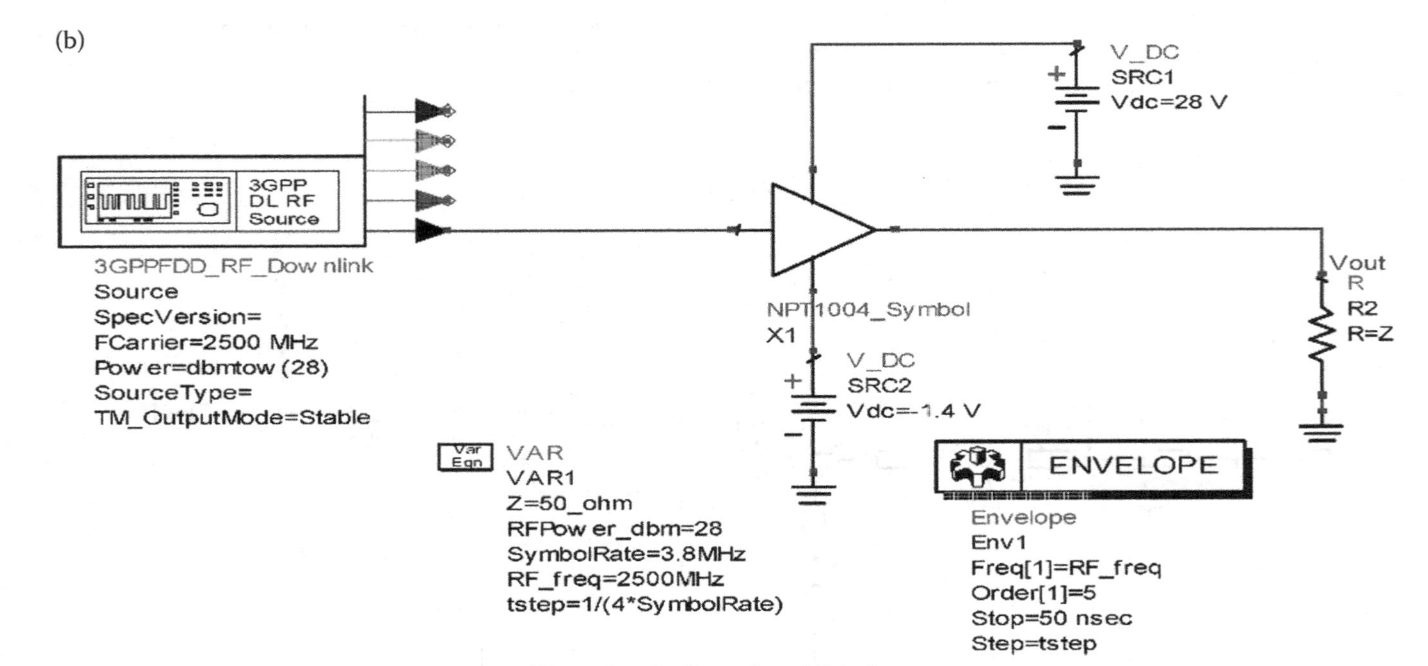

FIGURE 5.29 *Continued*

model includes 4 FIR filters, each containing 3 taps. A static AM/AM and AM/PM characteristic table is obtained by means of offline identification of predistorter parameters, and the model parameters are obtained by using the LMS error method.

Figure 5.30 shows the PA predistortion effects obtained by using the augmented Hammerstein model and improved Hammerstein model, respectively. As indicated by the figure, compared with the augmented Hammerstein model, the improved Hammerstein model produces the PA linearization effect that not only compensates the in-band distortion caused by STM but also inhibits the out-of-band spectral regrowth caused by LTM; in addition, the power spectral density curve computed by the improved Hammerstein model gets more approximate to the real one of PA and the improved Hammerstein model delivers higher modeling accuracy.

Figure 5.31 shows a comparison of AM/AM and AM/PM characteristics of the two models after predistortion. As indicated by the simulation results, both models after predistortion improve the PA linearization to some extent and improve the divergence state due to strong memory effects. However, the predistortion model established by the improved Hammerstein model delivers better AM/AM and AM/PM characteristics, compared with the augmented Hammerstein model.

Memory effects are considered as another important role after nonlinearity that restricts the development of predistortion technology. It is of great significance to investigate memory effects deeply. Following the modeling principle of Hammerstein model, this section focuses on the impact of memory effects on the entire modeling process. The memory effect subsystem in the Hammerstein model is divided into a weak memory effect module and a strong memory effect module, and the two modules are modeled respectively. In addition, an additional path is constructed to ensure the model accuracy, thereby enabling accurate modeling. Simulation results show that the improved Hammerstein model delivers superior performance in PA modeling and predistortion application. More investigations are still in progress regarding the behavioral modeling scheme of PA. Modeling schemes having higher accuracy will be proposed, especially for the long-term memory effects of PA, which is the development direction of PAbehavioral modeling in the future.

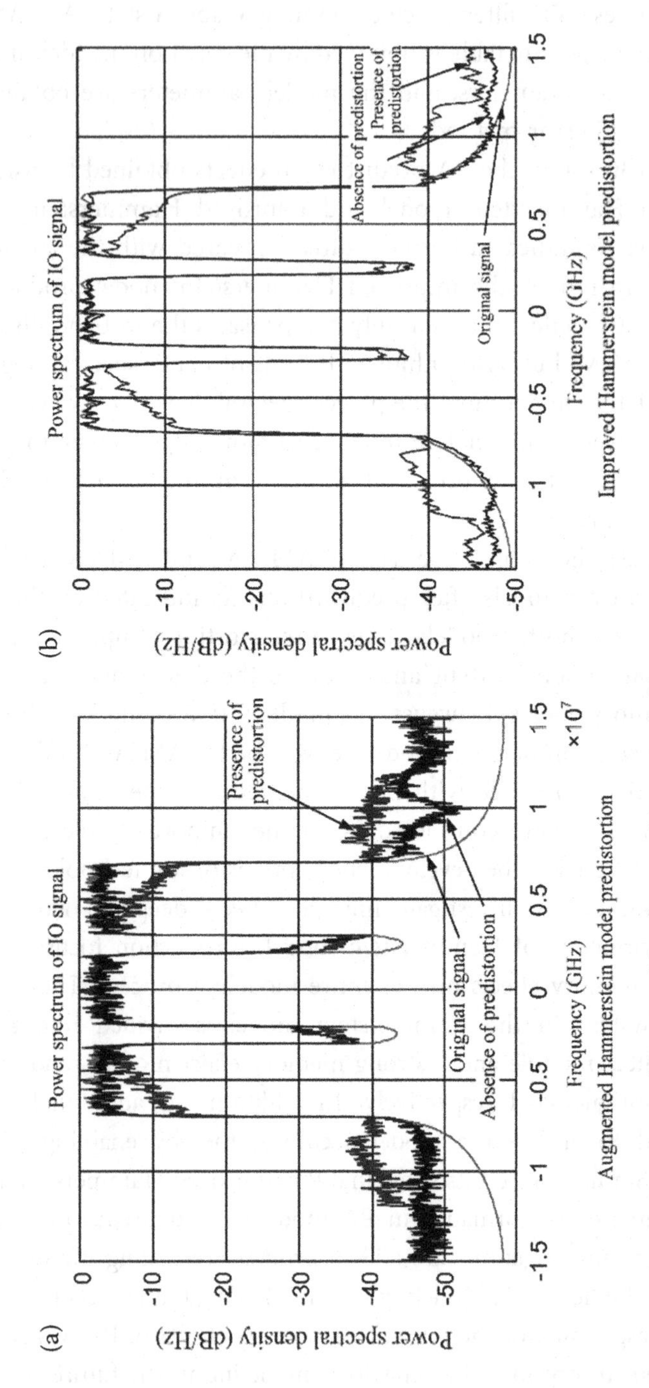

FIGURE 5.30 Comparison of two models after predistortion (a,b).

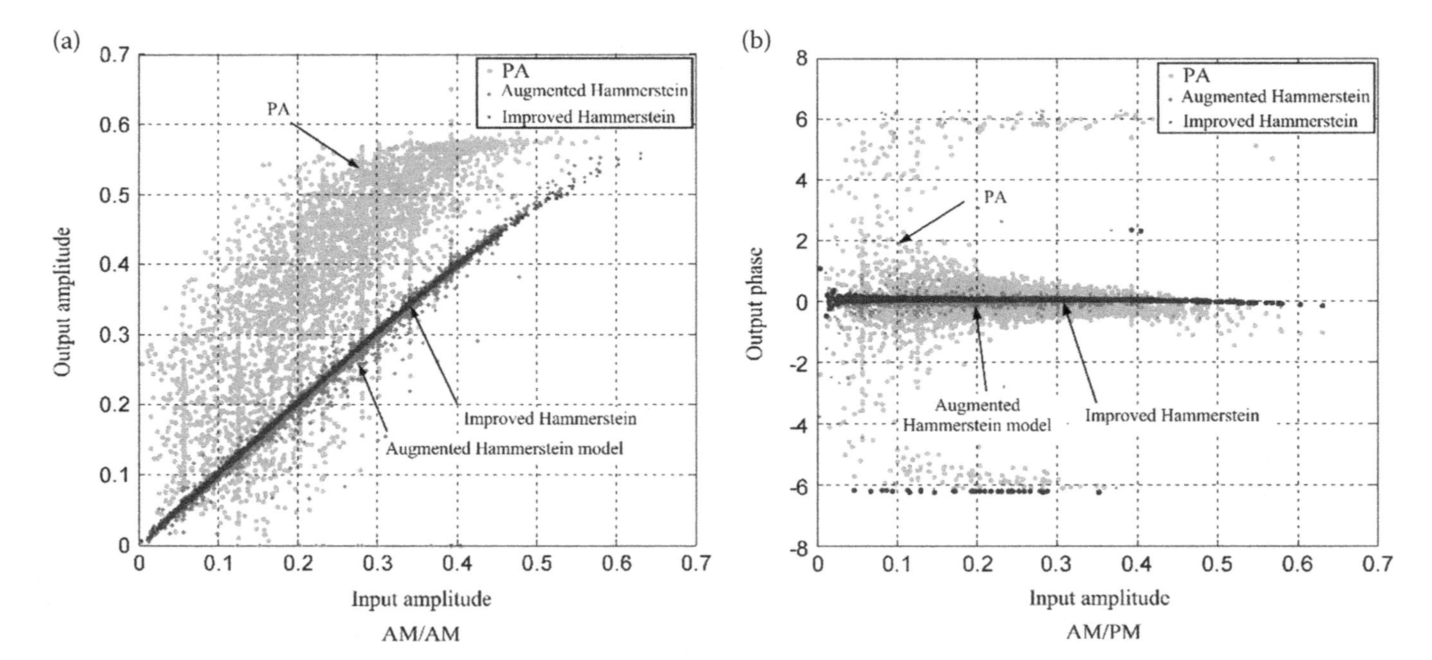

FIGURE 5.31 AM/AM and AM/PM characteristics of two models.

REFERENCES

1. HYUNCHUL KU, KENNEY J S. Behavioral Modeling of Nonlinear RF Power Amplifiers Considering Memory Effects[J]. IEEE Trans on Microwave Theory and Techniques, 2003, 51(12):2495–2503.
2. SILVEIRA D, GADRINGER M, ARTHABERL H. RF-Power Amplifier Characteristics Determination Using Parallel Cascade Wiener Models and Pseudo-Inverse Techniques. Asia-Pacific Microwave Conference, APMC2005 Proceeding, China: Suzhou, 2005[C]. Vol 2, 1–4.
3. ISAKSSON M, WISELL D, RONNOW D. A Comparative Analysis of Behavioral Models for RF Power Amplifiers[J]. IEEE Trans on microwave theory and techniques, 2006, 54(1):348–359.
4. ZHANG JING, HE SONG BAI, GAN LU. Design of a Memory Polynomial Predistorter for Wideband Envelope Tracking Amplifiers[J]. Journal of Systems Engineering and Electronics, 2011, 22(2):193–199.
5. HAMMI O, GHANNOUCHI F M, VASSILAKIS B. A Compact Envelope Memory Polynomial for RF Transmitters Modeling with Application to Baseband and RF-Digital Predistortion[J]. IEEE Trans on Microwave Wireless Component Letters, 2008, 18(5):359–361.
6. MORGAN D, MA ZHENG XIANG, KIM J, et al. A Generalized Memory Polynomial Model for Digital Predistortion of RF Power Amplifiers[J]. IEEE Trans on Signal Process, 2006, 54(10):3852–3860.
7. XIE ZHONG SHAN, LIU BING, LIU WEI, et al. A Novel Approach to Pruning Volterra Models with Memory Effects. Proc of International Conference on Information Technology, Computer Engineering and Management Sciences, Nanjing, Jiangsu, China, 2011[C]. Piscataway: IEEE, 59–61.
8. STAUDINGER J. DDR Volterra Series Behavioral Model with Fading Memory and Dynamics for High Power Infrastructure Amplifiers. Prec of IEEE Topical Conference on Power Amplifiers for Wireless and Radio Applications, Phoenix, AZ, USA, 2011[C]. Piscataway: IEEE, 61–64.
9. GADRINGER M E, SILVEIRA D, MAGERL G. Efficient Power Amplifier Identification Using Modified Parallel Cascade Hammerstein Models. Proc of Radio Wireless Symposium, Long Beach, CA, USA, 2007[C]. Piscataway: IEEE, 305–308.
10. MAYADA Y, FADHEL M G. An Accurate Predistorter Based on a Feedforward Hammerstein Structure[J]. IEEE Trans on Broadcasting, 2012, 58(3):254–460.
11. HAMMI O, GHANNOUCHI F M. Twin Nonlinear Two-Box Models for Power Amplifiers and tRansmitters Exhibiting Memory Effects with Application to Digital Predistortion[J]. IEEE Trans on Microwave Wireless Component Letters, 2009, 19(8):530–532.
12. SILVA C P, CLARK C J, MOULTHROP A A, et al. Optimal-Filter Approach for Nonlinear Power Amplifier Modeling and Equalization. Proc

of IEEE MTT-S International Microwave Symposium, Boston, MA, USA, 2000[C]. Piscataway: IEEE, 437–440.

13. SINGERL P, KUBIN G. Chebyshev Approximation Of Baseband Volterra Series for Wideband RF Power Amplifiers. IEEE International Symposium on Circuit and System, ISCAS 2005, Japan, 2005[C]. Piscataway: IEEE, Vol. 3, 2655–2658.
14. ZHU ANDING, BRAZIL T J. RF Power Amplifier Behavioral Modeling Using Volterra Expansion with Laguerre Functions. Microwave Symposium Digest, 2005 IEEE MTT-S International, Long Beach, CA, USA, 2005[C]. Piscataway: IEEE, 963–966.
15. Freescale Simeconductor, Inc. Freescale Device Data-Wireless RF Product. 2005, 1:495–499.
16. DU TIANJIAO, YU CUIPING, LIU YUANAN. A New Accurate Volterra-Based Model for Behavioral Modeling and Digital Predistortion of RF Power Amplifiers[EB/OL]. 2012.
17. 都天骄,于翠屏,刘元安,等.一种有效的基于宽带功率放大器强记忆效应特性的 PMEC 预失真方法[J]. 电子与信息学报, 2012, 34(2):440–445.
DU T, YU C, LIU Y, et al. An Effective PMEC Predistortion Method Based on the Strong Memory Effect Characteristics of Broadband Power Amplifier[J]. Journal of Electronics and Information Technology, 2012, 34(2):440–445.
18. 南敬昌,李诗雨,汪赫瑜,等.一种新型射频功放建模结构LMEC研究及预失真应用[J].计算机应用研究, 2013, 30(08):2447–2450.
NAN J, LI S, WANG H, et al. New LMEC Model for Behavior Modeling and Predistortion for RF Power Amplifiers[J]. Application Research of Computers, 2013, 30(08):2447–2450.
19. 常铁原,王月娟.一种具有快速跟踪能力的改进 RLS 算法研究[J]. 计算机工程与应用, 2011, 47(23):147–149+227.
CHANG T, WANG Y. Modified RLS Algorithm with Fast Tracking Capability[J]. Computer Engineering and Applications, 2011, 47(23): 147–149+227.
20. 郑晓林.宽带通信系统中功率放大器的数字预失真技术研究[D].广州:广东工业大学, 2011.
ZHENG X. Study on Digital Predistortion Techniques for Power Amplifiers in Broadband Communication Systems[D].Guangzhou: Guangdong University of Technology, 2011.

CHAPTER 6

Power Amplifier Modeling Based on Neural Network

6.1 RESEARCH ON BEHAVIORAL MODEL OF RF POWER AMPLIFIER BASED ON RBF NEURAL NETWORK

It is extremely important to construct an accurate behavioral model of PA in system-level simulation, for the purpose of design and optimization in a high-linearity PA or communication subsystem. This section applies the real transistor testing board of PA, collects massive input-output (IO) data of PA from ADS simulation and constructs a behavioral model based on the RBF neural network [1]. In addition, this section describes the structure design of RBF neural network, K-means clustering algorithm and conjugate gradient optimization algorithm, and also tests the model. Results show the PA behavioral model based on the RBF neural network delivers high accuracy, and the model has a better performance in accurate approximation and fast computation compared with the BP neural network model.

6.1.1. RBF Neural Network Structure and Learning Algorithm

The RBF neural network is a three-layer feedforward network with a single hidden layer. It was proposed by Moody J. and Darken C. in the late 1980s, which is a kind of local approximation network.

Owing to the advantages in nonlinear modeling, the RBF neural network has been expanding its application field. In recent years, experts and scholars worldwide have also deeply explored the RBF neural network, such as adding a compensation network structurally, adding an intra-class competition layer between hidden layer and output layer or introducing a feedback network. Regarding the algorithm, an evolutionary algorithm is adopted to optimize the RBF neural network. Especially, the combination of variable length chromosome genetic algorithm and least square method enables the structure and parameters of RBF neural network to be simultaneously determined. Based on the optimized approach, the thermal load forecasting model established for a thermal power plant has achieved an ideal effect. In addition, the combination of RBF neural network and other artificial intelligence techniques such as expert system or fuzzy system also exhibits an optimum performance.

The basic idea of RBF neural network is to use RBF as the "basis" of the hidden layer unit for constituting the hidden layer space, and to enable the hidden layer to transform an input vector, so that the low-dimensional pattern input data is transformed to the high-dimensional space. In this way, linearly inseparable problems in the low-dimensional space are turned into linearly separable ones in the high-dimensional space.

1. RBF neural network structure

An RBF network is structurally a three-layer forward network with a single hidden layer. The first layer is an input layer consisting of signal source nodes; the second layer is a hidden layer (i.e., radial base layer); the third layer is an output layer, which is a linear layer. The RBF neural network structure is outlined in Figure 6.1.

The nonlinear transfer function for the hidden layer is

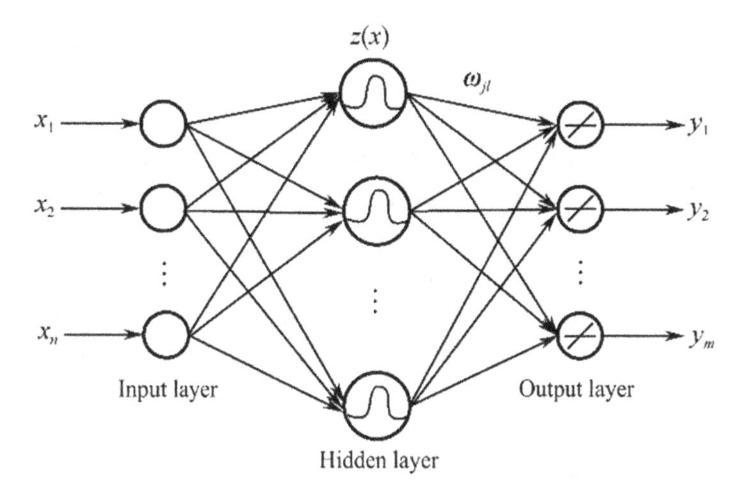

FIGURE 6.1 RBF neural network structure.

$$\{z_l(\boldsymbol{x}) = G(\|\boldsymbol{x} - \boldsymbol{c}_l\|),\ l = 1, 2, \cdots, L\}$$

It is expressed by a Green's function [2], where L is the number of hidden layer units. The Green's function is defined as

$$G(\|\boldsymbol{x} - \boldsymbol{c}_l\|) = \exp\left(-\frac{\|\boldsymbol{x} - \boldsymbol{c}_l\|^2}{2\sigma_l^2}\right) \tag{6.1}$$

This is a multivariate Gaussian function with the variance σ_l^2, where $\boldsymbol{x}$ is the input vector; $\{\boldsymbol{c}_l | l = 1, 2, \cdots, L\}$ is the center point of $G(\cdot)$, that is, the center point of the lth hidden node; σ_l is the field width of the lth hidden node, equivalent to the standard deviation. The output of the jth output node is

$$y_j(\boldsymbol{x}) = \sum_{l=1}^{L} \omega_{jl} G(\|\boldsymbol{x} - \boldsymbol{c}_l\|) + b_j \quad (j = 1, 2, \cdots m) \tag{6.2}$$

where ω_{jl} is weight, that is, the connection weight between the l th neuron in the hidden layer and the j th neuron in the output layer; b_j is threshold. The approximate smoothness is determined by σ_l^2.

2. Learning algorithm of RBF neural network

The design of RBF neural network requires not only the network structure but also its parameters. The parameters to be learned by the RBF network include the transfer function center $\boldsymbol{c}_l$, field width σ_l, weight ω_{jl}, and the number of hidden layer nodes L. The K-means clustering algorithm is applied to determine the transfer function center and field width of the training sample, and the conjugate gradient optimization algorithm is used to adjust the connection weight between output layer and hidden layer of the network. In this way, the input–output nonlinear relationship model of PA is established.

The K-means clustering algorithm is one of the most widely used clustering algorithms. Herein, it is assumed that L (its value determined by prior knowledge) clustering centers are available. Suppose $\boldsymbol{c}_l(\tau)(l = 1, 2, \cdots, L)$ is the transfer function center for the τth iteration. The steps to determine the transfer function center are as follows.

Step 1: Initialize the clustering center. Based on experience, randomly select L different samples from the training sample set as the initial center $\boldsymbol{c}_l(0)$ $(l = 1, 2, \cdots, L)$. Set the number of iterations.

Step 2: Randomly input the training sample $\boldsymbol{X}_k$ of group k.

Step 3: Select the transfer function center that is nearest to the training sample $\boldsymbol{X}_k$. That is, select $l(\boldsymbol{X}_k)$ that satisfies:

$$l(\boldsymbol{X}_k) = \underset{k}{\operatorname{argmin}} \|\boldsymbol{X}_k - \boldsymbol{c}_l(\tau)\|, \quad l = 1, 2, \cdots, L \tag{6.3}$$

where $\boldsymbol{c}_l(\tau)$ is the lth center of the transfer function for the τth iteration.

Step 4: Adjust the center. Adjust the center of the transfer function with the following equation:

$$\boldsymbol{c}_l(\tau + 1) = \begin{cases} \boldsymbol{c}_l(\tau) + \eta[\boldsymbol{X}_k(\tau) - \boldsymbol{c}_l(\tau)], & l = l(\boldsymbol{X}_k) \\ \boldsymbol{c}_l(\tau), & \text{otherwise} \end{cases} \tag{6.4}$$

where η is the learning step size and $0 < \eta < 1$.

Step 5: Determine whether all training samples have been learned and the distribution of centers does not vary any more. If so, terminate the algorithm; otherwise, τ is incremented by 1 and then go to Step 2.

The final $c_l\,(l = 1, 2, \cdots, L)$ obtained is the transfer function center for the RBF network.

The above describes how to apply the K-means clustering algorithm to determine the transfer function center. Next, the field width needs to be determined. Since the hidden layer of RBF network is expressed as a Gaussian function, which is

$$G(\|\boldsymbol{X}_k - \boldsymbol{c}_l\|) = \exp\left(-\frac{\|\boldsymbol{X}_k - \boldsymbol{c}_l\|^2}{2\sigma_l^2}\right), \quad l = 1, 2, \cdots, L \tag{6.5}$$

Therefore, the field width is calculated as follows:

$$\sigma_1 = \sigma_2 = \cdots = \sigma_l = \frac{d_{\max}}{\sqrt{2L}} \tag{6.6}$$

where L is the number of hidden layer units and $d_{\max}$ is the maximum distance between the selected centers. Now, both the transfer function center and the field width have been determined.

The conjugate gradient optimization algorithm serves as an intermediate method between the steepest descent method and Newton's method. This algorithm is regarded as one of the most effective algorithms to solve large nonlinear optimization problems. Each search direction of the algorithm is conjugate with each other, and these search directions are only a combination of the negative gradient direction and the search direction of the previous iteration. Therefore, the algorithm features less storage and convenient computation. The purpose is to obtain a set of weight values $\boldsymbol{\omega}$ minimizing the error between the output y of the neural network and the desired output y_0 of the system; that is, $\min\limits_{\omega} E(\omega)$, where the following equation applies:

$$E = \frac{1}{2}\sum_{k\in T_r}\sum_{j=1}^{m}(y_j(\boldsymbol{X}_k, \omega) - y_0)^2 \tag{6.7}$$

This is the expression of the error sum of square, which should be minimized in the process of neural network training.

In the algorithm, the gradient is recorded as $g_k = \nabla f(\omega_k) = \nabla E|_{\omega_k}$ and the search direction is recorded as $\boldsymbol{d}$. The iterative formula is expressed as follows:

$$\omega_{k+1} = \omega_k + \alpha_k \boldsymbol{d}_k \tag{6.8}$$

And

$$\boldsymbol{d}_k = \begin{cases} -g_k, & k = 1 \\ -g_k + \beta_k \boldsymbol{d}_{k-1}, & k \geq 2 \end{cases} \tag{6.9}$$

where α_k is the parameter controlling the step size, and β_k is a parameter specific to a conjugate gradient method.

The specific steps of conjugate gradient optimization algorithm to solve the weight values are as follows.

Preset step: Determine the initial weight value ω_1. Calculate the gradient $g_1 = \nabla E|_{\omega_1}$. Determine the initial search direction $\boldsymbol{d}_1 = -g_1$, that is, the negative gradient direction [3]. Set $k = 1$ and enter the main steps.

Main steps:

1. Regulate α_k to minimize $E(\omega_k + \alpha_k \boldsymbol{d}_k)$. Calculate $\omega_{k+1} = \omega_k + \alpha_{\min} \boldsymbol{d}_k$.
2. Check whether the error function E produces the minimum value. If so, terminate the algorithm and generate the calculation result ω_{k+1}; otherwise, turn to (3).
3. Calculate the new gradient g_{k+1}. Calculate β_k as follows:

$$\beta_k = \frac{g_{k+1}^T (g_{k+1} - g_k)}{\boldsymbol{d}_k^T (g_{k+1} - g_k)} \tag{6.10}$$

4. (Set $k = k + 1$ and turn to (1) to repeat the calculation process.

6.1.2. Power Amplifier Modeling Based on RBF Neural Network

The testing board of the MRF6S19060N transistor made by Freescale applied to 1930–1990 MHz base stations is used as the PA [4,5] for behavioral modeling. The operating voltage is 28 V, the average output power is 12 W (31.08 dBm), and the gain is 16 dB. The transistor is an LDMOS power transistor featuring high gain, high output power, favorable cost efficiency and high reliability. It performs quite well in communication subsystem simulation or linearization techniques.

The underlying circuit of amplifier is created in ADS, and the existing device model symbols are used to characterize the established amplifier schematic diagram. Its peripheral circuit is connected, with the corresponding parameters of components in the circuit specified appropriately. The signal source input is set as a sinusoidal signal with an amplitude of 0.5 V and a frequency of 1960 MHz. The transient simulation stop time is set to 3 ns and the time interval is set to 0.05 ns. The amplitude and frequency of the sinusoidal signal can be regulated as per circuit requirements. The schematic diagram and simulation results of the PA transient simulation are shown in Figures 6.2 and 6.3, respectively.

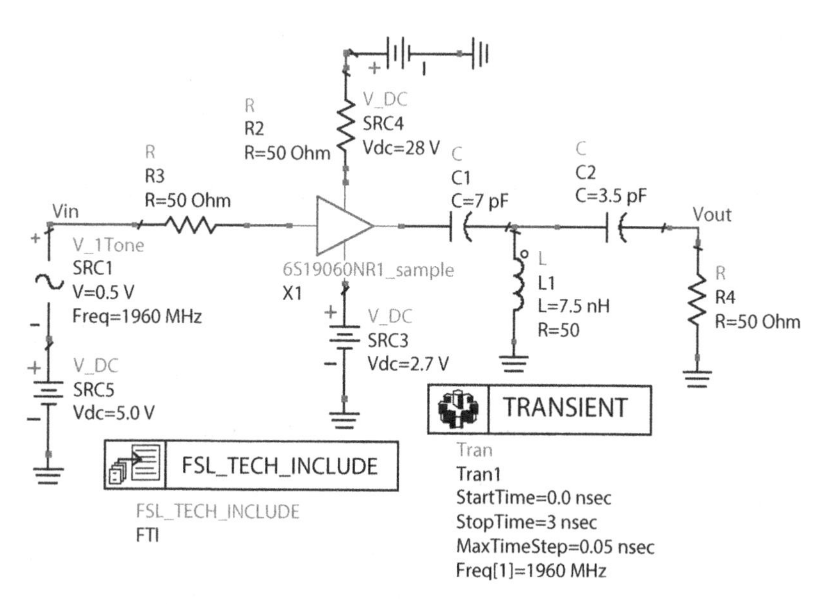

FIGURE 6.2 Schematic diagram of PA transient simulation.

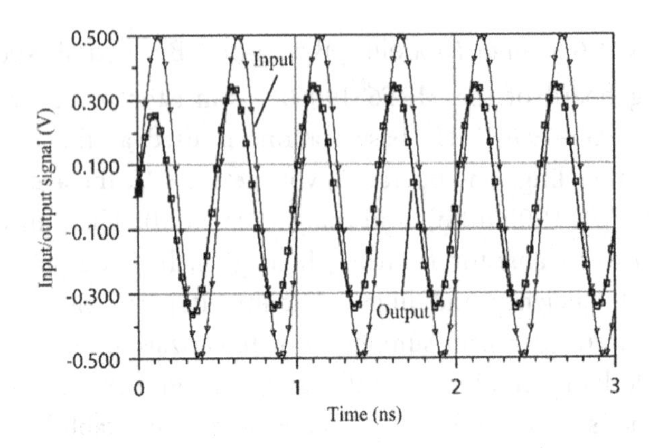

FIGURE 6.3 Simulation results of amplifier IO signals.

Due to the inherent PA nonlinearities, the signal obtained at the output end is distorted to a certain extent in amplitude and phase, compared with the original input signal. Based on the circuit setup and simulation as described above, the IO data of the trained neural network can be collected. That is, the input signal of amplifier and the sampling interval are used as the input to the neural network, and the output signal of amplifier can be used as the desired output of the neural network.

Assume n and m represent the number of input and output neurons of RBF neural network, respectively; $\boldsymbol{x}$ represents an n-order vector containing input information; $\boldsymbol{y}$ represents an m-order vector containing output information; $\boldsymbol{\omega}$ represents a vector containing all weight values. The input and output vectors $\boldsymbol{x}$ and $\boldsymbol{y}$ are respectively defined as

$$\boldsymbol{x} = [U_{\mathrm{n}}, f_{\mathrm{req}}, t_{\mathrm{stop}}, t_{\mathrm{step}}]^{\mathrm{T}}$$

$$\boldsymbol{y} = [y_1, y_2, \cdots, y_m]^{\mathrm{T}}$$

where $n = 4$; U_{n} is the voltage value of signal amplitude; f_{req} is the signal frequency; t_{stop} is the simulation cut-off time; t_{step} is the sampling interval. In this training process, their values are 0.5 V, 1960 MHz, 3 ns and 0.05 ns, respectively. In the output vectors, $y_1, y_2 \cdots, y_m$ correspond to the output signal values of amplifier at sampling points, respectively [6]. The collected input and output data of amplifier are used as the input and desired output training data of RBF neural network, respectively. The behavioral model of PA is established by adjusting the weight values. The output results of RBF neural network are shown in Figure 6.4.

As indicated by Figure 6.4, the output results generated by the trained RBF network are in good agreement with the real output results of PA. This demonstrates the RBF neural network can sufficiently approximate the nonlinearities of PA, and furthermore, the K-means clustering algorithm and the conjugate gradient optimization algorithm are proved to train the network parameters well.

The performance of the established model is tested by means of an evaluation on the error value between the real PA output and the output of RBF neural network. The smaller the error between the two, the better the performance of the model. The specific error value is shown in Figure 6.5. It can be seen that the error fluctuates with changes of the two waveforms but generally remains within a small range of 0.006. Therefore, the PA behavioral model established by RBF neural network has produced a favorable effect.

To further demonstrate the accuracy of the PA behavioral model established by RBF network, the same IO data of PA are used to train the BP neural network and the performance of established network model is observed. See Figure 6.6. As indicated by Figure 6.6, the output results of

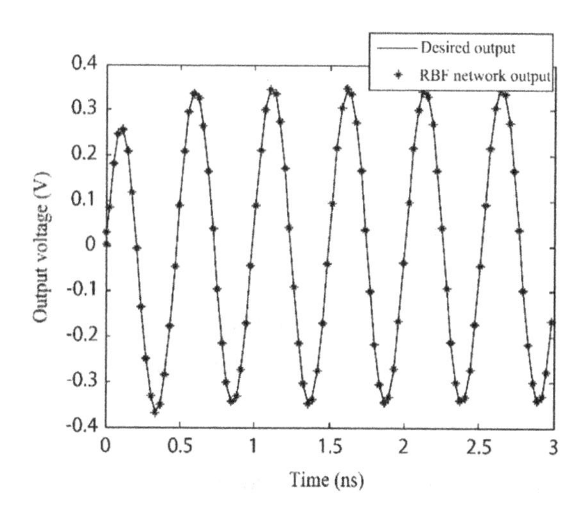

FIGURE 6.4 RBF network-based output and desired output.

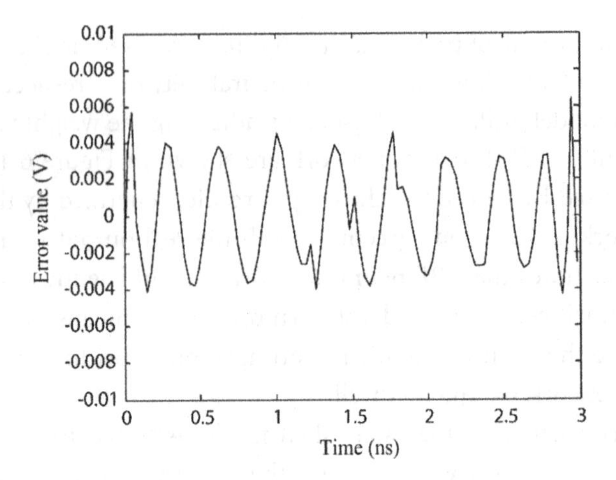

FIGURE 6.5 Error value between RBF network-based output and desired output.

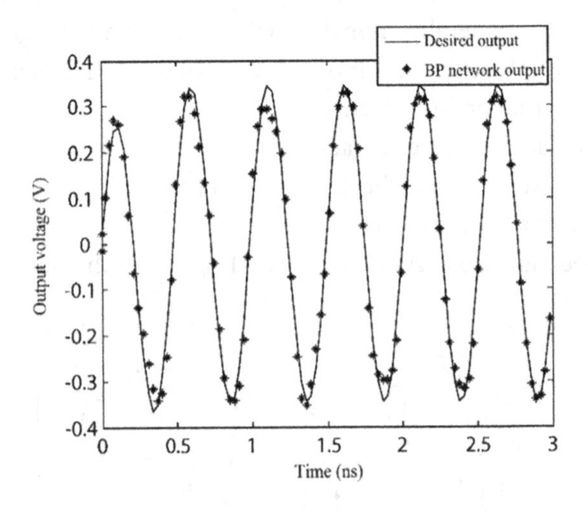

FIGURE 6.6 BP network-based output and desired output.

the trained BP network and the desired results do not produce a favorable fitting effect, especially at peaks or troughs. It is concluded that the RBF neural network outperforms the BP neural network, regarding the approximation simulation of transient nonlinearities of PA. The RBF neural network delivers a better performance in the established IO relationship model of PA and has higher model accuracy, compared with the BP neural network.

6.2. RESEARCH ON BEHAVIORAL MODEL OF RF POWER AMPLIFIER BASED ON BP-RBF NEURAL NETWORK

In this section, the PA circuit design is based on the MRF5P21180 (Freescale semiconductor chip) in ADS. The BP model, RBF model, and cascaded BP-RBF model are used for MATLAB® fitting simulation of the voltage data extracted from the designed PA. This section validates the model accuracy based on the comparison of the root mean square error (RMSE) of voltage data, and compares the model training speeds based on the number of training and the length of convergence time. Eventually, this section comes to a conclusion that the BP-RBF model delivers a better fitting performance compared with the other two models.

6.2.1. Theoretical Analysis of Three Models

1. BP Model

The BP network is a multilayer feedforward network trained according to the error back propagation algorithm. Figure 6.7 shows the topology of a three-layer BP neural network. It consists of two processes: forward calculation (positive propagation) of data streams and back propagation of the error signal. In positive propagation, the input signal originates from the input layer and is propagated to the hidden layer and finally to the output layer; the status of neurons at each layer only affects the neurons at the next layer. If the desired output is not obtained at the output layer, back propagation of the error signal is enabled. The weights and thresholds of the network are continuously regulated by means of back propagation, thereby minimizing the error sum of square of the network. These two processes are alternated, dynamically searching for a set of weight vectors in an iterative manner to minimize the value of network error function. In this way, the process of information extraction and memory is complete [7].

In Figure 6.7, suppose the input layer of BP network has n nodes; the hidden layer has L nodes; the output layer has m nodes; the weight value

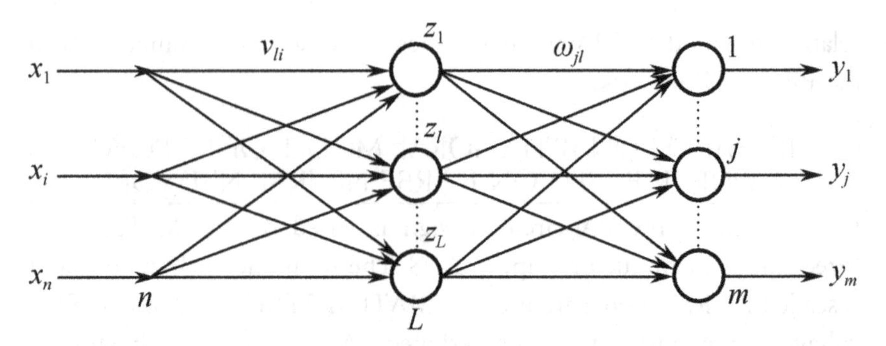

FIGURE 6.7 Topology of three-layer BP neural network.

between the input layer and the hidden layer is v_{li}; the weight value between the hidden layer and output layer is ω_{jl}. When the transfer function of the hidden layer is $f_1(\cdot)$ and the transfer function of the output layer is $f_2(\cdot)$, the output of a hidden layer node is (threshold value included into the summation term)

$$z_l = f_1\left(\sum_{i=1}^{n} v_{li}x_i\right) \quad l = 1, 2, \cdots, L \tag{6.11}$$

The output of the output layer node is

$$y_j = f_2\left(\sum_{l=1}^{L} \omega_{jl}z_l\right) \quad j = 1, 2, \cdots, m \tag{6.12}$$

Now, an approximate mapping of the n-dimensional space vector to the m-dimensional space is complete for the BP network.

If the actual output at the output layer does not match the desired output, the back propagation phase of the error signal is enabled. The output error will be propagated from the output layer back to the input layer via the hidden layer in a layer-by-layer manner, and the error will be allocated to each hidden layer to obtain the error signal of each layer unit, which is used as the basis for modifying the weight value of each unit. The modification process of weight values means the learning process of the BP network, until the accuracy of network output meets the requirements. The back propagation process is as follows. The error function is defined firstly, and then the weight values of the output layer and hidden layer are regulated to make the weight values an appropriate matrix, thereby obtaining the ideal output signal.

2. RBF Model

The RBF model structure is outlined in Figure 6.1. The learning process of RBF network is to automatically generate one RBF neuron at a time and regulate the corresponding network weight. In this way, the number of RBF neurons continuously increases with the weight being regulated, until the required error indicator and the maximum number of training steps are reached. The RBF model structure is shown in Figure 6.1. For an RBF network with n inputs and m outputs, the IO mapping relationship is

$$y_j(\boldsymbol{x}) = f_j = b_j + \sum_{l=1}^{L} \omega_{jl} G(||\boldsymbol{x} - \boldsymbol{c}_l||) \quad (j = 1, 2, \cdots m) \tag{6.13}$$

where b_j is the threshold, $G(\cdot)$ is the hidden layer transfer function, and $\boldsymbol{c}_l$ is the center point of the l th function.

The hidden layer neurons come with the local response characteristic for the input $\boldsymbol{x} = (x_1, x_2, \cdots, x_n)$:

$$G(||\boldsymbol{x} - \boldsymbol{c}_l||) = \exp\left(-\frac{||\boldsymbol{x} - \boldsymbol{c}_l||^2}{2\sigma_l^2}\right) \tag{6.14}$$

where σ_l^2 is the variance determining the width of the odd function around the center point. As $||\boldsymbol{x} - \boldsymbol{c}_l||$ increases, $G(||\boldsymbol{x} - \boldsymbol{c}_l||)$ rapidly decays to 0. For a given input $\boldsymbol{x}$, only a few processing units with their centers close to $\boldsymbol{x}$ are activated, which realize local approximation. In this way, the nonlinear transformation of input quantities can be realized from the input layer to the hidden layer, and the output layer neurons combine the output of the hidden layer. Therefore, the network output is

$$y_j(\boldsymbol{x}) = \sum_{l=1}^{L} \omega_{jl} G(||\boldsymbol{x} - \boldsymbol{c}_l||) + b_j \tag{6.15}$$

Since the output of RBF model is a linear weighted sum of the output of hidden layer units, the learning speed of RBF model is fast.

3. BP-RBF Model

BP model is known to be simple, easy to implement, less computationally intensive and have a strong parallelism, especially for its multi-dimensional nonlinear mapping capabilities. In combination with other methods, the BP model can overcome its vulnerability to local minima. In contrast, RBF network is able to approximate any nonlinear mapping. The algorithm is simple and practical, featuring a rapid training speed. An approach is to cascade the two models into one new model as BP-RBF. Structurally, the BP-RBF model takes the output of BP as the input to RBF, and then generates the network output (hidden layer neurons of BP and RBF summed up as the neurons of original BP). The structure of BP-RBF model is shown in Figure 6.8.

BP-RBF model combines the advantages of BP and RBF models and weakens their respective drawbacks, enabling the simulation results to get more approximate to the real data.

6.2.2. 3G Power Amplifier Design and Data Extraction

For PA modeling with the three models, the first step is to extract the IO data of PA. The MRF5P21180, which is made by Freescale suitable for PAs used in CDMA, TDMA or multi-carrier base stations, is selected for the PA circuit design. Its operating frequency band is 2110–21170 MHz, with the static operating points $V_{DS} = 28$ V, $I_D = 800$ mA, $V_{GS} = 4$ V. The matching circuit design adopts the load-pull technique, with the bias circuit added. The complete circuit schematic diagram is shown in Figure 6.9.

A CDMA2000 signal source is added to the above PA circuit for envelope simulation. The simulated input and output voltage amplitudes are shown in Figure 6.10.

BP → RBF

FIGURE 6.8 Block diagram of BP-RBF model.

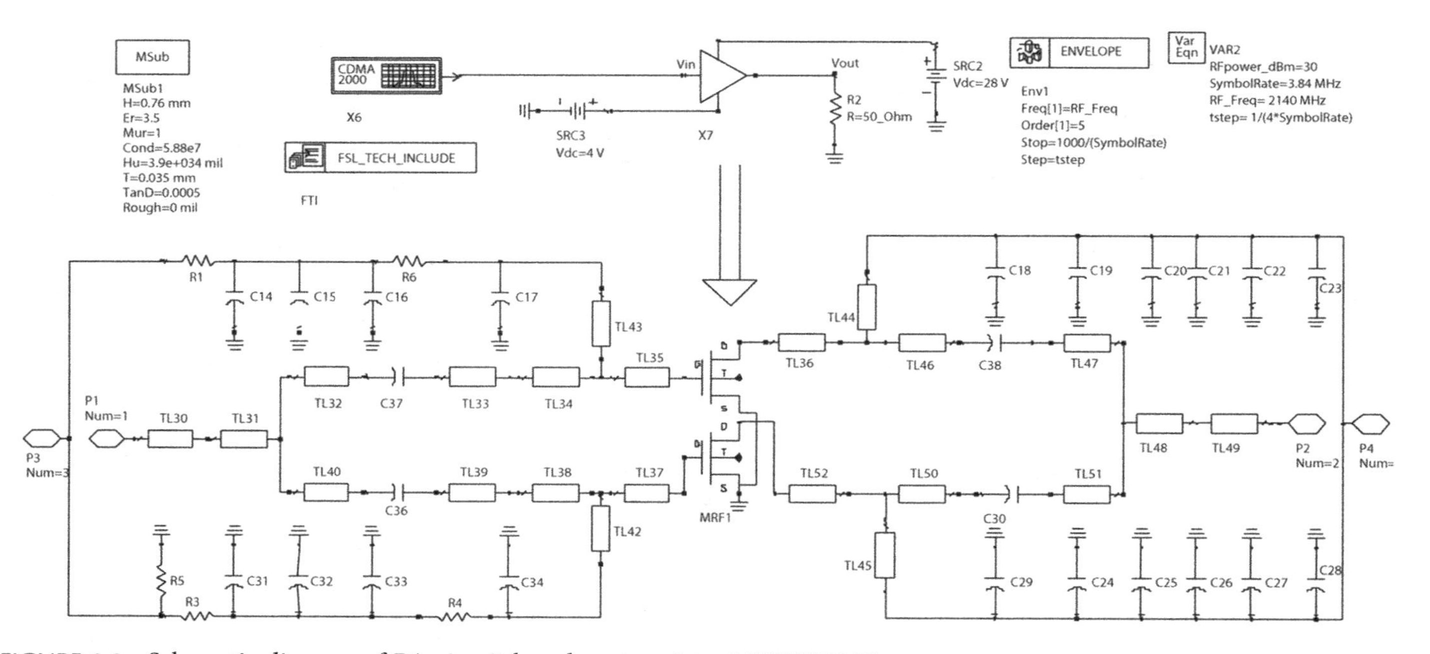

FIGURE 6.9 Schematic diagram of PA circuit based on transistor MRF5P21180.

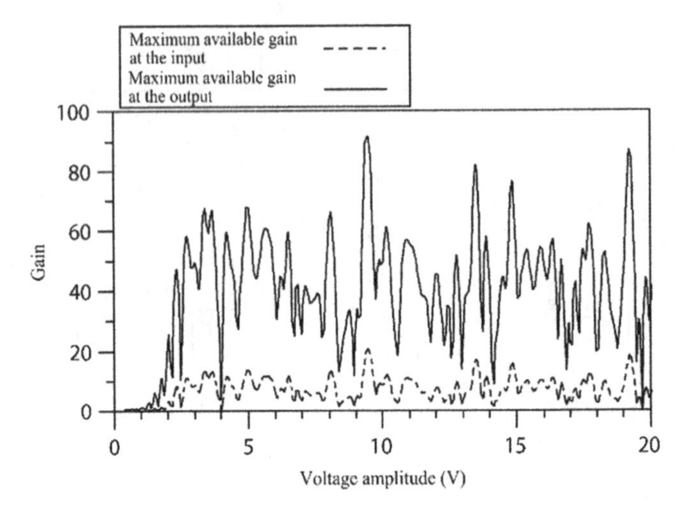

FIGURE 6.10 Input and output voltages of PA circuit based on MRF5P21180.

As indicated by Figure 6.10, the signal obtained at the output end is distorted to a certain extent in amplitude compared with the original input signal, due to the inherent PA nonlinearities. In the following neural network training process, the input voltage amplitude of the amplifier serves as the input to the neural network, and the output amplitude of the amplifier serves as the desired output of the neural network.

6.2.3. Simulation Experiment of Three Models

Firstly, 400 sets of input and output voltage amplitudes are extracted from the designed 3G PA circuit. Then, BP, RBF and BP-RBF models are used for PA modeling and simulation based on the selected IO data, and the simulation results are analyzed to further compare the fitting degrees of the three models to the input and output voltage amplitudes.

For the establishment of BP model, 3 nodes are selected on the input layer, 1 node on the output layer, and 5 nodes on the hidden layer; the tansig function is selected as the transfer function for the hidden layer; the purlin function is selected as the transfer function for the output layer [8]. Generally, a neural network is trained first until it becomes a stabilized network good enough, as proved by further data testing to produce desired outputs, to be used for PA modeling. The test and training data come from odd- and even-numbered sets of the 400 sets of

PA voltage values, respectively. The fitting simulation result of MRF5P21180 voltage amplitude based on the BP model is shown in Figure 6.11.

As indicated by Figure 6.11, the real voltage output of PA is basically consistent with the output of the established model, with a small error maintained at about 0.0022 on average.

The RBF network is established by use of newrbe and newrb. When the data error between training and testing is zero, the former is selected for training and testing. When the data error between training and testing is not zero, the latter is selected. The data is adjusted based on its own characteristics and network structure. The RBF model uses the same trained data as that for the BP model to predict the network performance, thus determining the data fitting effect of the RBF model. The fitting simulation result of MRF5P21180 voltage amplitude based on the RBF model is shown in Figure 6.12.

As indicated by Figure 6.12, the error between the real voltage output of PA and the output of the established model is slightly larger, which is maintained at about 0.0070 on the whole.

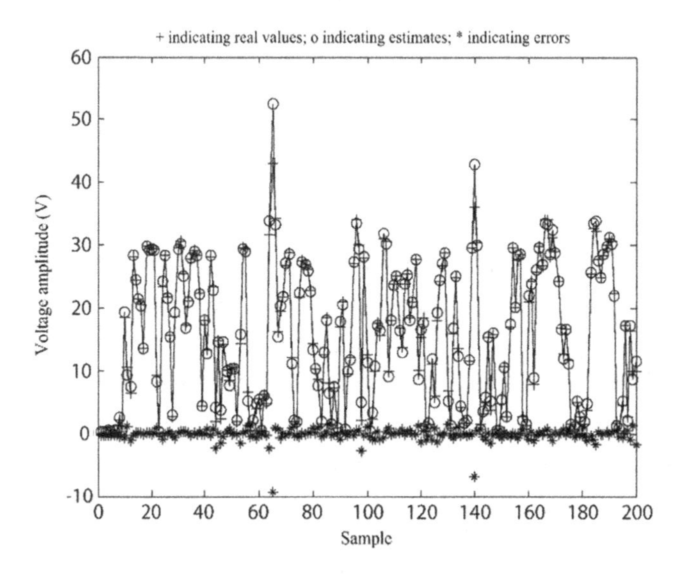

FIGURE 6.11 Fitting simulation result of MRF5P21180 voltage amplitude based on BP model.

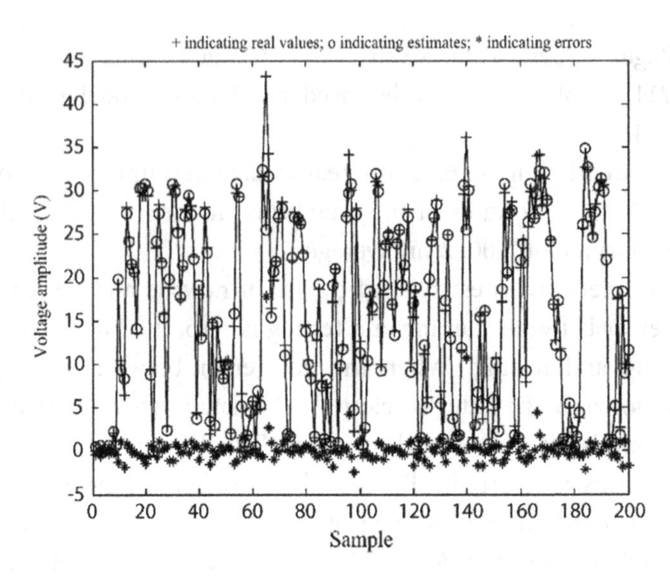

FIGURE 6.12 Fitting simulation result of MRF5P21180 voltage amplitude based on RBF model.

The specific BP and RBF parameters in the BP-RBF model are specified as above, and the cascaded structure is fitted with the same set of voltage data. The result is shown in Figure 6.13.

The cascaded BP-RBF model not only overcomes the drawback of low training efficiency in the BP network, but also makes full use of the RBF network's advantages such as fast training speed and not falling into local minima. This makes the whole BP-RBF network combine an accurate fitting effect of BP model and fast learning speed in the simulation process. Repeated simulations have shown that the mean square error is maintained at about 0.0011, as shown in Figure 6.13.

Based on the comparison of Figures 6.11, 6.12, and 6.13, it can be concluded that the trained BP-RBF model delivers the best fitting effect among the three models and more accurately simulates the established PA model. Secondarily, the BP model is better than the RBF model. In this subsection, the average error between the real PA output and the respective output of the above three neural network models is evaluated to test the performance of established models. The network training speed of each model is measured by the number of voltage data training and the length of convergence time. Specific simulation data is listed in

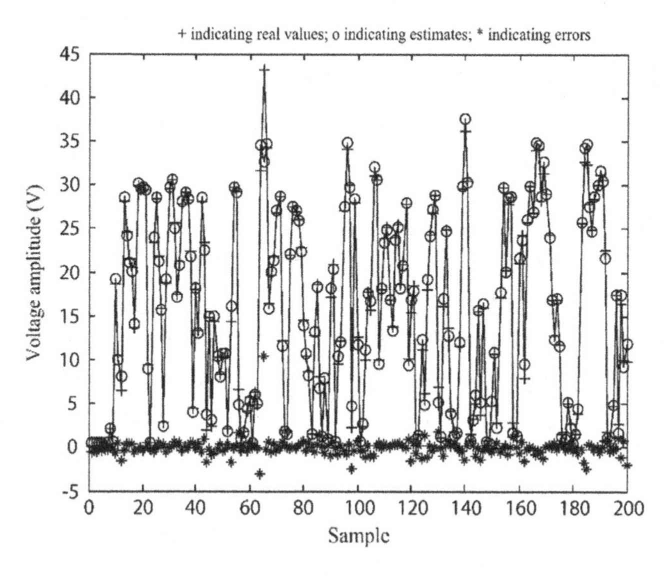

FIGURE 6.13 Fitting simulation result of MRF5P21180 voltage amplitude based on BP-RBF model.

Table 6.1. Obviously, the error produced by the BP-RBF model is kept within a small numerical range and the convergence time is maintained at a very short period of time.

The behavioral model of PA simulates the characteristics of a real amplifier, playing an important role in simulations of communication subsystems and PA linearization systems. Comparison of data fitting results indicates the superiorities of BP-RBF neural network, such as optimal approximation function, excellent performance in interpolation of discrete data, low computational intensity and fast training speed. Therefore, the BP-RBF neural network model is more suitable for the establishment of a PA behavioral model delivering better performance,

TABLE 6.1 MATLAB® Simulation Results of Voltage Amplitudes Using Three Models

Name	BP	RBF	BP-RBF
Mean square error	0.0022	0.0070	0.0011
Number of iterations	1	12	3
Convergence time (s)	0.94	1.96	1.19

higher accuracy and faster training speed. Furthermore, when applied to a PA linearization system, the BP-RBF model can help increase the working efficiency, shorten the development cycle and improve the communication quality.

6.3. FUZZY NEURAL NETWORK MODELING WITH IMPROVED SIMPLIFIED PARTICLE SWARM OPTIMIZATION

Fuzzy neural network has attracted extensive attention in different fields owing to its favorable nonlinear function approximation ability, learning adaptability and parallel information processing ability. As a kind of fuzzy neural network, an Adaptive Neural Fuzzy Inference System (ANFIS) combines the learning mechanism of neural networks and the linguistic reasoning ability of fuzzy systems, and is also capable of self-organization, self-learning and logical reasoning. ANFIS is applied to nonlinear system modeling. Literature [9] proposes a method for device and circuit modeling based on the fuzzy logic, with parameter adjustment by means of the least square method and BP algorithm. Literature [10] proposes to apply a fuzzy neural network to PA predistortion. Literatures [11] and [12] propose a method for RF PA behavioral modeling based on an adaptive fuzzy neural network, but its applications are limited due to disadvantages such as complex structure and parameter learning, slow convergence speed, high risk of falling into local optimum, etc.

Particle Swarm Optimization (PSO), a stochastic optimization algorithm based on swarm intelligence theory, has been widely used in optimization problems such as neural networks and fuzzy control. To improve the local optimization ability of PSO algorithm and increase the convergence speed, a variety of improved PSO methods have been proposed, such as PSO based on the inertia weight, a combination of genetic algorithm and PSO [13,14], improved momentum PSO [15], etc. Literatures [16] and [17] use the PSO algorithm for behavioral modeling of RF PA based on the BP neural network and SVM, respectively. Both approaches prove to have the disadvantages of slow convergence speed and low accuracy.

To characterize an RF PA with memory effects more accurately, this section proposes an improved simplified PSO algorithm and also establishes a fuzzy neural network PA model by combing the proposed algorithm and ANFIS. The improved simplified PSO algorithm only retains the position terms of particles and adds the influence of the random individual optimal candidate solution. The position term of a particle is collectively determined by the particle's current position, individual optimal solution, global optimal solution, and random individual optimal candidate solution. The proposed algorithm adopts the linear decreasing inertia weight and the asynchronous dynamic learning factor, and introduces the Laplace coefficient, thereby increasing the population diversity, accelerating the convergence speed and avoiding falling into local optimum. The simulation results demonstrate that the model established by the method proposed in this section features small error, high accuracy, fast convergence speed and few iterations, and can favorably characterize a PA.

6.3.1. Power Amplifier Model Based on Fuzzy Neural Network

Nonlinearity of a PA means a behavior in which the output signal strength does not vary in direct proportion to the input signal strength, since the output signal reaches saturation while the input signal amplitude is gradually increased. The discretized output expression of a nonlinear PA behavioral model is as follows:

$$v_{\text{out}}(k) = f(|v_{\text{in}}(k)|)\exp\{j[g(|v_{\text{in}}(k)|) + \arg[v_{\text{in}}(k)]]\} \tag{6.16}$$

where v_{in} and v_{out} are the input and output voltages of PA, respectively; $f(\cdot)$ and $g(\cdot)$ are the nonlinear distortion functions of AM/AM and AM/PM, respectively.

An RF PA with memory effects is usually a nonlinear dynamic system, whose output depends both on the current input and on the previous inputs. Its dynamic nonlinearities can be expressed as

$$\begin{cases} |v_{\text{out}}(k)| = f(|v_{\text{in}}(k)|, |v_{\text{in}}(k-1)|, \cdots, |v_{\text{in}}(k-q)|) \\ \arg[v_{\text{out}}(k)] - \arg[v_{\text{in}}(k)] = g(|v_{\text{in}}(k)|, \cdots, |v_{\text{in}}(k-q)|) \end{cases} \tag{6.17}$$

where q is the memory depth; $f(\cdot)$ and $g(\cdot)$ are multivariable functions. Therefore, the problem is equivalent to an approximation problem of static function, which means establishing a dynamic IO characteristic model of the circuit by a numerical method. The circuit model established by this method not only has a solid theoretical foundation but also uses a single algorithm ultimately in the entire modeling process, allowing different circuits or devices to use the same modeling method. The fuzzy neural network modeling method proposed in this section is one of these modeling methods.

A simplified structure of first-order T-S type ANFIS adopting the multiple-input single-output (MISO) design is shown in Figure 6.14. There are $r\{x_1, x_2, \cdots x_r\}$ input parameters and one output parameter y. The T-S type ANFIS consists of an antecedent part and a conclusion part that are used to match the model's fuzzy rules. The model is structurally divided into five layers, the first three layers constitute the antecedent part and the second two layers constitute the conclusion part. The square nodes in the figure require parameter learning.

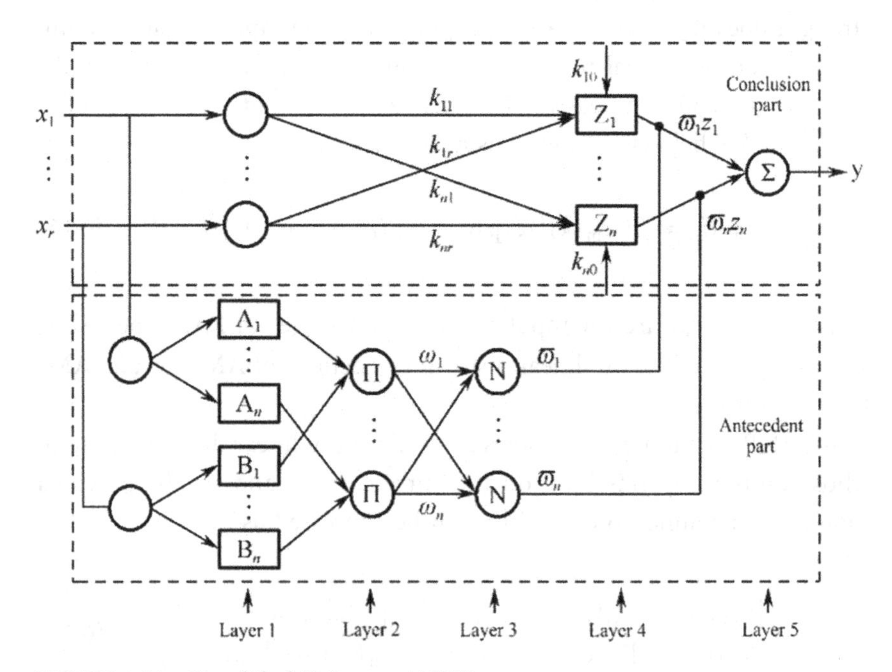

FIGURE 6.14 Simplified T-S type ANFIS.

Details about each layer and applied formulas are as follows.

Layer 1: Membership function layer, where the input parameters $\{x_1, x_2, \cdots x_r\}$ are made fuzzy. Each input parameter is mapped to fuzzy sets: $A_1 \cdots A_n$, $B_1 \cdots B_n$. The matching degrees of input variables are calculated, and each node represents a membership function. Assuming a Gaussian function, the output is expressed as

$$\mathrm{MF}_{ij}(x_i) = \exp\left[-\frac{(x_i - c_{ij})^2}{\sigma_{ij}^2}\right], i = 1, 2, \cdots, r, j = 1, 2, \cdots, n \tag{6.18}$$

where r is the number of input variables and n is the number of membership functions.

Layer 2: Fuzzy inference layer, where the incentive intensity of the current input to each rule is calculated, and the product operation on the membership of each fuzzy variable is performed. The number of nodes is equal to the number of fuzzy rules. The output of this layer is

$$\omega_j = \exp\left[-\frac{\sum_{i=1}^{r}(x_i - c_{ij})^2}{\sigma_{ij}^2}\right] = \exp\left[-\frac{\|\boldsymbol{x} - \boldsymbol{c}_j\|^2}{\sigma_j^2}\right] \tag{6.19}$$

where $\boldsymbol{x} = [x_1, x_2, \cdots, x_r]^{\mathrm{T}}$, $\boldsymbol{c}_j = [c_{1j}, c_{2j}, \cdots, c_{rj}]^{\mathrm{T}}$.

Layer 3: Normalization layer, where each node of the fuzzy inference layer is normalized. The number of nodes is equal to the number of rules. The output of this layer is

$$\varpi_j = \frac{\omega_j}{\sum_{j=1}^{n}\omega_j} \tag{6.20}$$

Layer 4: Rule output layer. The output of each node is the product of the incentive intensity of each rule and the conclusion part. The output of this layer is

$$W_j = z_j \cdot \varpi_j \tag{6.21}$$

For the T-S type fuzzy inference system, its weight values are linear, and the fuzzy rule is "if x_1 is A, and $\cdots$ and $\cdots$, x_r is B, then $z_j = k_{j0} + k_{j1}x_1 + \cdots + k_{jr}x_r$", where k_{ji} is a real number.

Layer 5: Network output layer. It sums the inferred outputs of all rules to obtain the final result for the whole system.

$$y(x) = \sum_{j=1}^{n} W_j \tag{6.22}$$

A mapping from input to output is defined in the above fuzzy logic inference system. The block diagram of the simplified T-S type ANFIS shows that the network output can be converted into linear product form.

This section proposes a simplified T-S type ANFIS for AM/AM and AM/PM function modeling of PA with memory effects. The number of fuzzy rules in a traditional fuzzy system is determined artificially, which may be random or unreasonable. In addition, the fuzzy rules increase rapidly as the input entries grow, resulting in dimension curse and an extremely complex model structure. In contrast, the subtractive clustering algorithm does not require the predefined number of clusters, and can automatically cluster the data into several categories based on the characteristics of data samples and quickly determine the cluster center, so that the number of fuzzy rules and the network structure are more reasonable without considering the dimensions of problems. Therefore, this section adopts the online subtractive clustering algorithm to determine the structure and initial parameters of a fuzzy neural network.

A better division of the input space can be obtained by online dynamic adjustment of the cluster center and the number of clusters. The structure of ANFIS is identified by means of the result of subtractive clustering. The number of fuzzy rules is determined by the number of cluster centers, the initial value of membership function center is determined by each cluster center, and the initial value of membership function width is determined according to the Euclidean distance formula.

6.3.2. Improved Particle Swarm Optimization

1. Simplified Particle Swarm Optimization

The traditional PSO provides low convergence accuracy due to its slow convergence speed in the later stage of evolution and high risk of falling into local optimum. Therefore, scholars have proposed improved methods. Literature [18] proposes a simplified PSO algorithm from the perspective of model and proves that the PSO evolution process is independent of particle velocity. The simplified PSO algorithm is

$$x_{id}^{\tau+1} = wx_{id}^{\tau} + c_1 r_1 (p_{id}^{\tau} - x_{id}^{\tau}) + c_2 r_2 (p_{gd}^{\tau} - x_{id}^{\tau}) \tag{6.23}$$

where w is the inertia factor; r_1 and r_2 are random numbers in [0, 1] that follow a uniform distribution; c_1 and c_2 are known as learning factors; x_{id}^{τ} is the current position of particle i in the dth-dimensional subspace at the τ th iteration. In the iteration process, particles update their positions by tracking the individual extremum p_i and the population extremum p_g.

The first part on the right-hand side of Equation (6.23) is the influence term of the position at the previous moment, which balances the global and local search abilities by tuning the inertia factor. The second part is the "self-learning part", in which particles weigh the positions they have experienced to determine the influence of historical trajectories on the current particle movement trend and have a strong local search ability. The third part is the "mutual learning part", in which particles consider the search ability among particles and exhibit the information sharing among particles.

The simplified algorithm removes the particle velocity term and can find the optimal solution iteratively with only the particle position term, so that the second-order optimization equation is changed to a first-order one. Consequently, the algorithm becomes simpler and more efficient. On this basis, literature [19] proposes a simplified PSO algorithm with adaptive extension. This algorithm replaces the individual extremum of each particle with the mean of individual extrema of all particles. It makes full use of useful information about all particles and effectively avoids the premature convergence problem, significantly improving the global convergence performance and further optimizing the algorithm performance.

2. Improved Simplified Particle Swarm Optimization

An improved simplified PSO algorithm is proposed to increase the population diversity, speed up the convergence and avoid the premature phenomenon, with the interaction between individual particles taken into account. This algorithm introduces the influence of random individual optimal candidate solutions to express the information sharing among individual particles. In addition, the algorithm uses the Laplace coefficient, asynchronous inertia weight factor and dynamic learning factor to increase the population diversity, accelerate the convergence speed and diversify the search direction for an improvement in the training efficiency. In this algorithm, the position term of particle swarm is collectively determined by the current positions of particles, individual optimal solution p_{id}^{τ}, global optimal solution p_{gd}^{τ} and random individual optimal candidate solution p_{jd}^{τ}, where $i \neq j$ to cover the entire search space. Based on literatures [18] and [19], this subsection proposes an improved simplified PSO algorithm expressed as follows

$$x_{id}^{\tau+1} = wx_{id}^{\tau} + c_1 r_1 (p_{id}^{\tau} - x_{id}^{\tau}) + c_2 r_2 (p_{gd}^{\tau} - x_{id}^{\tau}) + \lambda \xi [p_{jd}^{\tau} - x_{id}^{\tau}] \quad (6.24)$$

The search step size of the new influence term is adaptively determined by the random number with Laplace distribution, which also determines the influence of random individual optimal candidate solutions on particle positions. The Laplace function is

$$F(x) = \begin{cases} \frac{1}{2} \exp\left(\frac{x-\eta}{g}\right), & x < \eta \\ 1 - \frac{1}{2} \exp\left(\frac{\eta - x}{g}\right), & x \geq \eta \end{cases} \quad (6.25)$$

where η and g denote the position parameter and scale parameter rate of the Laplace distribution function, respectively. Assume u is a random number in [0, 1] that follows a uniform distribution and $\eta = 0$ applies, and the random number λ that follows a Laplace distribution is

$$\lambda = \begin{cases} g \ln(2u), & u < 0.5 \\ -g \ln(2 - 2u), & u \geq 0.5 \end{cases} \tag{6.26}$$

To ensure the algorithm convergence, a constraint function ξ is introduced to mitigate the influence of Laplace coefficient λ.

$$\xi = \left(1 - \frac{\tau}{\tau_{max}}\right)^b \tag{6.27}$$

where τ and τ_{max} are the current number of iterations and the maximum number of iterations, respectively; $b \in [0.5, 5]$ is an intensity factor.

The algorithm adopts the strategy of linear decreasing inertia weight. The global convergence ability is stronger in the early stage and the local convergence ability is stronger in the later stage with the increase of iterations. The associated expression is

$$w(\tau) = w_{max} - \frac{w_{max} - w_{min}}{\tau_{max}} \cdot \tau \tag{6.28}$$

where τ_{max} is the maximum number of iterations; w_{max} is the weight of maximum inertia; w_{min} is the weight of minimum inertia; τ is the current number of particle iterations.

Learning factors c_1 and c_2 control the self-learning part and the mutual learning part of a particle position, respectively. c_1 and c_2 are dynamically modified by means of asynchronous changes. The self-learning part basically considers the particle itself and dominates in the early stage of iteration; the mutual learning part takes more account of the global optimum and dominates in the later stage of iteration. The whole process is beneficial to the convergence of the global optimal solution and the convergence accuracy. The dynamic learning factor formula is

$$\begin{cases} c_1^{(\tau)} = a - \lambda \cdot \tau \\ c_2^{(\tau)} = b + \lambda \cdot \tau \end{cases} \tag{6.29}$$

where the relevant parameters are set to $a = 2.5$, $b = 0.5$ and $\lambda = 2 \times 10^{-2}$, considering the convergence and accuracy of the algorithm.

3. Improved PSO Applied to Optimize the ANFIS Modeling Process

In the ANFIS, the parameters of the antecedent part are nonlinear and those of the conclusion part are linear. A typical optimization process adopts a hybrid algorithm combining gradient descent method and least square method to regulate parameters, but the process is easy to fall into local optimum and the convergence speed is slow. To address this drawback, this section proposes an improved simplified PSO algorithm to optimize the ANFIS for PA modeling, thereby improving the accuracy and efficiency of modeling.

The PA modeling process for the whole network is shown in Figure 6.15.

6.3.3. Power Amplifier Modeling Simulation Analysis

For PA modeling, the error cost function is

$$E = \frac{1}{2} \sum_{j=1}^{n} (\hat{y}_j - y_j)^2 \tag{6.30}$$

where $\hat{y}_j$ is the desired network output value; y_j is the real network output value; E is the output error.

In simulation analysis, fitness of the error cost function is evaluated and parameters are fine-tuned with optimal values. By doing so, the proposed modeling method not only ensures the diversity of particle swarm and enhances the local optimization ability, but also improves the network generalization ability.

The MRF6S21140 transistor device model made by Freescale is selected for PA circuit design. ADS is used to extract data from the circuit where the input is a WCDMA signal. Two hundred (200) sets of data are selected as the network training data and 100 sets of data are selected as the network test data. A two-input single-output model is adopted for PSO-ANFIS modeling and simulation of PA with memory effects. Its

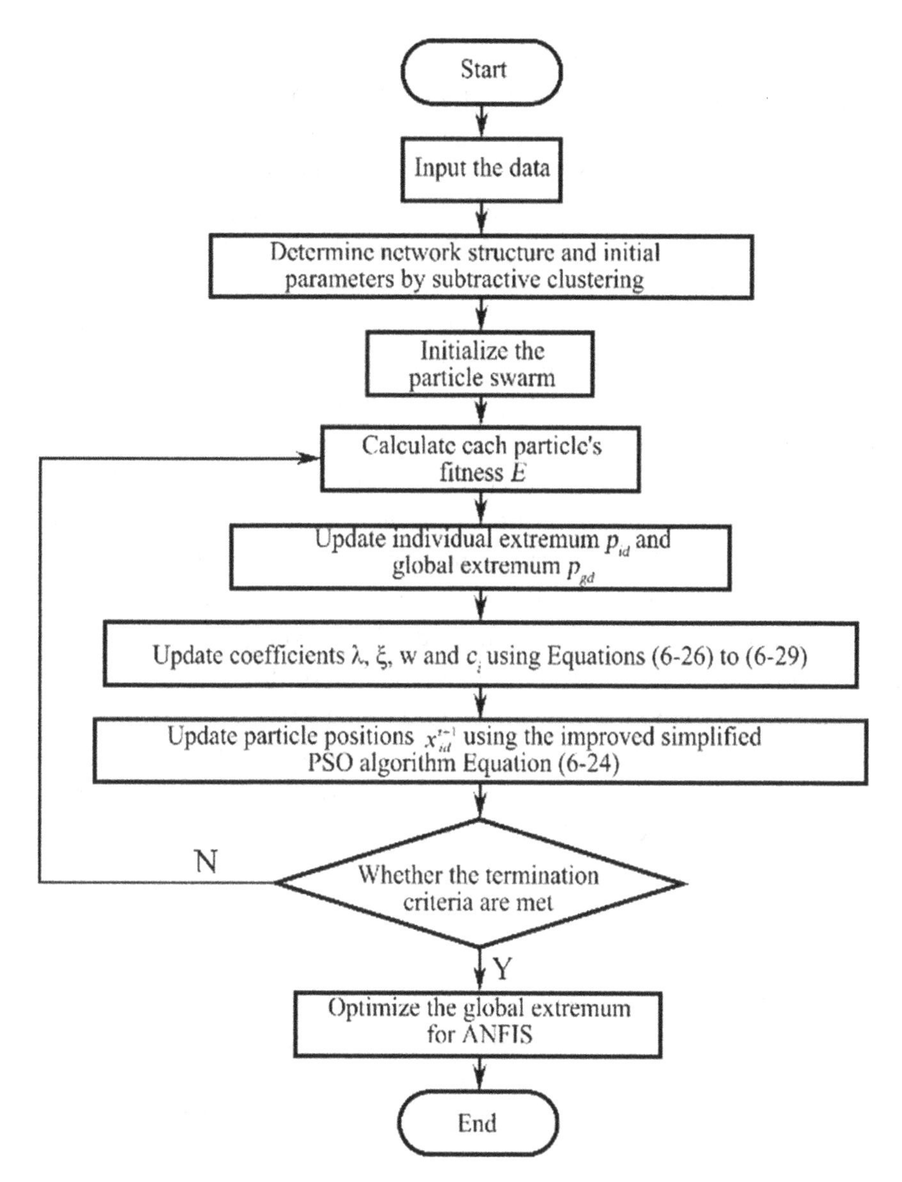

FIGURE 6.15 PA modeling process.

input variables are $v_{\text{in}}(k)$ and $v_{\text{in}}(k-1)$, and its output variable is $v_{\text{out}}(k)$. The modeling procedure shown in Figure 6.15 is used for an optimization design. The population size is set to 20 and the maximum number of iterations is set to 100. Parameters of the antecedent part and coefficients of the conclusion part are regulated by means of multiple iterations, and the PA model is simulated to validate the correctness of the proposed improved optimization algorithm and modeling method. For simplicity, only AM/AM characteristics of PA are investigated in this subsection.

The proposed PSO-ANFIS PA model with memory effects is simulated by MATLAB®, and the simulation results are compared with those generated by PSO-BP and ANFIS neural network PA models. Figures 6.16 and 6.17 illustrate the simulations of training samples and test samples of PSO-ANFIS model, respectively. As indicated by a comparison of voltage amplitude waveforms between real outputs and calculated outputs, the training and test samples of the PSO-ANFIS model are sufficiently close to the real outputs. This demonstrates the PSO-ANFIS model not only effectively fits the nonlinear curve of RF PA, but also favorably simulates the characteristics of RF PA. Figure 6.18 shows the AM/AM characteristic curve of PA, in which the simulation results are consistent with the real characteristics. The output signal is distorted to a certain extent compared with the original input signal, exhibiting the nonlinearity and memory effects of PA.

Figure 6.19 shows the comparison of training data errors based on the three models, and Table 6.2 lists the comparison of training and test data RMSEs based on the three models. Comparisons indicate favorable output fitting degrees of all the three models in RF PA modeling. It is

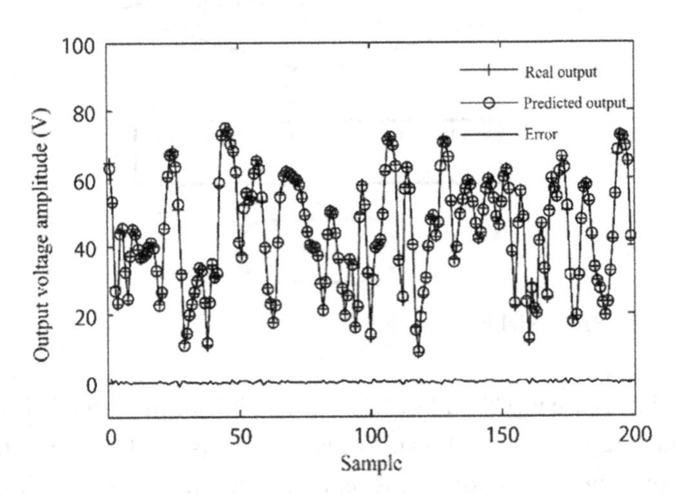

FIGURE 6.16 Simulation diagram of training samples based on PSO-ANFIS model.

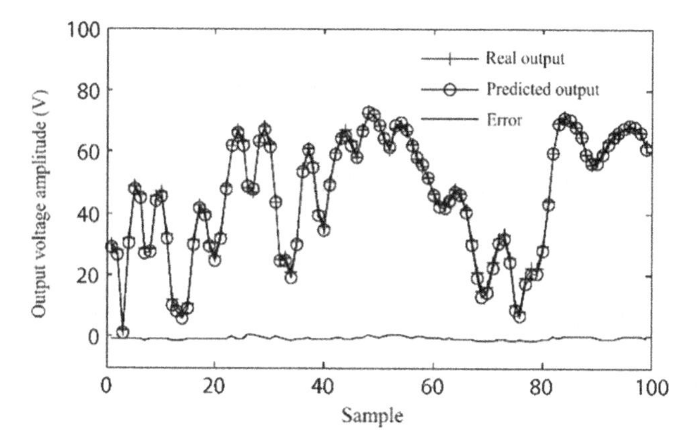

FIGURE 6.17 Simulation diagram of test samples based on PSO-ANFIS model.

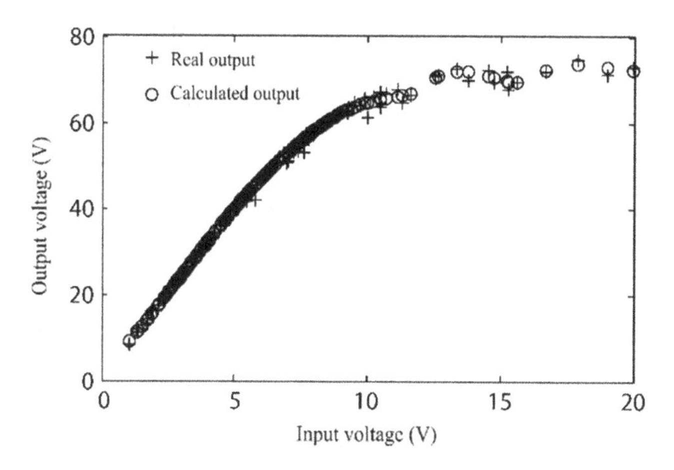

FIGURE 6.18 AM/AM characteristic curve of PA.

noted that the PSO-ANFIS model delivers the optimal degree of flatness and produces the smallest training and test RMSEs (0.5344 and 0.7905 respectively). The PSO-BP model produces the largest training and test RMSEs (0.8892 and 1.0628 respectively), and the ANFIS model produces the intermediate RMSEs (0.7635 and 0.9374, respectively). In addition, the PSO-ANFIS model delivers higher accuracy compared with the PSO-BP and ANFIS neural network models. Figure 6.20 shows the normalized mean square error (MSE) of the three models varying with the number of iterations, and Table 6.3 lists the performance analysis of the three

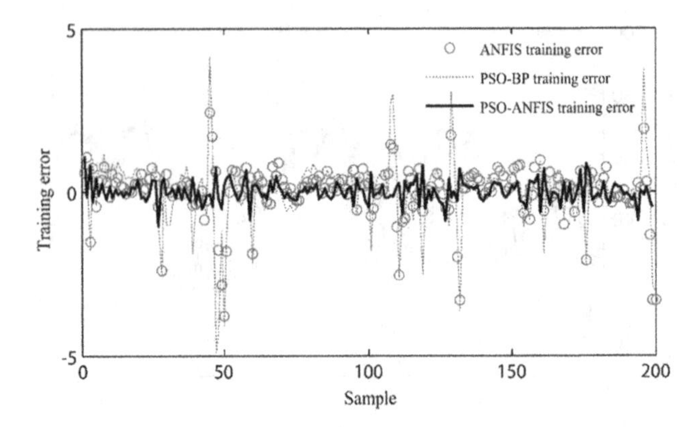

FIGURE 6.19 Comparison of training data errors based on three models.

TABLE 6.2 Comparison of Training and Test Data RMSEs Based on Three Models

	PSO-ANFIS	ANFIS	PSO-BP
Training sample	0.5344	0.7635	0.8892
Test sample	0.7905	0.9374	1.0628

models. As indicated by the figure and table, the PSO-ANFIS model reaches the desired MSE value (0.001) after 6 iterations and consumes the minimum running time, featuring fewer iterations and faster convergence speed. The above simulation results demonstrate the effectiveness and reliability of the modeling method proposed in this section.

Based on the adaptive fuzzy neural network, a nonlinear RF PA model with memory effects is established to accommodate the characteristics of RF PA. In this section, the subtractive clustering algorithm is adopted to determine the network structure by specifying the number of rules, and the improved simplified PSO algorithm is used to optimize the network parameters. The modeling method proposed in this section enhances the optimization ability of particle swarm, effectively improves the convergence speed and training efficiency, reduces the computational complexity, and enables easy implementation with fast speed and high accuracy. The simulation results have validated the effectiveness and reliability of the proposed method, which can be further applied to system-level simulation for effectively utilizing resources, shortening the development cycle and improving the communication quality.

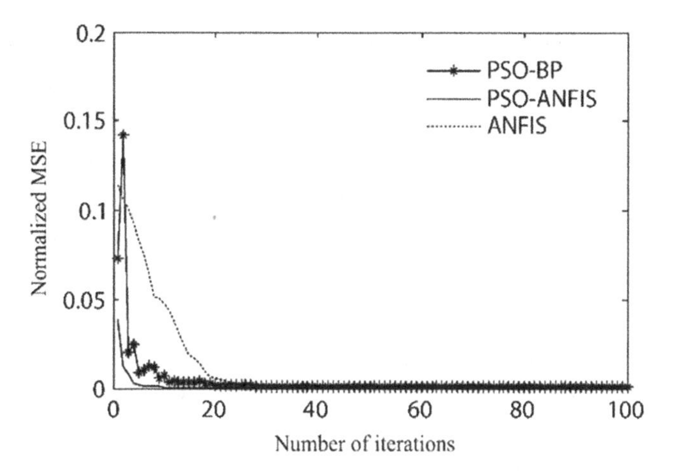

FIGURE 6.20 Normalized mean square error curves of three models.

TABLE 6.3 Performance Analysis of Three Models

Model	PSO-ANFIS	ANFIS	PSO-BP
Normalized MSE	0.001	0.001	0.001
Number of iterations	6	28	20
Running time (s)	14.50	20.47	42.35

6.4. FUZZY WAVELET NEURAL NETWORK MODELING BASED ON IMPROVED PARTICLE SWARM OPTIMIZATION

A fuzzy wavelet neural network (FWNN) combines wavelet technology with fuzzy methodology and neural network [20]. An FWNN integrates the reasoning ability of fuzzy methodology, learning ability of neural network as well as high-accuracy approximation feature of wavelet analysis. It has been applied to nonlinear system modeling and promises broad prospects. Literatures [11] and [12] propose a behavioral modeling method of RF PA based on the adaptive fuzzy neural network. Although the proposed method produces favorable modeling effects, its convergence speed is slow and it is easy to fall into local optimum. Based on the PA model in Literature [11], this section introduces the wavelet function into the rules of fuzzy neural network to establish an adaptive fuzzy wavelet neural network (AFWNN) PA model. For a further network optimization to obtain an accurate model, this section proposes an improved particle swarm optimization (IPSO) algorithm. Literatures [12] and [21] both adopt a simplified PSO algorithm and prove its correctness. Literature [22] adopts asynchronous dynamic learning factors and introduces the Laplace coefficient. However, these literatures ignore the influence of the worst search experience of particles on the search ability of particle swarm. Based on the findings of these literatures, this section introduces a factor for the individual worst position regarding the particle swarm velocity and position update and further simplifies the algorithm. In addition, the weight-related to the fitness function value is adopted to further improve the convergence and accuracy of particle swarm.

This section proposes an adaptive fuzzy wavelet neural network model and establishes a PA model with memory effects by using an IPSO algorithm. In the proposed method, the wavelet function is integrated into the fuzzy rules of the ANFIS to obtain a novel network model; an influencing factor for the worst position is introduced into the PSO algorithm to improve the search efficiency. This method further simplifies the algorithm by ignoring the particle velocity term. In addition, a dynamically changing inertia weight related to the fitness function value is adopted to speed up the convergence, avoiding the "premature" phenomenon. The simulation results show that the PA model established by this method delivers high accuracy with a small error, enabling the effective characterization of PA.

6.4.1. Adaptive Fuzzy Wavelet Neural Network

A fuzzy wavelet neural network (FWNN) integrates the advantages of fuzzy logic, neural network and wavelet function. It has attracted more and more attention. The wavelet function in combination with the ANFIS, a popular fuzzy neural network, constitutes an AFWNN. An AFWNN integrates the wavelet function into the fuzzy IF-THEN rules of the T-S type. Its expression form is:

$$\begin{aligned} &R^i \text{ if } x_1 \text{ is } A_1^j,\ x_2 \text{ is } A_2^j,\ \cdots,\ \text{and } x_r \text{ is } A_r^j, \\ &\text{then } y_j = \omega_j \sum_{i=1}^{r} \psi_{ji}(x_i) \end{aligned} \tag{6.31}$$

where x_i is the input of the jth rule of the neural network; $j = 1, 2, \cdots, n$ is the number of rules; $i = 1, 2, \cdots, r$ is the input dimensions; y_j and ω_j are the output and weight coefficient of the jth rule, respectively; A_i^j is the membership function of the input; ψ_{ji} is obtained by scaling and translating the mother wavelet function. The Mexican Hat mother wavelet function is

$$\psi(x) = (1 - x^2)\exp\left(\frac{-x^2}{2}\right) \tag{6.32}$$

After the mother wavelet function is scaled by b_{ji} times and translated by a_{ji}, its expression is given by Equation (6.35):

$$\psi_{ji}(x_i) = \psi\left(\frac{x_i - a_{ji}}{b_{ji}}\right) \tag{6.33}$$

Figure 6.21 shows the model structure of AFWNN. To develop a reduced-complexity model, a simplified ANFIS is applied. The AFWNN model is structurally divided into six layers incorporating the antecedent part and the conclusion part for matching the fuzzy rules of the model. The layers are described as follows:

Layer 1: The input layer, which introduces the input data into the network without information processing. The domain of discourse of each input is (0, 1);

Layer 2: The membership function layer, which computes the matching degrees of input variables. The number of nodes is obtained by the subtractive clustering algorithm, which is given by a Gaussian function as follows:

$$A_i^j(x_i) = \exp\left[-\frac{(x_i - c_{ji})^2}{2\sigma_{ji}^2}\right] \tag{6.34}$$

Layer 3: The fuzzy inference layer, which computes the incentive

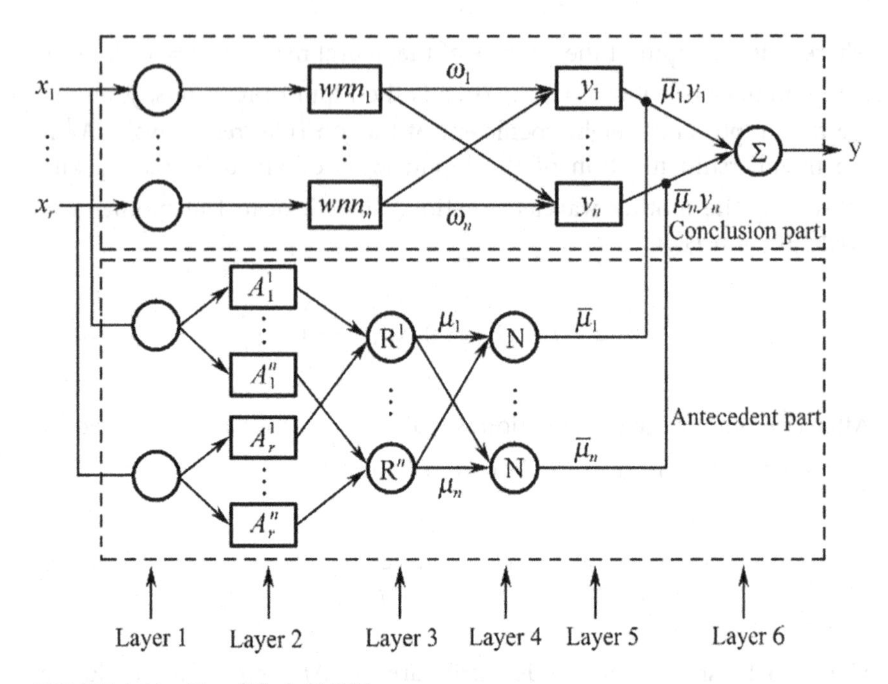

FIGURE 6.21 Simplified ANFIS structure.

intensity of each rule. For a simplified model, assume that the number of fuzzy rules equals the number of membership functions. The output of this layer is given by

$$\mu_j = \prod_{i=1}^{r} A_i^j (x_i) (j = 1, 2, \cdots n) \tag{6.35}$$

Layer 4: The normalization layer, which normalizes all the nodes of the previous layer. The output of this layer is given by

$$\bar{\mu}_j = \frac{\mu_j}{\sum_{j=1}^{n} \mu_j} \tag{6.36}$$

Layer 5: The rule output layer, whose output is

$$W_j = \bar{\mu}_j \cdot y_j \tag{6.37}$$

Layer 6: The network output layer. It sums the inferred outputs of all rules to obtain the final result for the whole system.

$$y = \sum_{j=1}^{n} W_j \tag{6.38}$$

6.4.2. Improved Particle Swarm Optimization

The PSO algorithm initializes a random particle swarm at the beginning, regulates the position of each particle according to its moving experience, and updates its position and velocity following the current individual extremum p_{best} and global extremum g_{best} until the optimal solution is found. Each particle has a fitness value determined by the optimization function. A fitness value indicates whether the particle is a good or bad one, and it also represents a potential solution to the optimized problem.

Assume that in a D-dimensional target search space, the dimensional factor is $d = j = 1, \cdots, n$ and the population size is N, $i = 1, \cdots, N$; $\boldsymbol{x}_i = (x_{i1}, x_{i2} \cdots, x_{in})$ and $\boldsymbol{v}_i = (v_{i1}, v_{i2}, \cdots, v_{in})$ are the current position and velocity of particle i, respectively; $\boldsymbol{p}_i = (p_{i1}, p_{i2}, \cdots, p_{in})$ is the individual extremum of particle i; $\boldsymbol{p}_g$ is the global extremum of all particles in the swarm. For a particle that is updated to the τth generation, the velocity and position of the particle are

$$v_{ij}(\tau + 1) = wv_{ij}(\tau) + c_1 r_1 [p_{ij}(\tau) - x_{ij}(\tau)] + c_2 r_2 [p_{gj}(\tau) - x_{ij}(\tau)] \tag{6.39}$$

$$x_{ij}(\tau + 1) = x_{ij}(\tau) + v_{ij}(\tau + 1) \tag{6.40}$$

where c_1 and c_2 are learning factors, which control the self-learning and mutual learning parts of a particle, respectively. In general, these learning factors take positive integers. r_1 and r_2 are random numbers in [0, 1] that follow a uniform distribution.

However, the search efficiency of PSO is affected by the randomly generated initial state, so the influence of the worst position is introduced into the update formula of particle velocity [23]. Since the particle remembers the worst position, the best search path can be determined more efficiently in the process of exploration. The velocity and position update formula are

$$v_{ij}(\tau + 1) = wv_{ij}(\tau) + c_1r_1[p_{ij}(\tau) - x_{ij}(\tau)] + c_2r_2[p_{gj}(\tau) - x_{ij}(\tau)] + c_3r_3[p_{wj}(\tau) - x_{ij}(\tau)]] \quad (6.41)$$

$$x_{ij}(\tau + 1) = x_{ij}(\tau) + v_{ij}(\tau + 1) \quad (6.42)$$

where c_3 is a learning factor; r_3 is a random number in [0, 1] that follows a uniform distribution; p_{wj} is the worst position. This formula comes with more influence factors and lower convergence speed, compared with the standard PSO algorithm. Therefore, we simplify Equations (6.41) and (6.42) and propose an IPSO algorithm. The updated formula is

$$x_{id}^{\tau+1} = wx_{ij}^{\tau} + c_1r_1(p_{ij}^{\tau} - x_{ij}^{\tau}) + c_2r_2(p_{gj}^{\tau} - x_{ij}^{\tau}) + c_3r_3(p_{wj}^{\tau} - x_{ij}^{\tau}) \quad (6.43)$$

This algorithm ignores the particle velocity term and finds the optimal solution iteratively with only the particle position term, so that the second-order optimization equation is changed to a first-order one. Consequently, the algorithm becomes simpler and more efficient.

Generally, a large value of inertia factor w enhances the global search ability of the algorithm and a small value improves the local search ability. To improve the global search performance of the system, we can allow a dynamic change of w. Considering the inertia weight w capable of affecting the global and local search abilities of PSO algorithm, an improved method is proposed in which the value of w is allowed to vary with the fitness value. That is, w is expressed as

$$w = w_{\min} + \frac{f(x)}{f(x)_{\max}}(w_{\max} - w_{\min}) \quad (6.44)$$

where $f(x)$ is the fitness value of the current function and $f(x)_{\max}$ is the maximum fitness value available for the whole particle swarm. The smaller the fitness value $f(x)$, the closer the current position is to the optimal position. In this sense, Equation (6.44) increases the value of inertia weight if the fitness value $f(x)$ of the current position is large (indicating a great deviation from the optimal position), thereby enhancing the global search ability. Oppositely, the value of inertia weight is reduced to increase the local search ability.

The flowchart of IPSO algorithm is depicted in Figure 6.22, where $\tau_{\max}$ denotes the maximum number of iterations.

6.4.3. Power Amplifier Modeling and Simulation

The AFWNN model and IPSO algorithm are used for RF PA modeling. Nonlinearities of RF PA with memory effects can be expressed as

$$I_{out}(n) = f\,[I(n), I(n-1), \cdots, I(n-q_1), \\ Q(n), Q(n-1), \cdots, Q(n-q_2)] \tag{6.45}$$

$$Q_{out}(n) = g\,[I(n), I(n-1), \cdots, I(n-q_1), \\ Q(n), Q(n-1), \cdots, Q(n-q_2)] \tag{6.46}$$

where $I(n)$ and $Q(n)$ are in-phase and quadrature components of the input signal, respectively; $I_{out}(n)$ and $Q_{out}(n)$ are in-phase and quadrature components of the output signal, respectively; $f(\cdot)$ and $g(\cdot)$ are nonlinear distortion functions of PA; q_1 and q_2 are memory depths.

In behavioral modeling of RF PA based on Equations (6.45) and (6.46), the input signal of RF PA is used as the input to the AFWNN and its output signal as the output of the network. To develop a reduced-complexity model, subtractive clustering is used for processing data and determining the number of fuzzy rules, thereby streamlining the network structure.

For RF PA modeling, the fitness function is

$$f = \frac{1}{2}\sum_{j=1}^{n}(\hat{y}_j - y_j)^2 \tag{6.47}$$

where y_j is the actual network output value; $\hat{y}_j$ is the desired network output value; f is the error function.

The simulation in this subsection is to validate the correctness of the proposed RF PA modeling method. The test signal uses a 16QAM signal with a bandwidth of 15 MHz. An RF PA with a center frequency of 1.96 GHz and

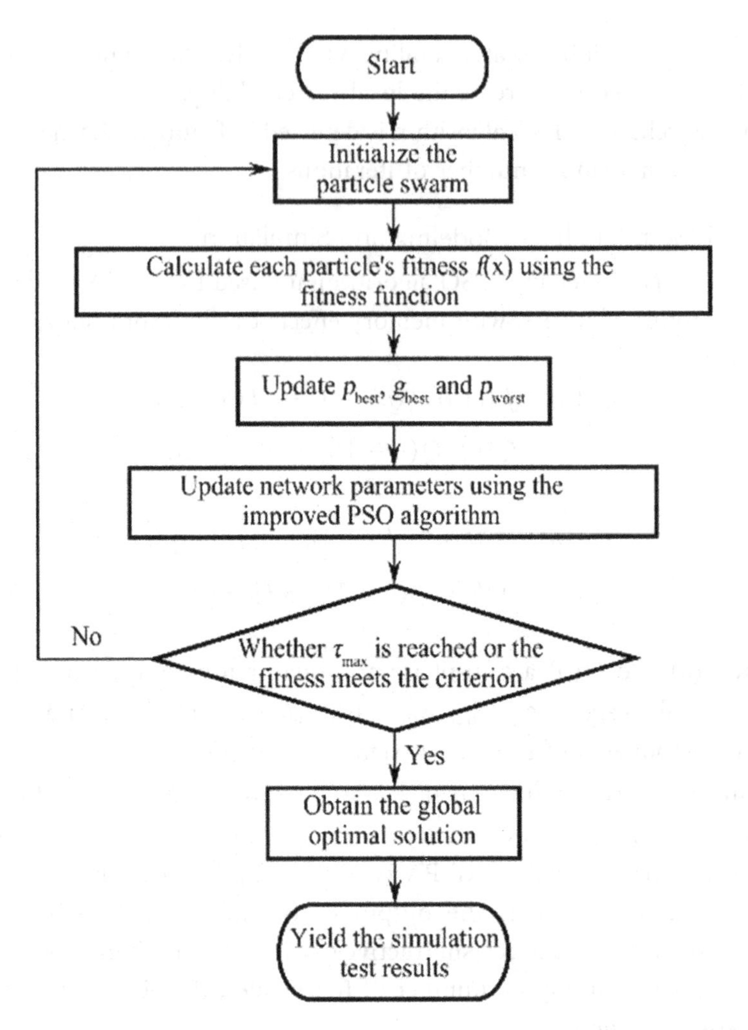

FIGURE 6.22 IPSO flowchart.

a gain of 50 dB is used to extract the IO data. Three hundred (300) sets of data are selected as the training data for the network model and 200 sets of data are selected as the test data. The optimization process depicted in Figure 6.23 is used for simulation on the MATLAB® platform. The population size is set to 30 and the maximum number of iterations is set to 100. Variable parameters of the model are regulated by means of iterations, and the RF PA model is simulated to validate the correctness of the proposed modeling method.

Figures 6.23 and 6.24 show the simulation results of the real part and imaginary part of the output voltage in PA modeling based on IPSO-AFWNN, respectively. As indicated by a comparison of waveforms between measured outputs and calculated outputs, the calculated outputs of the IPSO-AFWNN model are sufficiently close to the measured outputs. This demonstrates that the IPSO-AFWNN model can effectively fit the nonlinear curve of RF PA and simulate the characteristics of RF PA.

Table 6.4 lists the RMSEs of the real and imaginary parts based on IPSO-AFWNN, AFWNN and ANFIS models for PA modeling.

Comparisons indicate favorable output fitting degrees, regarding both the real and imaginary parts, of all the three models in RF PA modeling. It is noted that the IPSO-AFWNN model delivers the optimal degree of flatness and produces the smallest training RMSEs for both the real and imaginary parts (0.5528 and 0.5946 respectively).

Figures 6.25 and 6.26 show the power spectral densities and their error curves of proposed IPSO-AFWNN, AFWNN and ANFIS models, respectively. As indicated by the figures, the output of proposed IPSO-AFWNN model is closer to the measured result with a smaller error, compared to the other two models. Table 6.5 lists the performance analysis for the three models. As indicated by the table, the IPSO-AFWNN model reaches the desired MSE value (0.2) after 9 iterations, while the AFWNN and ANFIS models require 14 and 23 iterations,

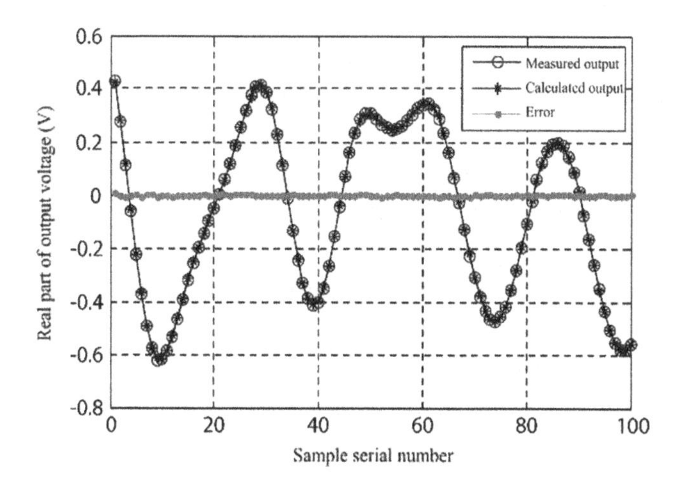

FIGURE 6.23 Real part simulation of PA output voltage.

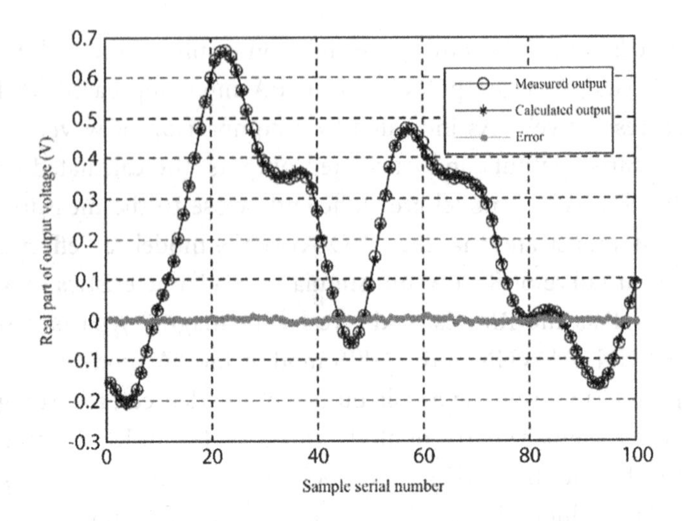

FIGURE 6.24 Imaginary part simulation of PA output voltage.

TABLE 6.4 RMSE Comparison of Real Part and Imaginary Part Based on Three Models

	IPSO-AFWNN	AFWNN	ANFIS
Real part	0.5528	0.8344	0.9295
Imaginary part	0.5946	0.7862	0.9628

respectively. It is noted that the IPSO-AFWNN model consumes the minimum running time, featuring faster convergence. The above simulation results demonstrate the effectiveness and reliability of the modeling method proposed in this section.

In this section, an AFWNN is utilized to establish a nonlinear RF PA model with memory effects. The proposed model adopts subtractive clustering to determine the number of fuzzy rules and IPSO algorithm to optimize the model parameters. As indicated by simulation results, the modeling method proposed in this section brings out the best of the nonlinear approximation ability in the network model, effectively improves the convergence speed and training efficiency, reduces the computational complexity, and enables easy implementation. The proposed model can be further applied to predistortion techniques and other fields.

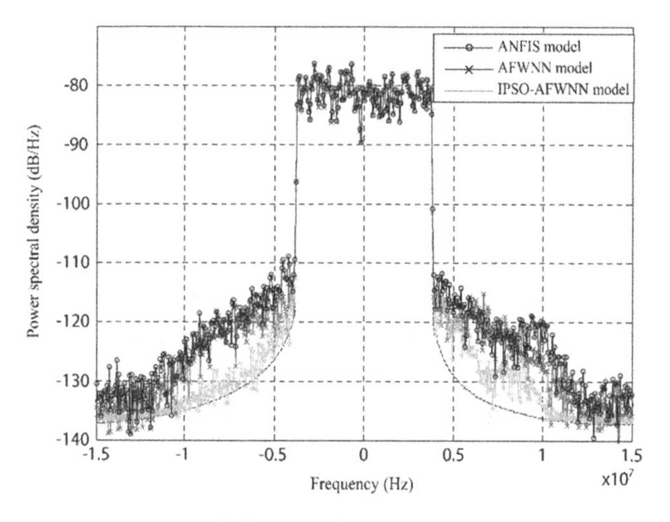

FIGURE 6.25 Power spectral density of PA model.

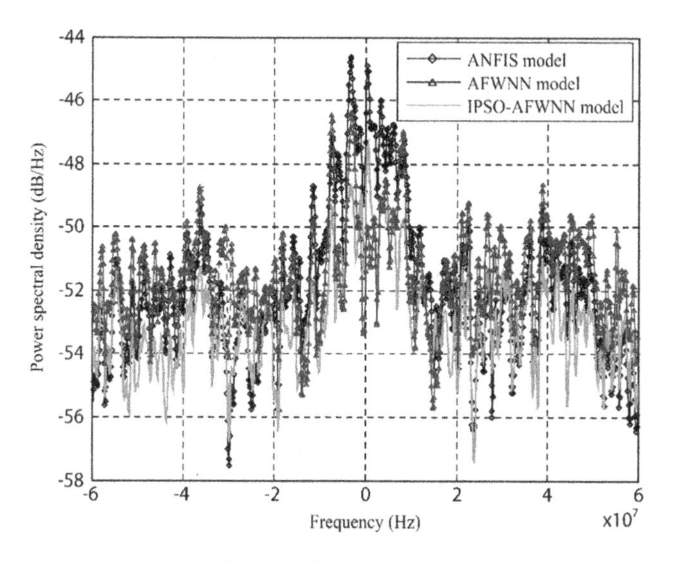

FIGURE 6.26 Error curve of PA model.

6.5. PSO-IOIF-ELMAN NEURAL NETWORK MODELING BASED ON ROUGH SET THEORY

Elman neural network (ENN) is one of the recurrent neural networks (RNNs). Compared with traditional feedforward neural networks such as BP or RBF, ENN has additional feedback connections and can simulate a

TABLE 6.5 Performance Analysis of Three Models

Model	IPSO-AFWNN	ANWNN	ANFIS
MSE	0.2	0.2	0.2
Number of iterations	9	14	23
Running time (s)	12.05	17.36	31.93

higher-order dynamic system [24]. There have been quite a few literatures about an improved ENN. Literature [25] selects the wavelet function with favorable time-frequency domain characteristics and the approximation characteristic as the activation function. Literature [26] introduces the states layer and sets self-feedback coefficients to enhance the adaptability of ENN. Literature [27] combines an adaptive vector quantization (AVQ) clustering method in feedforward network with OIF-Elman network. Literature [28] establishes a two-layer ENN structure. All these improvements aim to enhance the ability of ENN to process dynamic information, enabling the system to adapt to time-varying characteristics. However, these investigations do not consider the influence of input signal amplitude on memory effects and nonlinearities, and therefore the accuracy is not high enough.

To address the above problem and accommodate the advantages and disadvantages of PA modeling in system-level simulation, this section proposes a PA behavioral model based on an improved OIF-Elman neural network adopting the simplified PSO algorithm (PSO-IOIF-Elman). Considering the influences of both small signal and large signal on memory effects and nonlinearities of PA, the proposed model expresses the self-feedback coefficients of OIF-Elman neural network as normalized input and output voltages, in combination with AM/AM and AM/PM distortion. For the proposed model, the simplified PSO algorithm avoids falling into local optimum and the rough set theory corrects and compensates the predicted values of the model to improve the prediction accuracy. In a MATLAB® simulation comparison, it is noted that the proposed PSO-IOIF-Elman model based on the rough set theory produces a training error reduced by 9.53% and delivers a convergence speed increased by 11.31%, compared with the basic Elman (Belman) model. Therefore, the effectiveness and reliability of the proposed modeling method are validated.

6.5.1. OIF-Elman Neural Network Model

As a locally recursive dynamic RNN, ENN structurally consists of four layers: input layer, hidden layer, states layer and output layer. It introduces feedback branches to store internal states so that it can remember historical information. OIF-Elman neural network serves as an improvement on ENN. In the OIF-Elman neural network, outputs from the hidden layer and output layer are fed back to different states layers, respectively. Data from both the states layers and input layer serves as the input to the neural network, and self-feedback connections are added to the states layers (self-feedback coefficients α and β). See Figure 6.27. The closer α and β are to 1, the more historical input information the system contains, which means that a higher-order system can be simulated. When α and β are both 0, a Belman neural network is indicated.

Assume that the connection weight matrices from hidden layer to states layer 1, from input layer to hidden layer, from hidden layer to output layer and from output layer to states layer 2 are represented by $\boldsymbol{\omega}^1$, $\boldsymbol{\omega}^2$, $\boldsymbol{\omega}^3$ and $\boldsymbol{\omega}^4$, respectively. $u(k)$ is the input to input layer; $x_c(k)$, $y_c(k)$ and $x(k)$ are the outputs of states layer 1, states layer 2 and hidden layer, respectively; $y(k)$ is the output of output layer.

$$x(k) = f(\boldsymbol{\omega}^1 x_c(k) + \boldsymbol{\omega}^2 u(k) + \boldsymbol{\omega}^4 y_c(k)) \tag{6.48}$$

$$x_c(k) = \alpha x_c(k-1) + x(k-1) \tag{6.49}$$

$$y_c(k) = \beta y_c(k-1) + y(k-1) \tag{6.50}$$

$$y(k) = g(\boldsymbol{\omega}^3 x(k)) \tag{6.51}$$

where $f(\cdot)$ and $g(\cdot)$ are the excitation functions of hidden layer and output layer, respectively. Generally, sigmoid function and linear function are frequently adopted, in which $f(x) = \frac{1}{1+e^{-x}}$ and $g(x) = x$ apply.

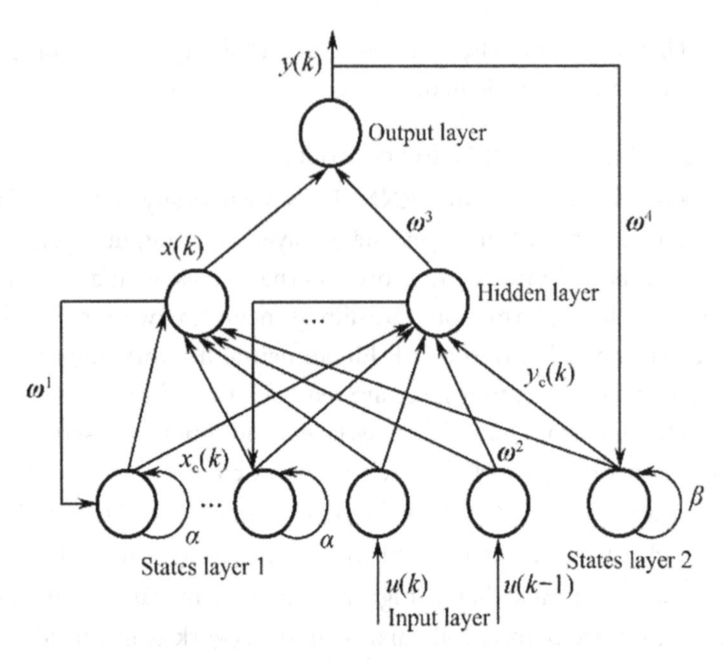

FIGURE 6.27 OIF-Elman neural network.

The AM/AM and AM/PM characteristic curves of PA exhibit scattering points. The smaller the signal, the larger the scattering range. This phenomenon indicates that the PA has remarkable memory effects under small signals, while the PA is less affected by historical information under large signals. On the other hand, when the saturation point is approached, the PA shows stronger static nonlinearities due to its compression performance; that is, large signals suffer more serious static nonlinearity distortion than small signals. It is quite different for small signals and large signals, to some extent, to influence static nonlinearities and memory effects of PA. To exhibit the accuracy and effectiveness of PA behavioral modeling, it is necessary to improve the OIF-Elman neural network model.

First, the input and output voltage data of PA are normalized according to the following equations:

$$\overline{x(k)} = \frac{x(k) - x_{\min}}{x_{\max} - x_{\min}} \tag{6.52}$$

$$\overline{y(k)} = \frac{y(k) - y_{\min}}{y_{\max} - y_{\min}} \tag{6.53}$$

where $x_{\min}$ and $y_{\min}$ are the minimum numbers in the sequences of collected PA IO data, respectively; $x_{\max}$ and $y_{\max}$ are the maximum numbers in the sequences, respectively. The self-feedback coefficients α and β in OIF-Elman neural network are expressed by normalized IO data. The equations are as follows:

$$\alpha(t) = \overline{x(k)}_{\max} - \frac{\overline{x(k)} - \overline{x(k)}_{\min}}{\tau_{\max}} \cdot \tau \tag{6.54}$$

$$\beta(t) = \overline{y(k)}_{\max} - \frac{\overline{y(k)} - \overline{y(k)}_{\min}}{\tau_{\max}} \cdot \tau \tag{6.55}$$

where τ is the number of iterations for the particle and $\tau_{\max}$ is its maximum value.

In this way, the influence of small signal and large signal on memory effects of PA can be differentially exhibited. Moreover, the self-feedback coefficients α and β are formulated to decline with an increase of the input signal, which means weakened memory effects. This algorithm also ensures a strong global convergence ability in the earlier stage and a strong local convergence ability in the later stage, so that the improved OIF-Elman neural network can converge to the target function in a faster and more accurate manner.

6.5.2. OIF-Elman Neural Network with Simplified PSO

The PSO algorithm has been described in Section 6.4.2. The flowchart of OIF-Elman network with simplified PSO is depicted in Figure 6.28.

6.5.3. Correction on Predicted Values of Power Amplifier Based on Rough Set Theory

When a neural network prediction model is used to predict the output voltage of PA, the prediction error is large near the voltage peak, leading to a fluctuation in the prediction accuracy. Therefore, the predicted

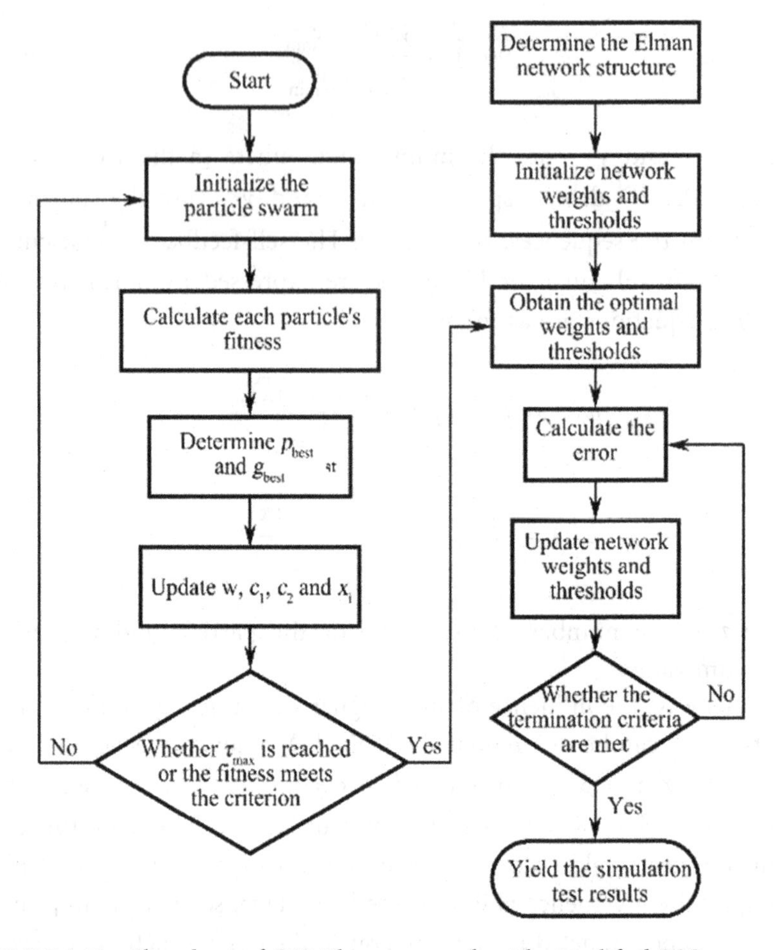

FIGURE 6.28 Flowchart of OIF-Elman network with simplified PSO.

voltage values near the peak are corrected and compensated on the basis of rough set theory.

The predicted output voltage of PA obtained by PSO-IOIF-Elman model simulation is corrected by the following equation:

$$\begin{cases} V'_{t+1} = V_{t+1} + s|p_{t+1} - p_t| \\ p_{t+1} = V_{t+2} - V_{t+1} \\ p_t = V_{t+1} - V_t \end{cases} \quad (6.56)$$

where V'_{t+1} is the predicted voltage value at moment $t + 1$ after correction; V_{t+2}, V_{t+1} and V_t are the predicted voltage values based on the PA model at moments $t + 2$, $t + 1$ and t before correction, respectively; p_{t+1} and p_t are slopes of the prediction function on both sides of moment $t + 1$, respectively; s is a scale factor.

In a given information system $K = (U, A, V, F)$, U is a non-empty set (called domain of discourse) of predicted output voltage values based on the PSO-IOIF-Elman neural network model. $A = C \cup D$ is a non-empty finite set of the information system, where the conditional attribute set C is a collection of input voltage data of PA and the decision attribute set is $D = \{d\}$. d is the solved scale factor s; that is, s is determined by establishing an information system based on the rough set theory. An information function $f(x, a) \in V$ applies where $V = \bigcup_{a \in A} V_a$, $\forall\, a \in A$ and $x \subset U$.

Assume $C = \{a, b, c\}$ and then we have

$$a = \frac{|p_{t+1} - p_t|}{V''_t} \tag{6.57}$$

$$b = \mathrm{sgn}(p_{t+1} - p_t) \tag{6.58}$$

$$c = \left| \frac{V_t}{M_{\max}} \right|_{t=1} \tag{6.59}$$

where sgn denotes a symbolic function; V''_t denotes the actual output voltage at moment t; M denotes the number of data points.

According to Equations (6.57) to (6.59), the conditional attribute set C is extracted from the predicted voltage values output from the improved PSO-IOIF-Elman neural network model. The decision attribute set D is determined according to expert experience. In this way, a complete information system takes its shape. The decision table is generated by means of equal frequency division and discrete processing for both C and D separately. Then, attribute reduction and value reduction are applied to the decision table to yield the minimum decision rule, thereby determining the scale factor s to be used for correcting the predicted output voltage values of PA model.

6.5.4. Power Amplifier Modeling Simulation and Results

Based on previous theoretical investigations, a WCDMA signal is used as the input to simulate the PA designed with the MRF6S21140 transistor made by Freescale. ADS software is used for the circuit design and data extraction for simulation. The equivalent circuit model is shown in Figure 6.29.

In MATLAB® simulation for Elman PA model with memory effects, 200 sets of data are selected as network training data and 100 sets of data are selected as network test data. A two-input single-output model is adopted for modeling and simulation of PA with memory effects. The number of hidden layer neurons is 6, input variables are $u(k)$ and $u(k-1)$, and output variable is $y(k)$. The mean square error (MSE) of test points is used as the error cost function. The expression is

$$\text{MSE} = \frac{1}{n}\sum_{k=1}^{n}(y_d(k) - y(k))^2 \tag{6.60}$$

where $y_d(k)$ is the actual output.

Figures 6.30 and 6.31 depict the training simulation of PSO-IOIF-Elman neural network model before and after rough set correction, respectively. As indicated by a comparison of voltage amplitude waveforms between real outputs and calculated outputs, simulation results of

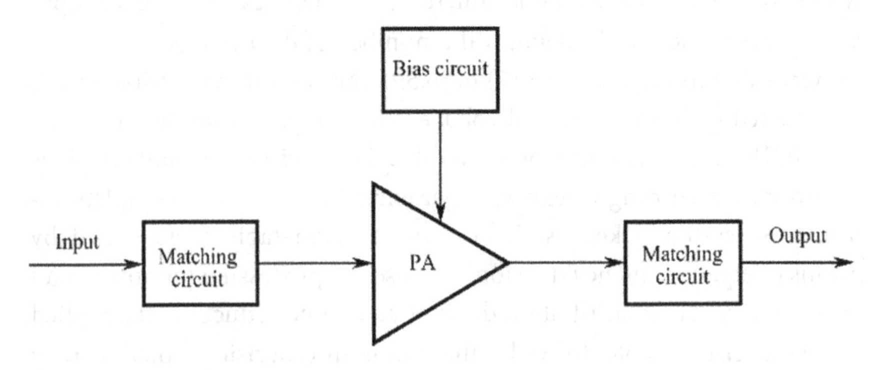

FIGURE 6.29 Equivalent PA circuit model.

training samples before rough set correction have large errors at peaks, while those after rough set correction are closer to real outputs and therefore better simulate the characteristics of RF PA. Figures 6.32 shows the test simulation of PSO-IOIF-Elman model based on the rough set theory. As indicated by a comparison of voltage amplitude waveforms between real outputs and calculated outputs, simulation results of test samples based on the model produce small errors, validating the accuracy of the proposed model.

Figure 6.33 shows the AM/AM characteristic curve of PA, in which the simulation results are basically consistent with the real characteristics. The output signal is distorted to a certain extent compared with the original input signal, accurately exhibiting the nonlinearity of PA.

Figure 6.34 shows a comparison of training data errors based on BP, Belman and PSO-IOIF-Elman after rough set correction. It is clearly observed that the proposed model in this section produces the smallest error.

Table 6.6 gives a comparison of training data error and convergence time based on the four models. It is clearly observed that the PSO-IOIF-Elman model based on the rough set theory produces the smallest training data error (0.5912), reduced by 9.53% compared to the Belman

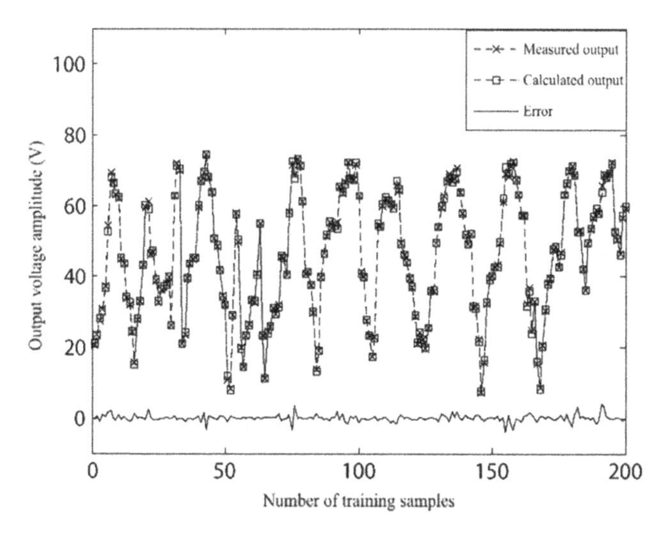

FIGURE 6.30 Training simulation of PSO-IOIF-Elman model before rough set correction.

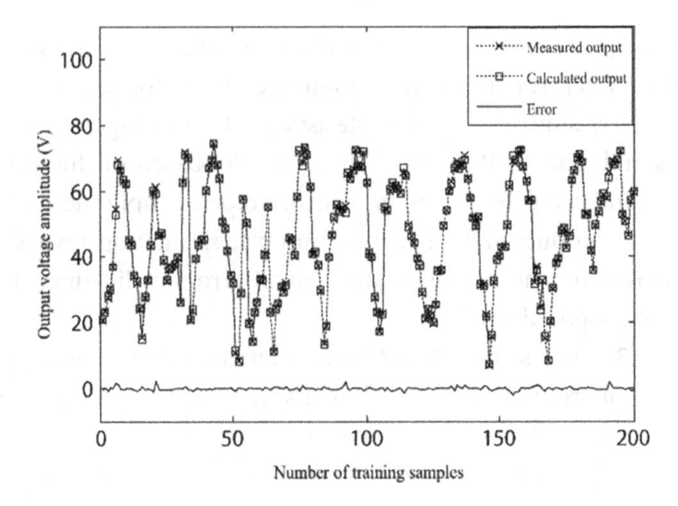

FIGURE 6.31 Training simulation of PSO-IOIF-Elman model after rough set correction.

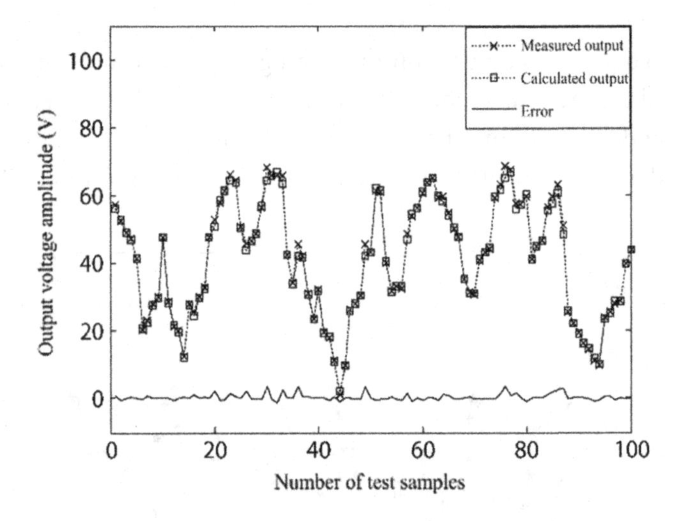

FIGURE 6.32 Test simulation of PSO-IOIF-Elman model based on rough set theory.

model. When the normalized MSE is the same (0.001) for all the models, the PSO-IOIF-Elman model before rough set correction requires a running time of 9.03s, and the PSO-IOIF-Elman model after rough set correction requires 9.25s. Although the proposed model has a slightly

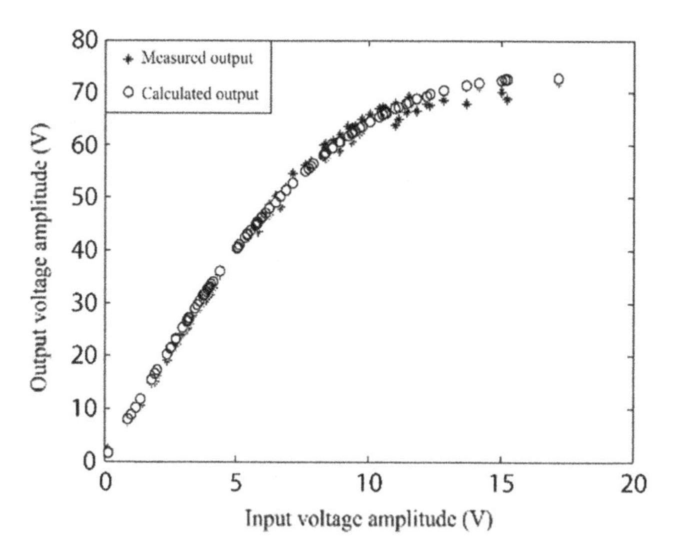

FIGURE 6.33 AM/AM characteristic curve of PA.

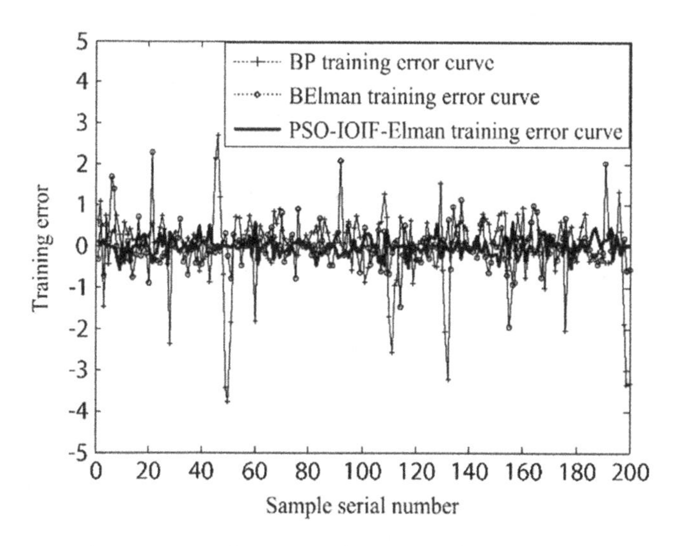

FIGURE 6.34 Comparison of training data errors based on three models.

longer running time, it delivers the optimal model performance on an overall basis.

Based on ENN, this section introduces the influence of input and output voltages expressed as the self-feedback coefficients of OIF-Elman network, and optimizes the weights and thresholds of IOIF-Elman

TABLE 6.6 Comparison of Training Data Error and Convergence Time Based on Four Models

Model	Training Data Error	Running Time (RMSE = 0.001)
BP	0.8344	13.96s
Belman	0.6535	10.43s
PSO-IOIF-Elman (before rough set correction)	0.6104	9.03s
PSO-IOIF-Elman (after rough set correction)	0.5912	9.25s

network by means of the simplified PSO algorithm, thus improving the optimization ability of particles. In this section, the PA behavioral model based on PSO-IOIF-Elman is established and simulated, and the predicted voltage values near the peak are corrected and compensated based on the rough set theory. Simulation results validate the effectiveness and reliability of the proposed model. Furthermore, it dynamically characterizes the nonlinearities and memory effects of PA more accurately, shortens the development cycle, and is proven to be a great help in the actual design.

6.6. NEURAL NETWORK INVERSE MODELING METHOD AND APPLICATIONS

Microwave device modeling based on neural network techniques is the process of neural network forward modeling. In nature, this process selects an appropriate neural network model to approximate or substitute a real device [29–31]. Usually, its inputs include geometrical parameters of the device, such as length and width, while its outputs include electrical parameters, such as scattering parameters, characteristic impedance, quality factor, etc. Microwave device design based on neural network techniques is the process of neural network inverse modeling, in which the inputs to the inverse model are electrical parameters and outputs are geometrical parameters. Both the forward model and inverse model come with the following advantages of neural networks. They have a strong ability to process complex nonlinear problems and allow different devices to be modeled with the same modeling technique. They adopt parallel data processing so that the neural network yields results in a very short time. They do not require prior knowledge about the input or output of the known device, which makes the modeling process simpler [31–34].

In the design of a real microwave device, it is usually the case that the geometrical parameters of the device are to be solved with the known target electrical parameters that the device can achieve. However, EM simulation software solves the electrical parameters after the geometrical parameters are determined. If the target electrical parameters are obtained by constantly changing the geometrical parameters of the device input to EM simulation software, long-term optimization simulation will be required, which is time-consuming. Therefore, it becomes more important to use a neural network inverse model to solve the device's geometrical parameters in an accurate and fast manner. A direct inverse modeling method is suitable for the device having a simple IO relationship. That is, a direct inverse model can be obtained simply by swapping the input and output data used to train the forward model, with the training algorithm identical to that for the forward model. However, the accuracy of a direct inverse model may fail to meet the requirements in the case of a complex IO relationship, basically because the inverse model cannot address the problem of multivalued solutions for parameters, i.e., correspondence between an electrical parameter and multiple values of a geometrical parameter. To address this problem, Humayun Kabir, Dr. K. Sri Rama Krishna and Padarthi Vijaya Kumar et al. proposed an inverse modeling method [35,36]. Firstly, a neural network forward model is trained. Next, the training data of inverse model is divided into groups according to certain rules, and each group of data is used to train a separate inverse submodel. Then, the forward model is used to combine all the inverse submodels into a complete model. If the combined inverse model reaches the desired accuracy, the procedure will stop; otherwise, the data will be further grouped and trained until the accuracy requirements of the inverse model are satisfied. Liu Zhao proposed an inverse modeling method optimized by back propagation (BP) algorithm in his doctoral thesis. This method uses the target electrical parameters as the input to the direct inverse model, and uses the geometrical parameters yielded by the direct inverse model as the output of the simulation module. The simulation module is regarded as an equivalent circuit or empirical formula for the device. Next, the electrical parameters yielded by the simulation module are compared with the target electrical parameters, and the error is propagated back to the direct inverse model. Then, the weight values of the inverse model are regulated accordingly, until the error between output electrical parameters and target parameters meets the requirements [37]. Heriberto Jose Delgado et al. also proposed an inverse modeling method by means of algorithm optimization [38]. These methods may establish inverse models meeting accuracy requirements, although at the cost of extremely complicated processes notably slowing down the modeling speed. This is because these methods establish both the inverse model and forward model or equivalent circuit and empirical formula, in the context of massive iterative calculations. Moreover, some of these methods are not applicable for the devices without equivalent circuit or empirical formula. In actual engineering, engineers pursue fast and accurate design. Simulation software can reach the desired accuracy but at a particularly low speed. This is not tolerable because neural network modeling techniques are practically expected to complete microwave device design in a quick fashion while meeting accuracy requirements.

Neural network inverse modeling techniques are frequently used for microwave device design. The direct inverse modeling method cannot reach the desired accuracy when a complicated IO relationship is observed for a device, while other inverse modeling methods that can reach the desired accuracy have complex structures and high computational intensities. Therefore, this section proposes a novel inverse modeling method for microwave device design. This method establishes only a neural network forward model and updates the input parameters of the forward model by adaptively tuning the learning rate of the steepest descent method with the weight values kept unchanged, thereby minimizing the error between the model output and ideal output and achieving the function of the inverse model. This method is much simpler than other methods because it achieves the function of an inverse model without the need to establish it separately. The introduction of adaptive learning rate further improves the speed and accuracy of the proposed model. In the application of impedance transformer design, the proposed method is demonstrated to reduce the running time by 7.49% and improve the MSEs of solved length and frequency by 99.95% and

98.81% for the transformer respectively, compared with the direct inverse modeling method. The proposed method can address the problem of multivalued solutions and be applied to the actual microwave device design.

6.6.1. Inverse Modeling Method

This subsection proposes a fast and accurate inverse modeling method, aiming to address the problem that the existing inverse modeling methods cannot combine modeling speed and accuracy. A neural network inverse model is designed to accurately compute the correlated geometrical parameters for given values of target electrical parameters of a device. A neural network forward model uses given geometrical parameters of a device as its input to compute the electrical parameters as its output. The inverse modeling method proposed in this section is to realize the function of an inverse model based on a forward model. That is, the forward model of a device is enabled to serve as the inverse model by means of an update algorithm, eliminating the need to establish the inverse model separately. In this manner, the modeling process is simplified. The specific scheme is summarized into 5 steps.

1. Extract the given device data and train the neural network forward model. The input to the forward model is a geometric parameter $\boldsymbol{X}$ of the device, the computed output of the forward model is an electrical parameter $\boldsymbol{Y}$, and the ideal output is an electrical parameter $\boldsymbol{D}$. Therefore, $\boldsymbol{Y}$ and $\boldsymbol{D}$ are expected to be equal or within a certain error range, ensuring the accuracy of the forward model.

2. Save the weight parameters of the established forward model and keep them unchanged.

3. Additionally, extract another set of geometric parameter $\boldsymbol{X}_1$ and target electrical parameter $\boldsymbol{D}_1$ of the device. Generate the difference between target electrical parameter $\boldsymbol{D}_1$ and computed output $\boldsymbol{Y}$ of the forward model, which is recorded as the error E.

4. Keep the network weight parameters unchanged. Update the input $\boldsymbol{X}$ of the forward model by adaptively tuning the learning rate of the steepest descent method to minimize the error.

5. Evaluate the performance of the model based on the fitting effect between the updated $\boldsymbol{X}$ and the ideal $\boldsymbol{X_1}$. The smaller the error between the two, the better the performance of this model.

The proposed method realizes the function of inverse model by updating the input data based on a forward model. Specifically, the given target electrical parameter $\boldsymbol{D_1}$ is substituted into the network and eventually the geometric parameter $\boldsymbol{X_1}$ is solved. Since the weight parameters are kept unchanged in the entire modeling process, the relationship between input and output parameters of the network does not change, thus avoiding the correspondence between an electrical parameter and multiple values of a geometrical parameter in the solution. Therefore, the geometrical parameter solution in the inverse process is the optimal solution at the moment, ensuring the accuracy of the solution. Moreover, the model has a significantly reduced complexity and accelerated modeling speed because only the neural network forward model is developed in the entire modeling process. The introduction of adaptive learning rate in the update algorithm enables the modeling process to be faster and more accurate.

6.6.2. Update Algorithm

While keeping the weights unchanged, the new inverse modeling method changes the input parameters of the neural network forward model to realize the function of the inverse model. The update algorithm is used to change the input parameter $\boldsymbol{X}$ of the forward model, as expressed in Equation (6.61). Compared to the BP algorithm with a variable learning rate, this update algorithm only alters the updated object but its meaning is quite different.

$$X'_i = X_i - \eta \frac{\partial E}{\partial X_i} \tag{6.61}$$

where X'_i is the ith input data after the update; X_i is the ith input data before the update; E is the sum of squares of the errors between the electrical parameter $\boldsymbol{Y}$ output by the forward model and the given target electrical parameter $\boldsymbol{D_1}$; η is the adaptive learning rate. The value of adaptive learning rate makes a great difference on the performance of the network and helps prevent the update algorithm from falling into the local minimum. Its expression is as follows:

$$\eta(n+1) = \eta(n)\lambda \tag{6.62}$$

where $\eta(n+1)$ and $\eta(n)$ represent the learning rates of the $n+1$ th iteration and the n th iteration, respectively;

$$\lambda = a^{-\text{sign}(\Delta E)} = \begin{cases} a, & \Delta E < 0 \\ 1/a, & \Delta E > 0 \\ 1, & \Delta E = 0 \end{cases}$$

where a is a constant greater than 1, usually specified as 2. ΔE is the difference between error sums of squares for the two iterations. By use of the introduced adaptive learning rate, the learning speed is accelerated in the same direction if $\Delta E < 0$ and is reduced with an altered direction if $\Delta E > 0$, thereby ensuring the accuracy.

Since a two-layer neural network can approximate any nonlinear functional relationship, the forward model adopts a two-layer BP neural network for the sake of simplicity. Its structure is depicted in Figure 6.35. Assume that the neural network has N inputs, the hidden layer has L neurons and the output layer has M neurons. The activation function $F(\cdot)$ for hidden layer neurons is a logsig function, and the activation function for output layer neurons is a purelin function.

Based on Figure 6.35, the output $Z(l,p)$ of the lth neuron in the hidden layer is expressed as

$$Z(l,p) = F\left(\sum_{i=1}^{N} v(l,i)X(i,p) + b_{1l}\right) \tag{6.63}$$

where p represents the data; $v(l,i)$ represents the weight from the ith input to the lth neuron in the hidden layer of the network; b_{1l} represents the threshold for the lth neuron in the hidden layer. The derivative for the $F(\cdot)$ function is $F' = F(1-F)$. Therefore, the derivative for the output $Z(l,p)$ of the lth neuron in the hidden layer to the ith input X_i can be expressed as

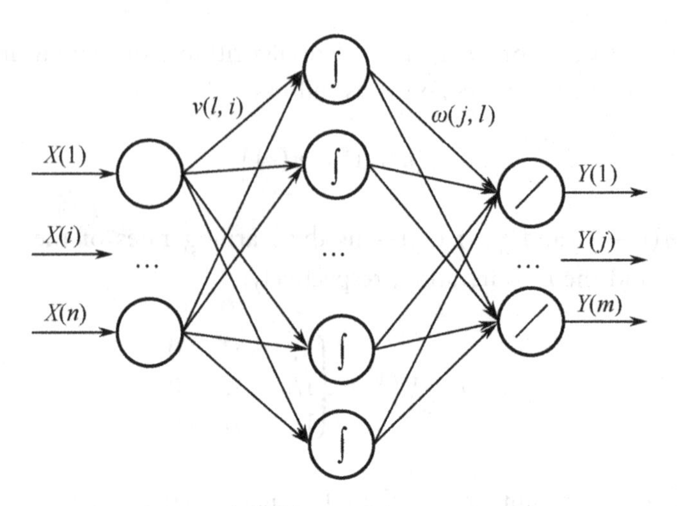

FIGURE 6.35 Neural network structure diagram.

$$\frac{\partial Z(l,p)}{\partial X_i} = v(l,i)Z(l,p)(1 - Z(l,p)) \tag{6.64}$$

The output $Y(j, p)$ of the jth neuron in the linear output layer of the network can be expressed as

$$Y(j,p) = \sum_{l=1}^{L} \omega(j,l)Z(l,p) + b_{2j} \tag{6.65}$$

where $\omega(j, l)$ is the weight parameter from the lth neuron in the hidden layer to the jth neuron in the output layer, and b_{2j} is the threshold for the jth neuron in the output layer.

The sum of squared error E between the output $\boldsymbol{Y}$ and the given target electrical parameter $\boldsymbol{D}_1$ of the neural network forward model can be expressed as

$$E = \sum_{p=1}^{P} E_p = \frac{1}{2}\sum_{p=1}^{P}\sum_{j=1}^{M}(Y(j,p) - D_1(j,p))^2 \tag{6.66}$$

where E_p represents the sum of squared error yielded by Group p.

The following equation is yielded by derivation:

$$\frac{\partial E}{\partial X_i} = \sum_{p=1}^{P} \sum_{l=1}^{L} \sum_{j=1}^{M} \left[(Y(j,p) - D_1(j,p))\omega(j,l) Z(l,p)(1 - Z(l,p)) v(l,i) \right] \tag{6.67}$$

Substituting Equation (6.67) into Equation (6.61) to solve the update algorithm for the input $\boldsymbol{X}$ in Equation (6.61), thereby further solving the input (geometrical parameter) to the forward model.

6.6.3. Application Examples and Simulation Analysis

This subsection uses the simplest quarter-wavelength impedance transformer as an example to validate the features of fast speed and high accuracy delivered by the new inverse modeling method. For validation of the accuracy in solving the transformer's geometrical parameters, the new inverse modeling method is compared with the direct inverse modeling method and ADS software, respectively. For validation of the solution speed, the new method is compared with the direct inverse modeling method.

1. Quarter-wavelength impedance transformer

A model for a quarter-wavelength impedance transformer is built in ADS software, as depicted in Figure 6.36.

As indicated in Figure 6.36, TL2 is a quarter-wavelength microstrip transmission line with a length defined by variable l. TL2 is designed to

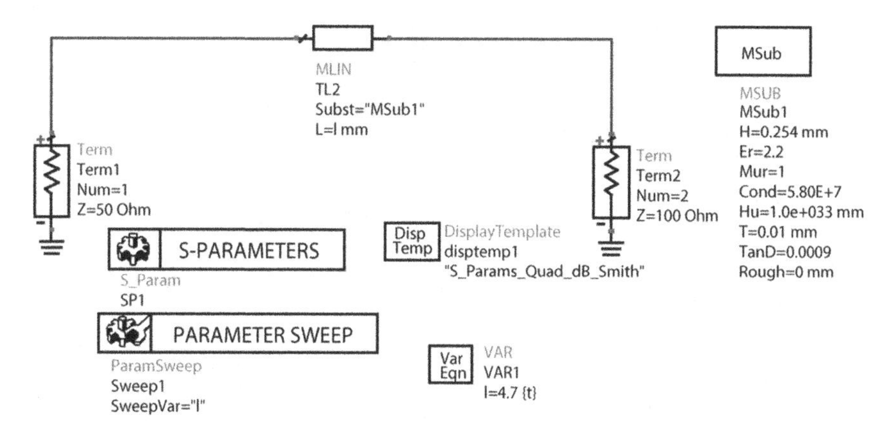

FIGURE 6.36 Model for quarter-wavelength impedance transformer in ADS.

achieve impedance matching between a 50 Ω port and a 100 Ω port. Theoretically, when l is equal to a quarter wavelength, the entire circuit is expected to resonate at the center frequency, where S_{11} reaches the minimum value. In practice, however, the microstrip line width varies and a frequency shift may occur, because the characteristic impedance of TL2 differs from the impedances of both ports. It is time-consuming to use simulation software to constantly regulate the values of l and f for a minimum S_{11}. Therefore, we adopt a neural network inverse model to design the impedance transformer and solve the values of l and f corresponding to S_{11}. We use l and f as inputs to a neural network forward model and S_{11} as its output. Extract more than 2,000 sets of IO data from the transformer, as shown in Figure 6.36, where 20 sets of l and 101 sets of f are extracted. Figure 6.37 illustrates S_{11} varying with different values of l and f. It is noted that one S_{11} corresponds to multiple l and f values, respectively. If a known S_{11} is taken as the input, it also corresponds to multiple l and f values.

Simulation results of the new inverse modeling method are compared to those of the direct inverse modeling method and software-based precise simulation respectively, for validation of the solution speed and accuracy delivered by the new method. For the direct inverse modeling method, the input is the target electrical parameter S_{11}, the outputs include the geometrical parameter l and frequency f, and 400 sets of

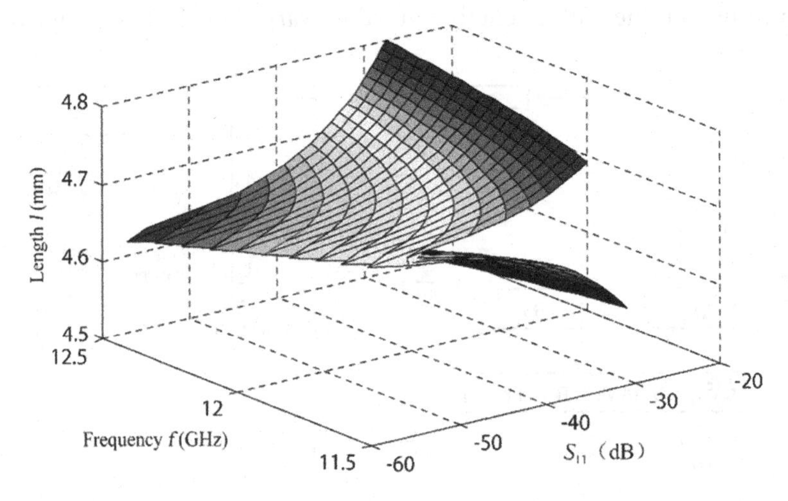

FIGURE 6.37 S_{11} variation curve.

training samples and another 75 sets of test samples are selected. For the new inverse modeling method, the neural network model is trained as per the modeling procedure in Section 6.6.2. Firstly, a neural network forward model is established. Four hundred (400) sets of training data and 75 sets of test data are used, which are the same as those for the direct inverse modeling method. The inputs are l and f, and the output is S_{11}. Then, the weight parameters and S_{11} for the 75 sets of test data of the trained forward model are extracted and substituted into the update algorithm in Section 6.6.3 to update the inputs l and f to the forward model. Software simulation uses a two-layer BP neural network depicted in Figure 6.36. The network model is trained and tested by means of MATLAB® programming. Based on the 75 S_{11} values of test data, the length value l of TL2 obtained by the direct inverse modeling method and new inverse modeling method and the l value corresponding to ADS simulation are shown in Figure 6.36; the frequency value f obtained by the direct inverse modeling method and new inverse modeling method and the f value corresponding to ADS simulation are shown in Figure 6.37.

In Figures 6.38 and 6.39, triangles denote l and f values corresponding to the 75 S_{11} values in ADS simulation; stars denote l and f values solved by the new inverse modeling method with the 75 S_{11} values as the inputs; circular points denote l and f values solved by the direct inverse modeling method with the 75 S_{11} values as the inputs. As indicated in Figures 6.38 and 6.39, the solutions yielded by the direct inverse modeling method cannot fit the geometrical parameters of the device at all. This is because the IO relationship of the transformer is complicated, as shown in Figure 6.37. An S_{11} corresponds to multiple l values and multiple f values, respectively. In this case, the direct inverse modeling method cannot yield multiple solutions simultaneously, let alone which solution is the optimal solution. Consequently, this method cannot accurately characterize such a device. In contrast, the new inverse modeling method accurately fits the device's l and f values yielded in ADS simulation. This is because the weight parameters of the neural network forward model do not vary all the time during the whole modeling process, indicating a fixed IO relationship for the network. In this way, the nonuniqueness of IO relationship is addressed, and the geometrical parameter solved in the

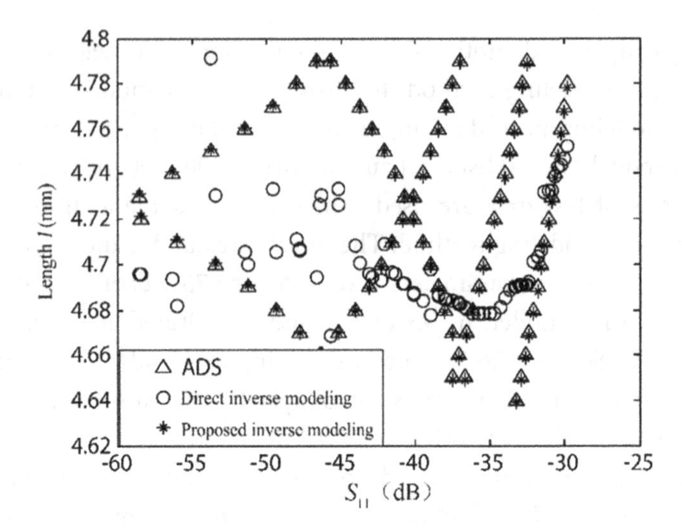

FIGURE 6.38 Comparison of TL2 lengths based on different methods.

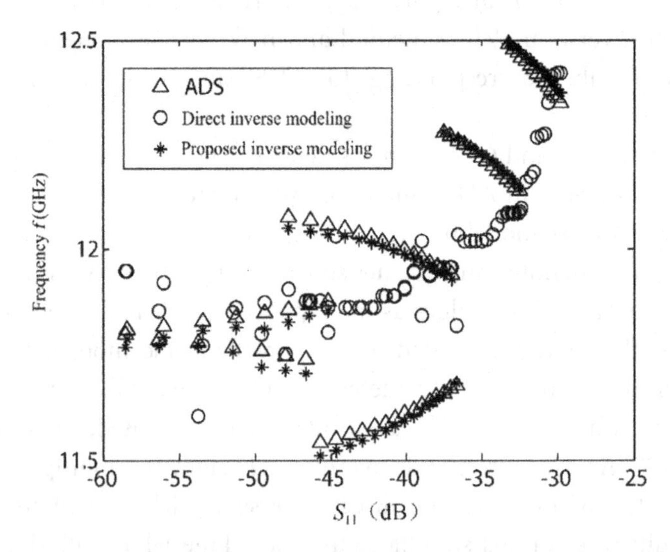

FIGURE 6.39 Comparison of frequencies based on different methods.

inverse process is the optimal solution at the moment. Therefore, the new inverse modeling method can fully meet the needs of practical device design, regarding the accuracy of inverse model.

The modeling method based on data grouping proposed by Literatures [29], [35] and [36] and the modeling method based on algorithm optimization proposed by Literatures [37,38] are both developed by means of multiple iterations on the basis of direct inverse modeling. Therefore, these methods have much lower modeling speeds compared to the direct inverse modeling method, and we only need to compare the direct inverse modeling method and the new inverse modeling method, regarding their modeling speeds. These two modeling methods have the same network complexity, so the network running time will not differ too much. The performance parameters of the direct inverse modeling method and the new inverse modeling method are listed in Table 6.7. The total running time of the network includes the training time and test time. The network structure is indicated by "number of model input parameters - number of hidden layer neurons of network - number of output layer neurons of network".

As indicated in Table 6.7, the new inverse modeling method improves the MSEs of solved length l and frequency f by 99.95% and 98.81% for the transformer respectively, compared with the direct inverse modeling method having the same network structure complexity. In addition, the new inverse modeling method is demonstrated to reduce the network running time by 7.49% compared with the direct inverse modeling method. In conclusion, the new inverse modeling method reduces the modeling complexity and accelerates the modeling speed, on the basis that its accuracy can meet the actual needs. The new method has proven to achieve an optimal trade-off between accuracy and speed in using neural networks to design microwave devices.

Neural networks support the same modeling technique applied to different devices. This advantage enables the new inverse modeling

TABLE 6.7 Performance Comparison of Different Modeling Methods

	Mean Square Error of Length l	**Mean Square Error of Frequency f**	**Total Network Running Time (s)**	**Neural Network Structure**
Direct inverse model	0.0024	0.0503	8.891878	1-27-2
Adaptive inverse model	0.0000013113	0.00059811	8.225913	2-27-1

method to solve geometrical parameters with given electrical parameters in design of any other active/passive device, provided that there is a certain mapping between electrical parameters and geometrical parameters.

This section proposes a new neural network inverse modeling method, which proves to accurately and quickly solve the geometrical parameters of microwave devices. This method updates the input parameters of the forward model by adaptively tuning the learning rate of the steepest descent method, thereby achieving the function of the inverse model. This method keeps the weight values of the forward model unchanged to guarantee the accuracy of inverse modeling, and enables an inverse model based on a forward model without the need to establish a separate inverse model, thereby accelerating the modeling speed. Although the efficiency of this method is validated only with a simple impedance transformer, this method can be used in design of any other active/passive device, since neural network modeling techniques do not require any prior knowledge of devices.

REFERENCES

1. NASKAS N， PAPANANOS Y. Neural-Network-Based Adaptive Baseband Predistortion Method for RF Power Amplifier[J]. IEEE Transaction on Circuits and Systems-II：Express Briefs， 2004， 51(11):619–623.
2. MAGNUS I， DAVID W， DANIEL R. Wide-Band Dynamic Modeling of Power Amplifiers Using Radial-Basis Function Neural Networks[J]. IEEE Transactions on Microwave Theory and Techniques， November 2005， 53(11):3422–3428.
3. 郑建国， 刘芳， 焦李成. 基于正交校正共扼梯度法的快速神经网络学习算法研究[J]. 电子与信息学报， 2002， 24(5):667–670.
4. ISAKSSON M， WISELL D， RONNOW D. Nonlinear Behavioral Modeling of Power Amplifiers Using Radial-Basis Function Neural Networks. IEEE MTT-S International Microwave Symposium Digest， Long Beach，CA，2005[C].Piscataway：IEEE，4.
5. ISAKSSON M， WISELL D， RONNOW D. A Comparative Analysis of Behavioral Models for RF Power Amplifiers[J]. IEEE Transactions on Microwave Theory and Techniques， 2006， 54(1):348–359.
6. 翟建锋， 周健义， 赵嘉宁， 等. 射频功率放大器前馈输入延迟神经网络模型（英文）[J]. Journal of Southeast University(English Edition)， 2008(01):6–9.

ZHAI J, ZHOU J, ZHAO J, et al. Behavioral Modeling of RF Power Amplifiers with Time-Delay Feed-Forward Neural Networks[J]. Journal of Southeast University (English Edition), 2008(01):6–9.

7. 于海雁， 林茂六， 许洪光. 基于 RBF 神经网络的射频功放器件大信号建模方法研究[J]. 哈尔滨工业大学学报， 2004, 16(4):89–92.
   YU H, LIN M, XU H. Large-Signal Modeling Method of RF Power Amplifier Based on RBFNN[J]. Journal of Harbin Institute of Technology, 2004, 16(4):89–92.
8. 孔范增， 季仲梅. 基于 BP 神经网络的视频H. 223协议标志符动态提取算法[J]. 计算机应用研究， 2010, 27(6):2113–2115.
   KONG F, JI Z. Dynamic Identify Algorithm for H.223 Protocol Identifier Based on Neural Network[J]. Application Research of Computers, 2010, 27(6):2113–2115.
9. 李郁松， 郭裕顺. 基于模糊逻辑的器件与电路建模技术[J]. 电路与系统学报， 2004， 9(01):59–63.
   LI Y, GUO Y. Circuits and Devices Modeling by Means of Fuzzy Logic[J]. Journal of Circuits and Systems, 2004, 9(01):59–63.
10. 邓洪敏， 何松柏， 虞厥邦. 基于模糊神经网络的自适应预失真功放[J]. 信号处理， 2003， 19(04):334–337.
    DENG H, HE S, YU J. Fuzzy Neural Network Based Adaptive Predistortion Power Amplifier[J]. Journal of Signal Processing, 2003, 19(04):334–337.
11. 翟建锋， 周健义， 洪伟， 等. 有记忆效应的功放实数延时模糊神经网络模型[J]. 微波学报， 2009， 25(05):41–44.
    ZHAI J, ZHOU J, HONG W, et al. Real-Valued Time-Delay Neuro-Fuzzy Model for Power Amplifier with Memory Effects[J]. Journal of Microwaves, 2009, 25(05):41–44.
12. 徐飞， 郭裕顺. 射频功率放大器的建模[J]. 电子器件， 2010， 33(03):384–387.
    XU F, GUO Y. The Modeling of RF Power Amplifier Circuit[J]. Chinese Journal of Electron Devices, 2010, 33(03):384–387.
13. 周燕， 刘培玉， 赵静， 等. 基于自适应惯性权重的混沌粒子群算法[J]. 山东大学学报（理学版), 2012, 47(3):27–32.
    ZHOU Y, LIU P, ZHAO J, et al. Chaos Particle Swarm Optimization Based on the Adaptive Inertia Weight[J]. Journal of Shandong University (Natural Science), 2012, 47(3):27–32.
14. CATALAO J P S， POUSINHO H M I， MENDES V M F. Hybrid Wavelet-PSO-ANFIS Approach for Short-Term Electricity Prices Forecasting[J]. Transactions on Power Systems， 2011， 26(1):137–144.
15. 黎林， 朱军， 刘颖， 等. 改进动量粒子群优化神经网络的语音端点检测[J]. 计算机工程与应用， 2013， 49(5):225–229.
    LI L, ZHU J, LIU Y, et al. Speech Endpoints Detection Based on BP Neural Network Optimized by Improved Momentum Particle Swarm

Optimization Algorithm[J]. Computer Engineering and Applications, 2013, 49(5):225–229.

16. 南敬昌， 任建伟， 张玉梅． 基于 PSO_BP 神经网络的射频功放行为模型[J]． 微电子学， 2011， 41(05):741–745.
NAN J, REN J, ZHANG Y. Study of Dynamic Behavioral Model for RF Power Amplifier Based on PSO BP Neural Network[J]. Microelectronics, 2011, 41(05):741–745.
17. 王瑞娜， 刘桂红， 南敬昌． 基于 PSO-SVM 的射频功率放大器模型研究[J]． 微电子学， 2013， 43(4):554–557.
WANG R, LIU G, NAN J. Modeling of RF Power Amplifiers Based on PSO_SVM[J]. Microelectronics, 2013, 43(4):554–557.
18. 胡旺， 李志蜀． 一种更简化而高效的粒子群优化算法[J]． 软件学报， 2007， 18(4):861–868.
HU W, LI Z. A Simpler and More Effective Particle Swarm Optimization Algorithm[J]. Journal of Software, 2007, 18(4):861–868.
19. 赵志刚， 张振文， 张福刚． 自适应扩展的简化粒子群优化算法[J]． 计算机工程与应用， 2011， 47(18):45–47.
ZHAO Z, ZHANG Z, ZHANG F. Simplified Particle Swarm Optimization Algorithm with Adaptive Extended Operator[J]. Computer Engineering and Applications, 2011, 47(18):45–47.
20. 孙娜， 张桂玲， 鄂明杰． 基于模糊小波神经网络的主机入侵预测[J]． 计算机工程， 2012， 38(8):89–91.
SUN N, ZHANG G, AO M. Host Intrusion Prediction Based on Fuzzy Wavelet Neural Network[J]. Computer Engineering, 2012, 38(8):89–91.
21. OYSAL Y， YILMAZ S. An Adaptive Fuzzy Wavelet Neural Network with Gradient Learning Algorithm for Nonlinear Function Approximation，Networking，Sensing and Control(ICNSC)，2013 10th IEEE International Conference on IEEE，Evry，France，2013[C].Piscataway: IEEE， 152–157.
22. 周丹， 南敬昌， 高明明． 改进的简化粒子群算法优化模糊神经网络建模[J]． 计算机应用研究， 2015， 32(4):1000–1003.
ZHOU D, NAN J, GAO M. Fuzzy Neural Network for Modeling Based on Improved Simplified Particle Swarm Optimization[J]. Application Research of Computers, 2015, 32(4):1000–1003.
23. EL SOUSY， F F M. Intelligent Optimal Recurrent Wavelet Elman Neural Network Control System for Permanent-Magnet Synchronous Motor Servo Drive[J]. Industrial Informatics，IEEE Transactions on， 2015， 9(4):1986–2003.
24. 杨春． 基于神经网络的非线性预测控制算法的研究[D]． 太原理工大学， 2012.
YANG C. Research of Nonlinear Model Predictive Control Algorithm Based on Neural Network[D]. Taiyuan University of Technology, 2012.
25. LI WANG， JIE SHAO， YAQIN ZHONG， et al. Modeling Based on Elman Wavelet Neural Network for Class-D Power Amplifiers[J]. Applied Mathematics & Information Sciences， 2013， 7(6):2445–2453.

26. 陈龙， 张可， 罗配明． 改进的 Elman 神经网络在 WSNs 距离预测中的应用[J]． 传感器与微系统， 2013， 32(1):149–152.
CHEN L, ZHANG K, LUO P. Application of Improved Elman Neural Network in Distance Prediction of WSNs[J]. Transducer and Microsystem Technologies, 2013, 32(1):149–152.
27. 朱小龙， 杨建国， 代贵松． 基于 AVQ 聚类和 OIF-Elman 神经网络的机床热误差建模[J]． 上海交通大学学报，2014，48(1):16–21.
ZHU X, YANG J, DAI G. AVQ Clustering Algorithm and OIF-Elman Neural Network for Machine Tool Thermal Error[J]. Journal of Shanghai Jiaotong University, 2014, 48(1):16–21.
28. 吴泽志， 傅佳． Elman 神经网络改进模型在脑膜炎诊断中的应用[J]． 计算机工程与应用， 2014， 50(3):221–226.
WU Z, FU J. Application of Improved Elman Neural Network Model in Diagnosis of Meningitis[J]. Computer Engineering and Applications, 2014, 50(3):221–226.
29. HUMAYUN KABIR， YING WANG， MING YU， et al. Neural Network Inverse Modeling and Applications to Microwave Filter Design [J]. IEEE Transactions on Microwave Theory and Techniques， 2008， 56(4):867–879.
30. 王亚静． 神经网络辨识及自适应逆控制研究[D]． 秦皇岛：燕山大学电气工程学院， 2010.
WANG Y. Study on Neural Network Identification and Adaptive Inverse Control[D]. Qinhuangdao: Institute of Electronic Engineering of Yanshan University, 2010.
31. HUMAYUN KABIR， MING YU， QI JUN ZHANG. Recent Advances of Neural Network-Based EM-CAD[J]. International Journal of RF and Microwave Computer- Aided Engineering， 2010， 20(5):502–511.
32. 张翼鹏， 陈亮， 郝欢． 一种改进的量子神经网络训练算法[J]．电子与信息学报， 2013， 35(7):1630–1635.
ZHANGY, CHEN L, HAO H. An Improved Training Algorithm for Quantum Neural Networks[J]. Journal of Electronics and Information Technology, 2013, 35(7):1630–1635.
33. JUAN PASCUAL GARCIA， FERNANDO QUESADA PEREIRA， DAVID CANETE REBENAQUE， et al. A Neural Network Method for the Analysis of Multilayered Shielded Microwave Circuits[J]. IEEE Transactions on Microwave Theory and Techniques， 2006， 54(1):309–320.
34. 张强， 许少华， 刘丽杰． 量子混合蛙跳算法在过程神经网络优化中的应用[J]． 信号处理， 2013， 29(8):1003–1011.
ZHANG Q, XU S, LIU L. Application of Quantum Shuffled Frog Leaping Algorithm in Process Neural Networks Optimization[J]. Journal of Signal Processing, 2013, 29(8):1003–1011.
35. SRI RAMA KRISHNA DR K， LAKSHMI NARAYANA J， PRATAP REDDY Dr L. A Neural Network Inverse Modeling Approach for the

Design of Spiral Inductor[J]. International Journal Of Computer Science & Engineering Technology， 2011，2(3):54–62.

36. PADARTHI VIJAYA KUMAR， DR K SRI RAMA KRISHNA，JLAKSHMI NARAYANA. A Novel Inverse Model Approach for the Design of the Spiral Inductor Using Artificial Neural Networks[J]. International Journal Of Advanced Engineering Sciences And Technologies， 2011，5(2):132–138.
37. 刘钊．微波神经网络技术研究[D]．天津：天津大学电子与信息工程学院，2004.
   LIU Z. Study of Microwave Neural Network[D]. Tianjin: School of Electronics and Information Engineering of Tianjin University, 2004.
38. HERIBERTO JOSE DELGADO, MICHAEL H， THURSBY. A Novel Neural Network Combined with FDTD for the Synthesis of a Printed Dipole Antenna[J]. IEEE Transactions on antennas and propagation， 2005， 53(7):2231–2236.

CHAPTER 7

Power Amplifier Modeling with X-Parameters

Modern communication systems are pursuing higher power, efficiency and linearity, which raises higher requirements on the performance of devices. RF nonlinearities become particularly significant in the context of large-signal excitation, which seriously affects system performance [1]. Consequently, nonlinear device modeling has become one of the most urgent problems in RF microwave circuit design. Linear S-parameters are not able to accurately characterize RF device behaviors, as RF devices often work in the nonlinear region or even in a saturation state. In recent years, scholars have proposed a theoretical modeling method that can characterize the nonlinearities of large signals, namely, PA modeling based on X-parameters.

7.1 DESIGN OF WIDEBAND POWER AMPLIFIER BASED ON X-PARAMETER TRANSISTOR MODEL

To accommodate RF nonlinearity modeling techniques, this section proposes a method extracting load-independent transistor X-parameters and uses the transistor model for PA design. This method introduces the load-pull technique into the traditional X-parameter extraction process and obtains the load-independent transistor X-parameter model and its optimal load resistance by means of iterations. Compared with the load-dependent X-parameter model [1], the proposed model is more suitable for analyzing and designing circuits, featuring a small amount of data, simple procedure and a short design period. The extracted transistor X-parameter model is used to design a wideband PA with a bandwidth ranging 1.7 to 2.7 GHz. Its matching circuit is designed based on a Chebyshev low-pass filter, and the bias circuit is designed with a double-sector open-circuit microstrip line structure. The comparison between measured data and simulated data indicates that, over the operating frequency range, the output power remains greater than 4 W with a power-added efficiency (PAE) higher than 45% and the power gain remains around 14 dB. The results validate the effectiveness of the proposed method.

Literature [2] proposes a measurement-based poly-harmonic distortion model to characterize large signals for the first time. Based on the findings, Professor Wang Jiali et al. proposed the concept and test method of nonlinear scattering function based on time-domain measurements [3]. In 2008, Agilent proposed the theory and test system of the X-parameter model based on frequency-domain measurements, which is an extension of S-parameters in the context of large-signal nonlinearity. In addition, the Cardiff model proposed by Professor TASKER P. J. from Cardiff University, United Kingdom, is also a nonlinear behavioral model based on IO waveforms. These novel nonlinear models are all black-box behavioral models based on measurements, which are designed to accurately characterize the nonlinearities of devices in the context of large signals. X-parameters have attracted extensive attention as a method to characterize, model, and design nonlinear devices. Literature [4] proposes an enhanced X-parameter model (i.e., load-dependent X-parameters) that extends the extraction of X-parameters from all impedance points over the entire Smith chart. Literature [5] proposes an analytical approach to obtain the optimum load impedance of X-parameter model. This approach optimizes the impedance value by means of mathematical operations to achieve the maximum power transmission. The approaches proposed in Literatures [4] and [5] require a fixed load impedance and are not suitable for analyzing or designing circuits.

7.1.1 Extraction of X-Parameters

Two approaches are available for the extraction of X-parameters. One is to use an NVNA and the other is the X-parameter generator in ADS simulation. The extracted X-parameters are imported into ADS and used in nonlinear simulation and design for circuits.

In the extraction of X-parameters, it is necessary to determine the basic elements in modeling, such as operating frequency, bias voltage and input power, and also to consider the matching state of the device. Usually, the matching load impedance is set to 50 Ω for extraction of parameters and modeling; according to Literature [6], simulation results in this context are in good agreement with measured results. However, the load impedance for a high-power transistor will be far away from 50 Ω and in a mismatched state when the ideal performance is achieved. In this case, the influence of load impedance should be considered. Literature [7] proposes a combination of the load-pull technique with NVNA in an impedance mismatching condition, introduces the source and impedance tuners, and extracts X-parameters for modeling based on given values of different load points. This method is referred to as a load-dependent X-parameter model. Literature [8] validates the feasibility of a method for extracting X-parameters by comparing the measurements of the X-parameter model against load-pull measurements. The model coefficient of the look-up function is A_{21}, but the look-up function is not suitable for an ideal circuit design process and a huge data volume is involved.

To address these defects in the aforementioned methods, this section proposes a new method to extract load-independent X-parameters, where each coefficient is independent of A_{21}. In the case of an impedance mismatch, the proposed method obtains effective X-parameters in the target area on the Smith chart based on the circuit structure and load-pull technique and extracts a transistor X-parameter model meeting the design needs. The method conducts the analysis based on X-parameters and has a faster convergence speed compared to the probability hypothesis density (PHD) model-assisted algorithm proposed in Literature [9]. The method retrieves the data of only one impedance point involving a minor data volume, compared to the method proposed in Literature [4].

Consequently, the method can be used to analyze an arbitrary load impedance of the nonlinear response at the output end of the device.

7.1.2 X-Parameter Model Description

X-parameter model generation involves the processing of the extracted nonlinear data, followed by simulation design and validation. X-parameter measurements apply to devices adopting any process technique. An X-parameter model has demonstrated its superiority over immature techniques that are not applicable to compact models or intellectual property (IP) protection. An X-parameter model allows model sharing without revealing internal circuits, which is an appealing feature to commercial organizations. In addition, an X-parameter model is easier to extract than a compact model [10]. Table 7.1 gives a simple comparison between an X-parameter model and a compact model.

7.1.3 Load-Independent X-Parameter Extraction Method

This method combines the traditional X-parameter extraction with the circuit load-pull technique by means of cross iteration to determine the reference impedance, extracts the X-parameter model meeting the design needs for a transistor in an impedance mismatching condition and identifies the optimal load impedance.

The accuracy of X-parameter model is subject to the effective range of extracted data. If the reference impedance Z_0 is determined on a Smith chart, the X-parameters are valid within a certain extended working area of the reference impedance Z_0. When the target impedance is far away

TABLE 7.1 Comparison Between X-Parameter Model and Compact Model

Method	Advantage	Disadvantage
Compact model	Scalability Noise model Formulated model Operable in multiple simulation modes	Inconvenient extraction Long development cycle
X-Parameter Model	Technical independence Simple and fast extraction Complete IP protection Quite accurate within the characterization range	Large file size NVNA bandwidth limitation

from the reference impedance, the extracted X-parameters will no longer be valid. Therefore, it is critical to determine the reference impedance of the target area to extract X-parameters.

In the beginning, the reference impedance is initialized. Assuming that the parameters of the nonlinear active device are unknown, both the initial reference impedance and the load impedance are set to 50 Ω, and the transistor X parameters are extracted in ADS. Next, data modeling is applied to circuit analysis and design (such as PA), and transistor load-pull simulation is performed to yield the maximum output power and load impedance. Whether the load impedance is close to the reference impedance is determined on the premise that the output power meets the requirements, and the optimal load impedance of the transistor is determined by means of iterations. If the optimal load impedance is close to 50 Ω, the model can be directly used for circuit analysis and design. If not, the reference impedance needs to be predicted again and new X-parameters are re-extracted. The aforementioned procedure is repeated. At each iteration, the weight function Equation (7.1) is applied, until the optimal load impedance (Z_{opt}) of the X-parameter model converges to the target load impedance (maximum power).

$$Z_0(i+1) = W(i) \cdot Z_0(i) + (1 - W(i)) \cdot Z_{opt}(i) \tag{7.1}$$

where W is the weight factor.

The flowchart of parameter extraction is depicted in Figure 7.1.

In the extraction of X-parameters, the re-selection of the reference impedance delivers higher accuracy in the optimized impedance region, which is also the purpose and advantage of the proposed method. Compared with the extraction of load-dependent X-parameters, the proposed method features a simpler design, less time consumption and higher modeling accuracy.

Use the method proposed in this subsection to extract data from the GaN HEMT transistor (NPTB00004), and establish an accurate X-parameter model used in the PA circuit design. Parameter settings for transistor modeling are: operating frequency range 1.7–2.7 GHz; bias voltages $V_{GS} = -1.4V$ and $V_{DS} = 28V$; and input power range

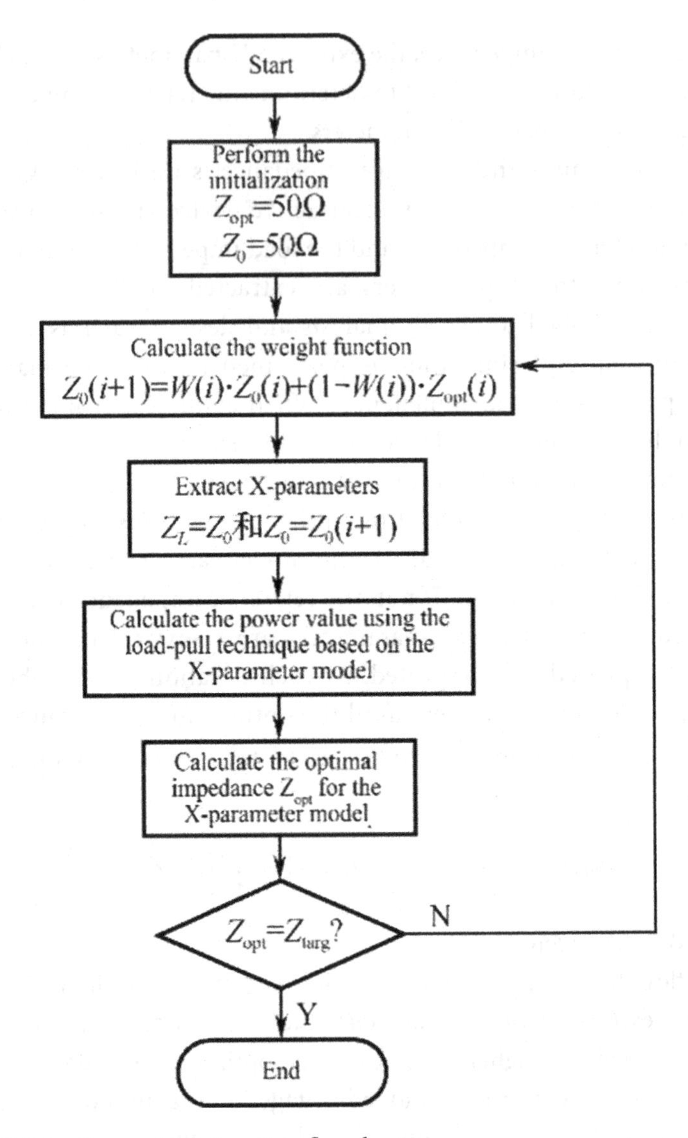

FIGURE 7.1 Parameter extraction flowchart.

0–30 dBm. Since the source impedance produces little influence on power conversion, the influence of load impedance is considered. The source reference impedance is set to a fixed value Z_0 = 50 Ω, the initial reference impedance at the output is set to Z_0 = 50 Ω and the extracted harmonic numbers *m*, *n* are both 3. Use harmonic balance simulation with the extracted model for analysis and prediction of load-pull behaviors, so that the target impedance Z_{targ} in the design can be yielded. The iterative process eventually yields the optimal load impedance $Z_{opt} = Z_{targ} = 17.2 - j35.5\ \Omega$, with the weight value set to 0.5.

7.1.4 Wideband Power Amplifier Design

This subsection uses an X-parameter model of the GaN HEMT transistor (NPTB00004) to analyze and design a high-efficiency wideband PA. Design objectives are: operating frequency range 1.7–2.7 GHz; power gain 14 dB; power gain flatness less than ±1 dB; and PAE in the operating frequency range greater than 45%.

In the design, the optimal output load impedance of the transistor has been determined by the extraction of X-parameters. The matching network is designed with an impedance transformer in the form of a Chebyshev low-pass filter. A combination of double-sector open-circuit microstrip line and filter circuit is adopted to broaden the bandwidth of high-input impedance and downsize the circuit. Based on optimization, the design of high-efficiency wideband PA is finally complete.

The output matching network structure produced by an impedance transformation design from real impedances to complex impedances is shown in Figure 7.2.

The wideband PA circuit after design, processing, assembly and debugging is shown in Figure 7.3.

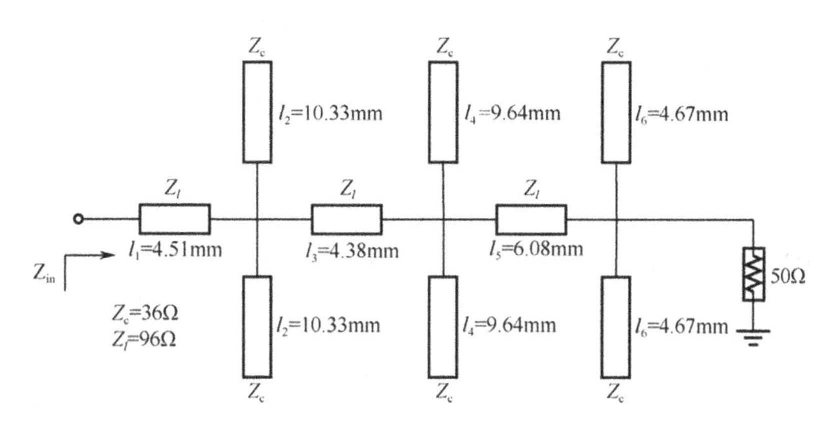

FIGURE 7.2 Output matching network structure.

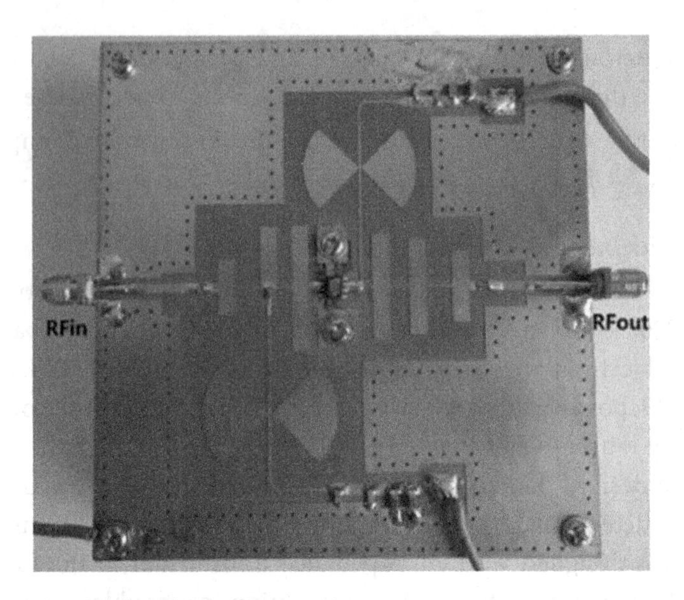

FIGURE 7.3 Wideband PA circuit.

7.1.5 Simulation and Testing

The PA operates in deep Class AB, with the drain voltage V_{DS} = 28V and gate voltage V_{GS} = −1.4V. The input signal uses a single-frequency continuous wave (CW) signal. The performance of the wideband PA is simulated and tested by means of frequency sweep and power sweep. Figure 7.4 depicts the power gain, output power and PAE of the tested PA when the input power is 22 dBm. In the operating frequency range of 1.7–2.7 GHz, the simulation gain varies from 14.1 to 14.9 dB, the output power from 36.1 to 36.9 dBm and the PAE from 52% to 60%. In the test gain range of 13.7–14.7 dB, the output power varies from 35.7 to 36.7 dBm and the PAE from 50% to 58%. Therefore, the simulation results are basically consistent with the test results.

Figure 7.5 shows the PA's harmonic spectrums of different orders. Over the entire operating frequency range, the second-order harmonic suppression ranges from −35 to −70 dBc and the third-order harmonic suppression ranges from −40 to −60 dBc. The harmonic suppression is demonstrated favorable and basically meets the design requirements. Figure 7.6 depicts the PA gain, output power and efficiency varying with input power at an operating frequency of 2 GHz. The PAE gradually

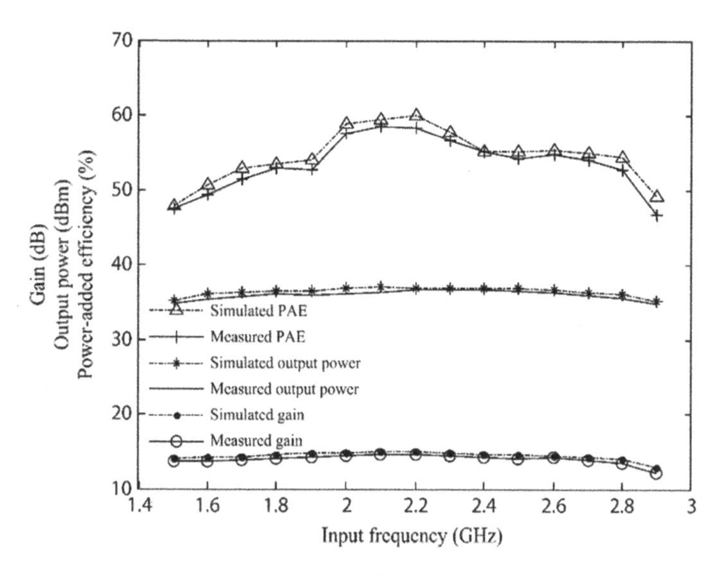

FIGURE 7.4 PA performance vs. input frequency.

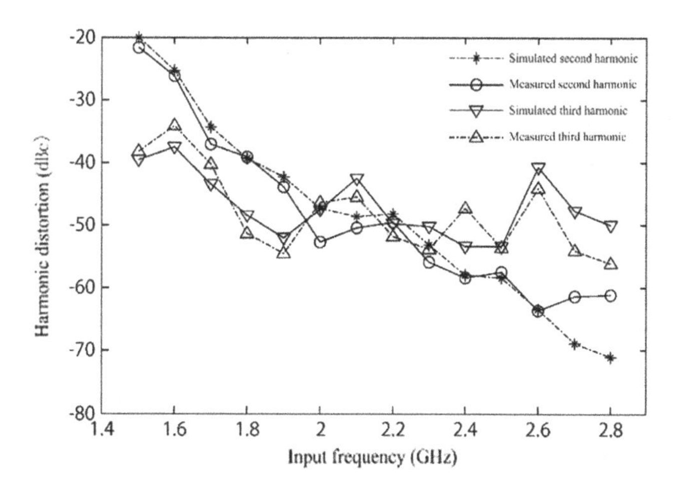

FIGURE 7.5 PA's harmonic spectrums of different orders.

increases with the growth of input power. When the input power reaches 22 dBm, the amplifier is basically in a saturated state. After saturation is reached, the nonlinearities of PA are slightly deviated, which also validates that X-parameters can characterize the large-signal nonlinearities. As indicated by the figures, the amplifier simulation results are basically consistent with the test results, proving the correctness and effectiveness

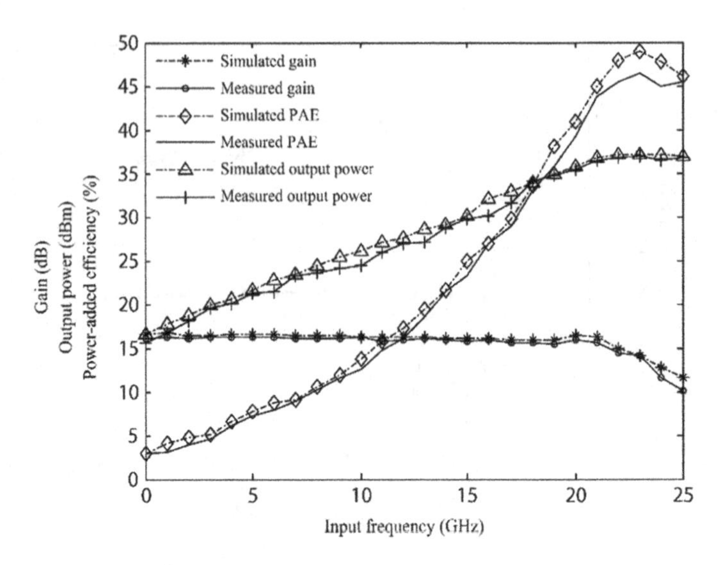

FIGURE 7.6 PA performance vs. input power.

of the design method. Furthermore, the X-parameter model has demonstrated its superiorities in less time consumption, high efficiency and reduced design time.

This section proposes a load-independent X-parameter extraction method, which is applicable for the analysis and design of large-signal nonlinear circuits. The proposed method yields an accurate X-parameter model by means of a load-pull simulation and the iterative method used to determine the optimal load impedance. Then, the extracted model is used for the wideband PA design, and the actual measurements and simulation results are compared to validate the correctness of the extraction method and the effectiveness of the established model. In conclusion, the established X-parameter model is demonstrated to accurately characterize the nonlinearities of the device, allow easy measurements and have a short design period. The model can be widely applied to microwave circuit design.

7.2 RESEARCH ON DYNAMIC X-PARAMETER MODEL BASED ON MEMORY EFFECTS OF POWER AMPLIFIER

A novel dynamic X-parameter PA model is proposed in this section to more accurately characterize the memory effects of PA. This model is considered an improvement on the dynamic X-parameter model. It uses the feed-forward (FF) modeling technique [11] to extract the kernel function characterizing memory effects of PA, and to replace the non-linear function in the X-parameter expression with the kernel function. The static part of the X-parameter expression is still the nonlinear function of PHD model [2], and the dynamic part is reasonably simplified. The proposed model comes with high simulation accuracy and fast simulation, compared to the traditional dynamic X-parameter model. Simulation and test results of NPT1004 PA with the proposed model demonstrate that its ACPR error is reduced by 3 dB.

Literature [12] proposes the original X-parameter model, that is, PHD model [13], and uses it to characterize the strong nonlinearities of PA. This model indicates a function of the amplitude of input large signals in the expression. The experiment has shown that this model favorably simulates the strong nonlinearities of PA. However, the PHD model can only indicate the static behaviors of PA but cannot characterize the dynamic part of PA, such as memory effects. Therefore, this model is often referred to as a static X-parameter model. Literature [16] proposes a dynamic X-parameter model, which is an improvement based on the original PHD model. The nonlinear function of the model is no longer merely a function of the current input signal amplitude, but a three-dimensional function of the current input signal amplitude, historical input signal amplitude and time. Although the model can accurately characterize the memory effects of PA, it is difficult to validate or apply the model due to the complexity of the three-dimensional kernel function. Literature [14] improves the dynamic X-parameter model based on the FF model proposed in Literature [15]. The improved model divides the memory effects of PA into short-term memory effects and long-term memory effects, regarding the generation mechanism. It is proved that this modeling scheme has improved the simulation speed to some extent but the simulation accuracy is unfavorable.

7.2.1 Dynamic X-Parameter Theory

As a natural extension of S-parameters in a nonlinear and large-signal context, X-parameters describe the mapping relationship between the outgoing wave and the incident wave at each port of a nonlinear device.

X-parameters are considered as a linear approximation of the scattering function under large-signal excitation conditions and come from the PHD framework. Dynamic X-parameters extend the application of X-parameters to memory effects [11]. The dynamic X-parameter model is expressed as follows:

$$B(t) = \left(F_{\mathrm{CW}}(|A(t)|) + \int_0^{\infty} G(|A(t)|, |A(t-u)|, u)\,du \right) \cdot \exp^{j\phi(A(t))} \tag{7.2}$$

where $B(t)$ is the superposition of $F_{\mathrm{CW}}(\cdot)$ (indicating the static nonlinear part) and $G(\cdot)$ (indicating the dynamic nonlinear part). $F_{\mathrm{CW}}(\cdot)$ and $G(\cdot)$ are both the functions of $A(t)$ (instantaneous amplitude). Compared with the traditional PHD model, the dynamic X-parameter model introduces the dynamic part, thereby characterizing the nonlinearities of PA more accurately. In addition, $G(\cdot)$ is a three-dimensional function of the current input signal amplitude, historical input signal amplitude and time, which can characterize the dynamic part of PA. However, the complexity of three-dimensional function makes it difficult to extract the kernel function. To address this problem, an improved FF dynamic X-parameter model has been proposed. The topology of FF model is outlined in Figure 7.7.

In Figure 7.7, $\hat{x}(t)$ denotes the input signal and $\hat{y}(t)$ denotes the output signal. Two paths are available between input and output, namely the short-term memory (STM) and the long-term memory (LTM). The expression for this model is

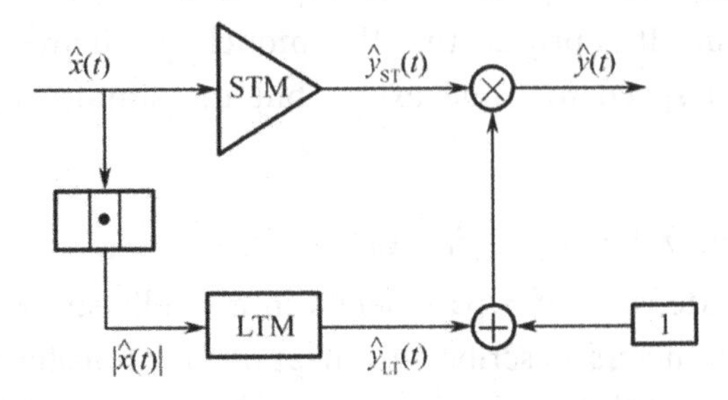

FIGURE 7.7 FF model topology.

$$\hat{y}(t) = (1 + \hat{y}_{\mathrm{LT}}(t)) \cdot \hat{y}_{\mathrm{ST}}(t) \tag{7.3}$$

where,

$$\hat{y}_{\mathrm{ST}}(t) = \int_0^{\infty} \hat{h}_{\mathrm{ST}}(|x(\hat{t} - \tau)|, \tau) \cdot x(\hat{t} - \tau) \cdot \mathrm{d}\tau \tag{7.4}$$

$$\hat{y}_{\mathrm{LT}}(t) = \int_0^{\infty} h_{\mathrm{LT}}(|x(\hat{t} - \tau)|, \tau) \cdot |x(\hat{t} - \tau)| \cdot \mathrm{d}\tau \tag{7.5}$$

Literature [15] has proved that this model can accurately simulate memory effects of PA, thereby establishing an effective memory effect mechanism of PA. Based on the memory effect mechanism, Literature [14] improves the dynamic X-parameter model. The improved dynamic X-parameter model is expressed as follows

$$\begin{aligned} B_{ik}(t) = {} & \mathrm{FB'}_{ik}(|A_{11}(t)|, t) \cdot P^k(t) + \cdots \\ & + \sum_{j \neq 1}^{N} \sum_{p \neq 1}^{\mathrm{NH}} S'_{ijkp}(|A_{11}(t)|, t) \cdot A_{jp}(t) \cdot P^{k-p}(t) + \cdots \\ & + \sum_{j \neq 1}^{N} \sum_{p \neq 1}^{\mathrm{NH}} T'_{ijkp}(|A_{11}(t)|, t) \cdot A^{*}_{jp}(t) \cdot P^{k+p}(t) \end{aligned} \tag{7.6}$$

where,

$$\mathrm{FB'}_{ik}(|A_{11}(t)|, t) = \mathrm{FB}_{ik}(|A_{11}(t)|) \cdot \mathrm{LF}_{\mathrm{FB}ik}(|A_{11}(t)|, t) \tag{7.7}$$

$$S'_{ijkp}(|A_{11}(t)|, t) = S_{ijkp}(|A_{11}(t)|) \cdot \mathrm{LF}_{Sijkp}(|A_{11}(t)|, t) \tag{7.8}$$

$$T'_{ijkp}(|A_{11}(t)|, t) = T_{ijkp}(|A_{11}(t)|) \cdot \mathrm{LF}_{Tijkp}(|A_{11}(t)|, t) \tag{7.9}$$

$$\mathrm{LF}_{\mathrm{FB}ik}(\cdots) = 1 + \int_0^{t} h_{\mathrm{FB}ik}(|A_{11}(t - \tau)|, \tau) \cdot A_{11}(t - \tau) \cdot \mathrm{d}\tau \tag{7.10}$$

$$\mathrm{LF}_{Sijkp}(\cdots) = 1 + \int_0^{t} h_{Sijkp}(|A_{11}(t - \tau)|, \tau) \cdot A_{11}(t - \tau) \cdot \mathrm{d}\tau \tag{7.11}$$

$$\mathrm{LF}_{Tijkp}(\cdots) = 1 + \int_0^{t} h_{Tijkp}(|A_{11}(t - \tau)|, \tau) \cdot A_{11}(t - \tau) \cdot \mathrm{d}\tau \tag{7.12}$$

In the improved dynamic X-parameter model, the kernel function is no longer three-dimensional but two-dimensional. The improved model characterizes the PA more accurately, and the kernel function is validated more easily. Compared to the traditional static X-parameter model, the improved model yields an ACPR value with the error reduced by 4 dB against the real PA value, thereby improving the accuracy of PA modeling. However, the in-depth study of memory effects has found that the accuracy of the improved model in characterizing memory effects of PA is not high, so a further improvement is required.

7.2.2 Improved Dynamic X-Parameter Model

The improved dynamic X-parameter model introduces a new variable, that is, the input signal envelope amplitude, to the LTM, preventing the FF model from generating an asymmetric equation when the LTM kernel function is solved. This subsection proposes an improved FF model (see Figure 7.8) for dynamic X-parameter PA modeling. The improved model can characterize the nonlinearities with high accuracy by using dynamic X parameters; in addition, it combines the mechanism of memory effects, especially the mechanism of LTM, thus improving the accuracy of the model and increasing the modeling speed. The STM of the improved dynamic X-parameter model adopts the kernel function of PHD model, ensuring the nonlinear characteristics. The kernel function of LTM is not only a function of the input signal amplitude but also a two-dimensional function of amplitude and phase. Consequently, the improved dynamic X-parameter model is

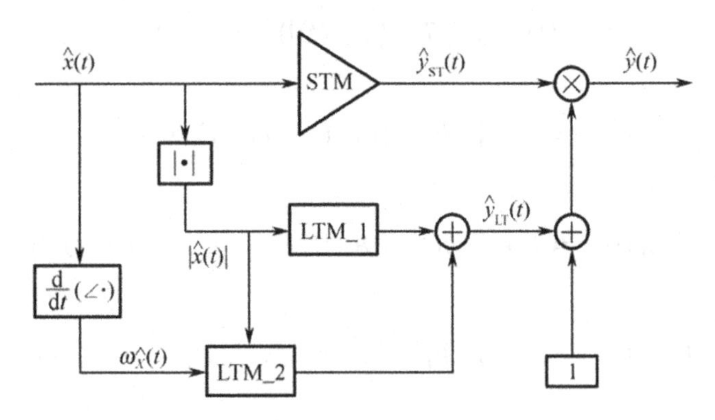

FIGURE 7.8 Improved FF model.

more concise than the model in Literature [13] and more accurate than the model in Literature [14].

The topology of the improved FF model is expressed as

$$\hat{y}(t) = (1 + \hat{y}_{\mathrm{LT}}(t)) \cdot \hat{y}_{\mathrm{ST}}(t) \tag{7.13}$$

where,

$$\hat{y}_{\mathrm{ST}}(t) = \int_0^{\infty} \hat{h}_{\mathrm{ST}}(|x(\hat{t} - \tau)|, \tau) \cdot x(\hat{t} - \tau) \cdot \mathrm{d}\tau \tag{7.14}$$

$$\begin{aligned} \hat{y}_{\mathrm{LT}}(t) = &\int_0^{\infty} \hat{h}_{\mathrm{LT1}}(|x(\hat{t} - \tau)|, \tau) \cdot |x(\hat{t} - \tau)| \cdot \mathrm{d}\tau \\ &+ \int_0^{\infty} \hat{h}_{\mathrm{LT2}}(|x(\hat{t} - \tau)|, \tau) \cdot \hat{\omega}_x(t - \tau) \cdot \mathrm{d}\tau \end{aligned} \tag{7.15}$$

The improved FF model is combined with X-parameters for PA modeling. The formula of the improved dynamic X-parameter model is as follows

$$\begin{aligned} B_{ik}(t) = &\mathrm{FB}''_{ik}(|A_{11}(t)|, t) \cdot P^k(t) + \cdots \\ &+ \sum_{j \neq 1}^{N} \sum_{p \neq 1}^{\mathrm{NH}} S''_{ijkp}(|A_{11}(t)|, t) \cdot A_{jp}(t) \cdot P^{k-p}(t) + \cdots \\ &+ \sum_{j \neq 1}^{N} \sum_{p \neq 1}^{\mathrm{NH}} T''_{ijkp}(|A_{11}(t)|, t) \cdot A^*_{jp}(t) \cdot P^{k+p}(t) \end{aligned} \tag{7.16}$$

where,

$$\mathrm{FB}''_{ik}(|A_{11}(t)|, t) = \mathrm{FB}_{ik}(|A_{11}(t)|) \cdot \mathrm{LF}_{\mathrm{FB}ik}(|A_{11}(t)|, t) \tag{7.17}$$

$$S''_{ijkp}(|A_{11}(t)|, t) = S_{ijkp}(|A_{11}(t)|) \cdot \mathrm{LF}_{Sijkp}(|A_{11}(t)|, t) \tag{7.18}$$

$$T''_{ijkp}(|A_{11}(t)|, t) = T_{ijkp}(|A_{11}(t)|) \cdot \mathrm{LF}_{Tijkp}(|A_{11}(t)|, t) \tag{7.19}$$

$$\begin{aligned} \mathrm{LF}_{\mathrm{FB}ik}(\cdots) = 1 &+ \int_0^{t} h_{\mathrm{FB}ik}(|A_{11}(t - \tau)|, \tau) \cdot A_{11}(t - \tau) \cdot \mathrm{d}\tau \\ &+ \int_0^{\infty} h'_{\mathrm{FB}ik}(|A_{11}(t - \tau)|, \tau) \cdot \omega_{A_{11}}(t - \tau) \cdot \mathrm{d}\tau \end{aligned} \tag{7.20}$$

$$\begin{aligned}\mathrm{LF}_{Sijkp}(\cdots) = 1 &+ \int_0^t h_{Sijkp}(|A_{11}(t-\tau)|, \tau)\cdot A_{11}(t-\tau)\cdot \mathrm{d}\tau \\ &+ \int_0^\infty h'_{Sijkp}(|A_{11}(t-\tau)|, \tau)\cdot \omega_{A_{11}}(t-\tau)\cdot \mathrm{d}\tau\end{aligned} \tag{7.21}$$

$$\begin{aligned}\mathrm{LF}_{TiJkP}(\cdots) = 1 &+ \int_0^t h_{Tijkp}(|A_{11}(t-\tau)|, \tau)\cdot A_{11}(t-\tau)\cdot \mathrm{d}\tau \\ &+ \int_0^\infty h'_{Tijkp}(|A_{11}(t-\tau)|, \tau)\cdot \omega_{A_{11}}(t-\tau)\cdot \mathrm{d}\tau\end{aligned} \tag{7.22}$$

$$\phi_{A_{11}}(t) = \tan^{-1}\left[-\mathrm{j}\frac{A_{11}(t) - A_{11}^*(t)}{A_{11}(t) + A_{11}^*(t)}\right] \tag{7.23}$$

$$\omega = \frac{\mathrm{d}\phi_{A_{11}}(t)}{\mathrm{d}t} \tag{7.24}$$

7.2.3 Kernel Function Extraction of New Model

The dynamic X-parameter model contains two parts of kernel functions, which are h_{ST}, h_{LT1} and h_{LT2}. h_{ST} can be regarded as a function of static X-parameters and therefore allows CW measurements. For the $h_{\mathrm{LT}}(\cdot)$ function, pulse measurements are used. Electromagnetic simulation software (ADS) is used to extract X-parameters, and the extracted data is used for envelope transient simulation of the circuit. A step simulation for the circuit yields all the coefficients in Equation (7.6), while the $h_{\mathrm{FB}}(\cdot)$ function is obtained by

$$h_{\mathrm{FB21}}(|A_{11}|, t) = \frac{1}{|A_{11}|}\frac{\partial}{\partial t}\left\{\frac{\mathrm{FB}''_{21}(|A_{11}|, t)}{\mathrm{FB}''_{21}(|A_{11}|, \infty)} - 1\right\} \tag{7.25}$$

where $\mathrm{FB}''_{21}(|A_{11}|, \infty)$ is the steady-state value of the coefficient $\mathrm{FB}''_{21}(|A_{11}|, t)$. This model can be implanted into the circuit or system simulation subject to the following rules: With the extracted $h_{\mathrm{FB21}}(|A_{11}|, t)$ coefficients, M orthogonal bias vectors $\Psi_{\mathrm{FB21},m}(|A_{11}|)$ are created by means of the singular value decomposition (SVD). Finally, the weight function is determined by using the least square method. The weight function represents the transfer function of a linear filter. It can be efficiently synthesized by classical poles/residues in way of time-domain vector fitting. In this way, Equation (7.16) changes from a convolution to a simple iterative algorithm, thus improving the modeling speed. The orthogonal basis vector $\Psi_{FB21,m}(|A_{11}|)$ is represented by a cubic spline.

7.2.4 Simulation and Data Analysis

The NPT1004 (GaN high-electron-mobility device made by Nitronex) is selected to design the PA. Its operating frequency band is 0–4 GHz, with the static operating points $V_{DS} = 28$ V, $I_{DS} = 350$ mA, $V_{GS} = -1.5$ V. The matching circuit design adopts the load-pull technique, and the designed circuit is used to extract X-parameters and test IMD3, respectively. See Figure 7.9.

As indicated in Figure 7.9, there is a serious asymmetry between the upper and lower sidebands in the PA output spectrum, which indicates that even the memory effects of the narrowband PA cannot be ignored.

In Figure 7.10, ① represents the measured values of PA, ② represents the power spectral density obtained with the improved dynamic X-parameter model and ③ represents the data obtained with the traditional X-parameter model. As indicated in the figure, the improved dynamic

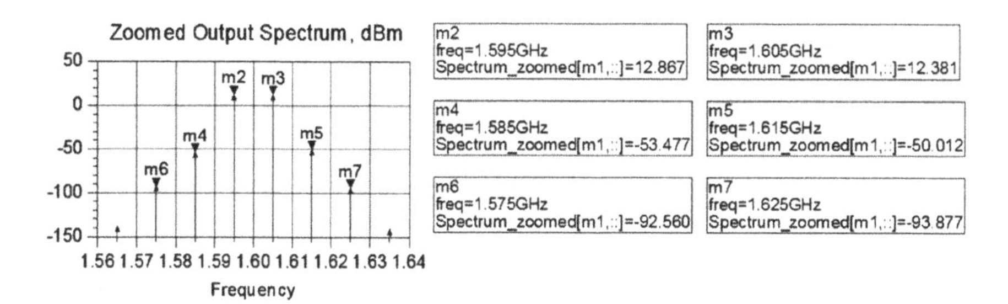

FIGURE 7.9 Spectrum diagram of dual-tone output signal.

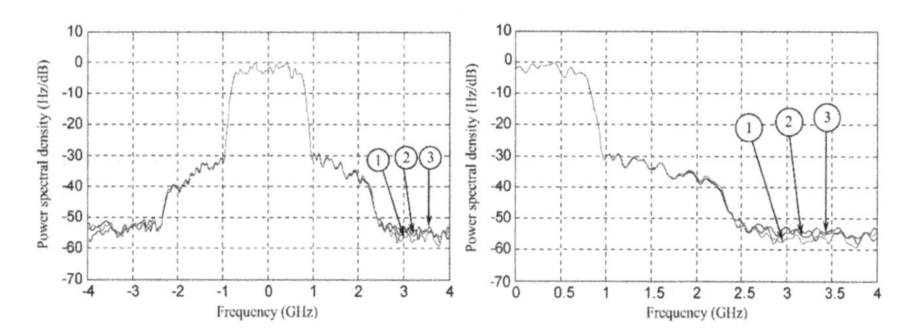

FIGURE 7.10 Power spectral density of output signal.

X-parameter model favorably predicts the characteristics of PA spectral regrowth and further improves the simulation accuracy, compared to the traditional X-parameter model. The improved model yields an ACPR value with the error reduced by approximately 3 dB compared to the traditional model. In addition, the improved model significantly reduces the simulation period.

Memory effects have become another major role after nonlinearity that hinders the linearization progress of PA. It is of great significance to investigate memory effects deeply. Dynamic X-parameters have proved to effectively characterize the strong nonlinearities of PA while setting up the memory effect mechanism of PA. Based on a combination of memory effects and dynamic X-parameters, this section investigates the behavioral model of PA. In this section, ADS software is used to extract the model coefficients and validate the model, and MATLAB® simulation and comparison are performed. Results demonstrate that the improved dynamic X-parameter model can characterize the nonlinearities and memory effects of RF PA more accurately.

REFERENCES

1. 华晓杰，林茂六，孙洪剑. 多谐波失真模型应用研究[J]. 电讯技术, 2007, 47(3):19–23.
   HAU X, LIN M, SUN H. Research on the Application of Polyharmonic Distortion (PHD) Model[J]. Telecommunication Engineering, 2007, 47(3):19–23.
2. JAN VERSPECHT, ROOT DAVID E.. Polyharmonic Distortion Modeling [J]. IEEE Microwave Magazine，2006，7(3):44–57.
3. 孙璐. 基于散射函数的一种微波非线性电路建模新方法[D]. 西安: 西安电子科技大学, 2010.
   SUN L. A New Method for Modeling Microwave Nonlinear Circuits Based on Scattering Function[D]. Xi'an: Xidian UniverSITY, 2010.
4. SIMPSON G, HORN J, GUNYAN D, et al. Load-pull+NVNA=Enhanced X-Parameters for PA Designs with High Mismatch and Technology-Independent Large-Signal Device Models: ARFTG Microwave Measurement Symposium，2008 72nd. IEEE Portland, OR, USA, 2008[C]. Piscataway: IEEE, 88–91.
5. ZARGAR H, BANAI A, CAI J, et al. An Analytical Approach to Obtain Optimum Load Impedance Using X-Parameters: Integrated Nonlinear Microwave and Millimetre-Wave Circuits(INMMIC), 2012 Workshop on IEEE，Dublin，Ireland，2012[C]. Piscataway: IEEE, 1–3.

6. 柳平，吕金凤，石中华. X 参数于负载牵引测量中的应用[J]. 信息通信，2011，6:45–46.
   LIU P, LV J, SHI Z. Application of X-Parameter in Load Pull Measurements[J]. Information & Communications, 2011, 6:45–46.
7. BESPALKO D T, BOUMAIZA S. X-Parameter Measurement Challenges for Unmatched Device Characterization, Microwave Measurements Conference (ARFTG), 2010 75th. ARFTG.IEEE, Anaheim, CA, USA, 2010[C]. Piscataway: IEEE, 1–4.
8. Gunyan D, Horn J, Xu J, et al. Nonlinear Validation of Arbitrary Load X-Parameter and Measurement-Based Device Model: Microwave Measurement Conference, ARFTG, 2009[C]. Boston: IEEE, 1–4.
9. LEONI R E, HARRIS S A, RIES D G. Active Simultaneous Harmonic Source and Load Pull Assisted by Local Polyharmonic Distortion Models: Microwave Symposium Digest(MTT)，2010 IEEE MTT-S International. IEEE, Anaheim, CA, USA, 2010[C]. Piscataway: IEEE，1166–1169.
10. ROOT D E. Future Device Modeling Trends[J]. IEEE Microwave Magazine, 2012, 13(6):45–59.
11. 何军. 宽带射频功率放大器记忆效应的研究[D]. 成都: 电子科技大学，2009.
    HE J. Research on Memory Effect of Broadband RF Power Amplifier[D]. Chengdu: University of Electronic Science and Technology of China, 2009.
12. CHIA SUNG CHIU, KUN-MING CHEN, GUO-WEI HUANG. Characterization of Annular-Structure RF LDMOS Transistors Using Polyharmonic Distortion Model[J]. IEEE Microwave Magazine, 2009: 977–980.
13. PATRICK ROBLIN, DAVID E ROOT, JAN VERSPECHT, et al. New Trends for the Nonlinear Measurement and Modeling of High-Power RF Transistors and Amplifiers With Memory Effects[J]. IEEE Transactions on Microwave Theory and Techniques, 2012, 60(6):1964–1978.
14. ARNAUD SOURY, EDOUARD NGOYA. Handling Long-Term Memory Effects in X-Parameter Model[J]. IEEE Microwave Magazine, 2012.
15. EDOUARD NGOYA, CHRISTOPHE QUINDROIT, JEAN MICHEL NÉBUS. On the Continuous-Time Model for Nonlinear-Memory Modeling of RF Power Amplifier[J]. IEEE Transactions on Microwave Theory and Techniques, 2009, 57(12):3278–3292.
16. ROBLIN P, ROOT D E, VERSPECHT J, et al. New Trends for the Nonlinear Measurement and Modeling of High-Power RF Transistors and Amplifiers with Memory Effects [J]. IEEE Transactions on Microwave Theory and Thechniques, 2012, 60(6):1964–1978.

CHAPTER 8

Other Power Amplifier Modeling

8.1 POWER AMPLIFIER MODEL BASED ON DYNAMIC RATIONAL FUNCTION AND PREDISTORTION APPLICATIONS

This section proposes a new dynamic rational function model with low complexity, which accommodates memory effects and more accurately characterizes the nonlinearities of RF PA in modern wireless communications. Based on an MP model, the proposed model is developed by a ratio of two polynomials. The numerator is an EMP containing memory effects, and the denominator consists of a memoryless polynomial without memory terms, thus reducing the number of coefficients to be identified. The numerator and denominator contain odd and even order polynomials, and the nonlinear order allows dynamic adjustment based on the changes in the static part and memory part. As validated by model simulation and predistortion application system, the dynamic rational function model requires 30.6% and 21.9% less coefficients and improves the ACPR by approximately 20 dB and 15 dB, compared to the MP model and rational function model, respectively, with similar model accuracy.

The rational function model[1–4] is developed by a ratio of two polynomials. Both the numerator and denominator contain memory effects and the numerator does not contain phase information. Plus, the model requires the numerator and denominator to have the same nonlinear order and memory depth. Therefore, the rational function model comes with strict conditions. To address the drawback of the rational function model, this section proposes a dynamic rational function model with low complexity.

8.1.1 Model Analysis

An MP model is essentially a simplified Volterra series model. It is considered as a trade-off between the complexity of the Volterra series model and the nonlinearity. Its expression is

$$y_{\mathrm{MP}}(n) = \sum_{p=0}^{P} \sum_{m=0}^{M} a_{pm} x(n-m) |x(n-m)|^{p} \tag{8.1}$$

where $x(n)$ and $y_{\mathrm{MP}}(n)$ are the input and output signals of the MP model, respectively; P and M are the nonlinear order and memory depth of the MP model, respectively; a_{pm} is the complex coefficient of the MP model.

Although the EMP model uses fewer coefficients compared to the Volterra series model, the coefficients involved in the model still need to be reduced. Therefore, Oualid Hammi proposed an EMP model[5], which is composed of the cross terms for the current moment and memory moments. Its expression is

$$y_{\mathrm{EMP}}(n) = \sum_{p=0}^{P} \sum_{m=0}^{M} a_{pm} x(n) |x(n-m)|^{p} \tag{8.2}$$

where $x(n)$ is the input signal, $y_{\mathrm{EMP}}(n)$ is the output signal, P is the nonlinear order, and M is the memory depth of the model.

The rational function is defined as a ratio of two power polynomials

$$y(n) = \frac{a_0 + a_1 x(n) + \cdots + a_J x^J(n)}{b_0 + b_1 x(n) + \cdots + b_K x^K(n)} = \frac{\sum_{i=0}^{J} a_i x^i(n)}{\sum_{j=0}^{K} b_j x^j(n)} \tag{8.3}$$

where $x(n)$ and $y(n)$ denote the input and output of the rational function at moment n, respectively.

According to Literatures [4] and [5], AM/AM and AM/PM characteristics are obtained by Equation (8.3), where $x(n)$ denotes the input signal amplitude of travel wave tube amplifier (TWTA) and solid-state power amplifier (SSPA). In this case, the PA is considered as a static nonlinear system where memory effects are not taken into account. Predistortion compensation is embodied in the simulation, SSPA performance is approximated by a cubic curve model, and TWTA is approximated by a Saleh model. Saleh model ignores the influence of measurement error and noise, which must be considered in practical applications.

Saleh model is used in TWTA and can be regarded as a special form of rational function model. A simpler rational function is expected to model the DPD. Therefore, an improved Saleh model is proposed for SSPA modeling, but it cannot compensate memory effects of PA either. To highlight memory effects, PA modeling adopts a rational function where memory effects are expressed as a complex number in the numerator and an absolute value in the denominator in form of time delay. The rational function proposed in Literature [4] is expressed as

$$y(n) = \frac{\sum_{i=0}^{K_n} \sum_{m_n=0}^{M_n} a_{i,m_n} x(n-m)|x(n-m)|^{2i}}{1 + \sum_{j=0}^{K_d} \sum_{m_d=0}^{M_d} b_{j,m_d} |x(n-m)|^{2j+1}} \tag{8.4}$$

where $x(n)$ and $y(n)$ denote the input and output of the baseband signal, respectively; K_n and K_d denote the nonlinear orders of the numerator and denominator, respectively; M_n and M_d denote the memory depths of the numerator and denominator, respectively.

As indicated by Literature [6], if $x(n)$ is very large, Equation (8.4) will converge to a finite value only when $K_n = K_d = K$ and $M_n = M_d = M$ apply. Expansion of the power series of Equation (8.4) yields

$$\begin{aligned} y(n) = & \left(1 - \sum_{j=0}^{K} \sum_{m_d}^{M} b_{j,m_d} |x(n-m_d)|^{2j+1} + \cdots \right) \\ & \times \sum_{i=0}^{K_n} \sum_{m_n=0}^{M_n} a_{i,m_n} x(n-m)|x(n-m)|^{2i} \end{aligned} \tag{8.5}$$

Equation (8.5) contains odd-order and even-order monomials, for example, $a_{i,m_n} b_{j,m_d} x(n-m)|x(n-m)|^{2i}\ |x(n-m_d)|^{2j+1}$, depending on the values of i and j.

As indicated in Equation (8.5), for an arbitrary nonlinear order K, the rational function model will have a nonlinear order $2(2K+1)$ after being expanded into a power series, and therefore it will become a highly nonlinear system. The characteristics of the rational function model are summarized as follows:

1. The numerator and denominator only contain odd-order polynomials.
2. The numerator and denominator have the same nonlinear order ($K_n = K_d$).
3. The numerator and denominator have the same memory depth ($M_n = M_d$).
4. The denominator of the model is in form of an absolute value and does not contain phase information of the signal.

Due to the aforementioned limitations, the rational function model does not deliver an ideal performance. Therefore, this section proposes a dynamic rational function model. The numerator of the model is an EMP that contains memory effects. The denominator is a memoryless polynomial without memory effects and its nonlinear order changes dynamically. The numerator and denominator contain odd- and even-order polynomials.

A dynamic rational function model is expressed as

$$y(n) = \frac{\sum_{j=0}^{N_n} \sum_{m=0}^{M} a_{j,m} x(n)|x(n-m)|^j}{1 + \sum_{i=0}^{N_d} b_i x(n)|x(n)|^i} \tag{8.6}$$

where $x(n)$ and $y(n)$ denote the input and output complex signals of PA at the sampling moment n, respectively; M denotes the memory depth; N_n and N_d denote the nonlinear orders of polynomials in the numerator and denominator, respectively; $a_{j,m}$ and b_i denote the complex coefficients of the numerator and denominator, respectively.

Expansion of the power series of Equation (8.6) yields

$$y(n) = \sum_{j=0}^{N_n} \sum_{m=0}^{M} a_{j,m} x(n)|x(n-m)|^j \times \left(1 - \sum_{i=0}^{N_d} b_i x(n)|x(n)|^i + \left(\sum_{i=0}^{N_d} b_i x(n)|x(n)|^i\right)^2 + \cdots\right) \quad (8.7)$$

As indicated in Equation (8.7), the nonlinear orders of the memory part and static polynomial part are N_n and $N_d + N_n$, respectively. In this way, the nonlinear orders can be dynamically regulated according to the static part and memory part. The denominator does not contain memory terms, thus reducing the number of coefficients to be identified in the model.

8.1.2 Model Determination and Coefficient Extraction

Usually, the model for DUT is determined by means of an iterative sweeping method, which is implemented in the way of offline training in the DSP. This method is used to perform three sweeps for the nonlinear order and memory depth of the numerator and denominator. Different dimension values (N_n, N_d, M) are selected in the model with their NMSE values compared. The initial value of each dimension is set to 0 and incremented in steps of 1 to improve the model performance. If the dimension value continues to grow but the model performance remains unchanged or deteriorates, the current dimension value will be the best one. The determined dimension values of the model remain unchanged for any PA. If the PA behavioral model fluctuates due to changes in the environment, adaptive PA matching will be performed by updating the model coefficients without changing the dimension values of the model.

Extraction of model coefficients uses the least square method. The dynamic rational function model expressed by Equation (8.6) can be written as

$$y(n) = -y(n) \sum_{i=0}^{N_d} b_i x(n)|x(n)|^i + \sum_{j=0}^{N_n} \sum_{m=0}^{M} a_{j,m} x(n)|x(n-m)|^j \quad (8.8)$$

In the extraction of model coefficients, the signal can be expressed as a matrix form

$$y = A\Phi_{\text{DRF-MFOD}} \tag{8.9}$$

where $\boldsymbol{y}$ is a $L \times 1$-dimensional output vector, L is the training data, and $\Phi_{\text{DRF-MFOD}}$ is a coefficient vector.

$$\Phi_{\text{DRF-MFOD}} = \left[b_0, \cdots, b_{N_d}, a_{00}, \cdots, a_{N_n 0}, a_{01}, \cdots, a_{N_n 1}, \cdots, a_{N_n M}\right]^{\mathrm{T}} \tag{8.10}$$

$$U_1(n) = \begin{bmatrix} -y(n)x(n) & \cdots & -y(n)x(n)|x(n)|^{N_d} \\ -y(n-1)x(n-1) & \cdots & -y(n-1)x(n-1)|x(n-1)|^{N_d} \\ \vdots & \vdots & \vdots \\ -y(n-L)x(n-L) & \cdots & -y(n-L)x(n-L)|x(n-L)|^{N_d} \end{bmatrix} \tag{8.11}$$

$$U_2(n) = \begin{bmatrix} x(n) & \cdots & x(n)|x(n)|^{N_n} & \cdots & x(n-M) & \cdots & x(n-M)|x(n-M)|^{N_n} \\ x(n-1) & \cdots & x(n-1)|x(n-1)|^{N_n} & \cdots & x(n-1-M) & \cdots & x(n-1-M)|x(n-1-M)|^{N_n} \\ \vdots & \vdots & \vdots & \vdots & \vdots & \vdots & \vdots \\ x(n-L) & \cdots & x(n-L)|x(n-L)|^{N_n} & \cdots & x(n-L-M) & \cdots & x(n-L-M)|x(n-L-M)|^{N_n} \end{bmatrix} \tag{8.12}$$

In Equation (8.9), matrix $\boldsymbol{A}$ is defined as $\boldsymbol{A} = [\boldsymbol{U}_1\boldsymbol{U}_2]$, where $\boldsymbol{U}_1$ and $\boldsymbol{U}_2$ are both Vandermonde matrices. Vandermonde matrix $\boldsymbol{U}_2$ is similar to an MP model.

As indicated from Equations (8.9) to (8.12), the coefficients of Equation (8.11) are linear. Use the least square method to yield the following equation

$$\Phi_{\text{DRF-MFOD}} = (A^{\mathrm{T}}A)^{-1}A^{\mathrm{T}}U_2 \tag{8.13}$$

After the coefficients are solved from Equation (8.13), the output can be calculated by means of Equation (8.9).

8.1.3 Model Performance Evaluation

This subsection compares the dynamic rational function model (Dominant Resource Fairness-Matrix Fractional Order Differentiators (DRF-MFOD)) with the MP model and rational function model (ADRF) to validate the accuracy of the dynamic rational function model proposed in this section.

A total of 9000 signal data points are used for modeling and simulation, of which 4000 sets of data are used for extraction and establishment of model parameters and the other 5000 sets are used for validation of model performance. The test signal is a 16 QAM signal with a chip rate of 15 Mcps and a signal bandwidth of 15 MHz. An LDMOS Doherty PA is driven with a gain of 50 dB and a center frequency of 1.96 GHz.

In the absence of memory effects, the MP model is a function of the nonlinear order p, the nonlinear order of the rational function model is a function of K, and the dynamic rational function model requires two-dimensional optimal nonlinear order values N_n and N_d. When the dimension values of the model are determined, the time delay is introduced to represent the memory depth. The variation of NMSE performance is observed by increasing the delay, and the memory depth of the model is determined when NMSE performance no longer improves.

Figure 8.1(a) depicts the comparison of DRF-MFOD model, MP model, and Aerosol Direct Radiative Forcing (ADRF) model. In the figure, the abscissa represents the nonlinear orders p and K of MP model and rational function model, respectively. For the dynamic rational function model, the abscissa represents the nonlinear order N_n. Figure 8.1(b) shows a local zoom-in part of Figure 8.1(a). For all models, the best memory depth is 3. The variation of correlated NMSE

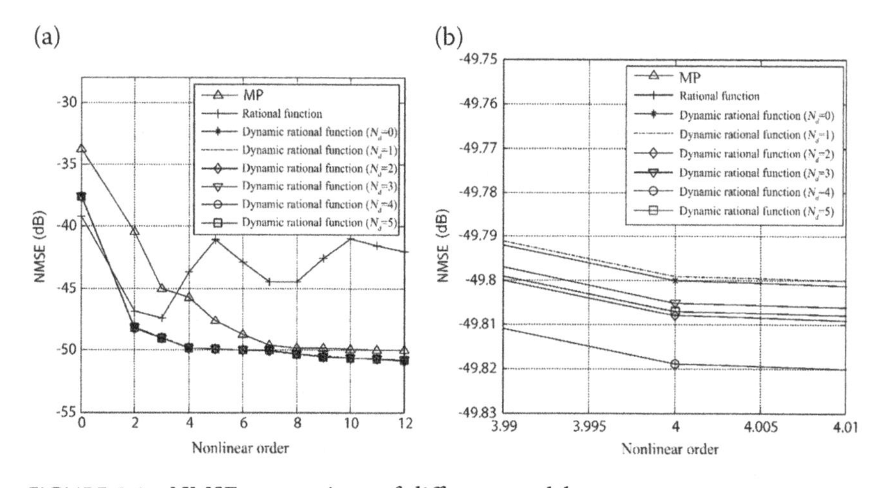

FIGURE 8.1 NMSE comparison of different models

performance is observed by increasing the memory depth, and the best memory depth is determined when the performance no longer improves.

As indicated by the trend of the NMSE curve in Figure 8.1, the rational function model delivers fast convergence and favorable performance when the nonlinear order is small. With the increase of nonlinear order, the performance of the rational function model deteriorates obviously, while the dynamic rational function model delivers much faster convergence and uses fewer coefficients compared to the MP model. This is because the denominator of the dynamic rational function model does not contain a time delay term, reducing model coefficients. For example, the MP model converges at $N = 8$ when a total of $9 \times 4 = 36$ coefficients are used; in contrast, the dynamic rational function model achieves a similar effect at $N_n = 4$, $N_d = 5$ when a total of $4 \times 5 + 5 = 25$ coefficients are used.

Table 8.1 gives a comparison of MP, ADRF and DRF-MFOD models in terms of model size, total coefficients and NMSE. Compared with the MP model, the DRF-MFOD model proposed in this section achieves similar accuracy with 30.6% less coefficients. Compared with the rational function model, the proposed model improves the accuracy by 2.4 dB with 21.9% less coefficients. Results demonstrate that the dynamic rational function model better simulates the nonlinearities of PA and reduces the complexity of real circuits.

Figures 8.2 (a), (b) and (c) show the spectra of calculated outputs based on the MP model, ADRF model and DRF-MFOD model against the real PA outputs plus the errors, respectively. As indicated by the figure, the rational function model is closer to the spectrum of real PA than the MP model, while the dynamic rational function model outperforms the rational function model and better simulates the characteristics of PA.

TABLE 8.1 Performance Comparison of MP, ADRF and DRF-MFOD Models

Model	**Model Size**	**NMSE (dB)**	**Total Coefficients**
MP	$(P, Q)(8, 3)$	–49.82	36
ADRF	$(K, M)(3, 3)$	–47.41	32
DRF-MFOD	$(M, N_n, N_d)(3, 4, 5)$	–49.81	25

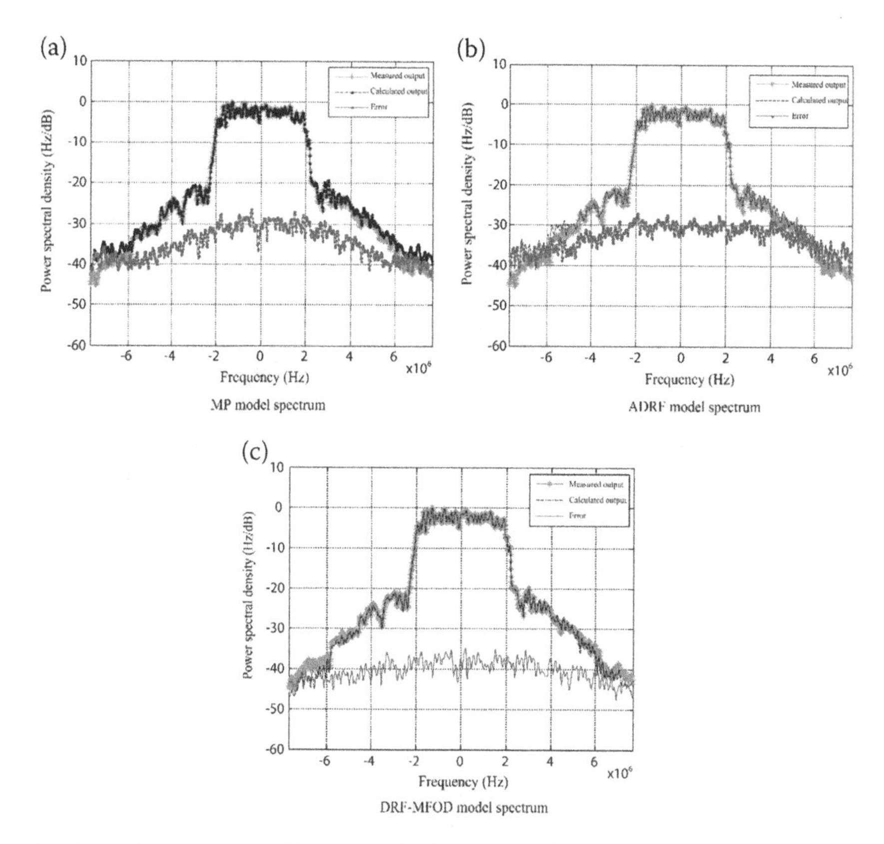

FIGURE 8.2 Spectrum diagrams of different models

8.1.4 Predistortion Application

The behavioral model of PA also serves as a predistorter model. Note that the input and output signals need to be swapped in the extraction of coefficients, and the extracted model coefficients are the predistorter coefficients. The dynamic rational function model, MP model and rational function model are applied to the predistortion system for simulation validation. Model performance is compared in terms of spectrum diagram and constellation diagram.

Figure 8.3 shows the power spectral densities of predistorted outputs based on different models. It is seen that out-of-band suppression performance and model size of different predistorter models are consistent with Table 8.1. As indicated by the figure, spectral regrowth is correspondingly suppressed in the outputs for all models when predistortion

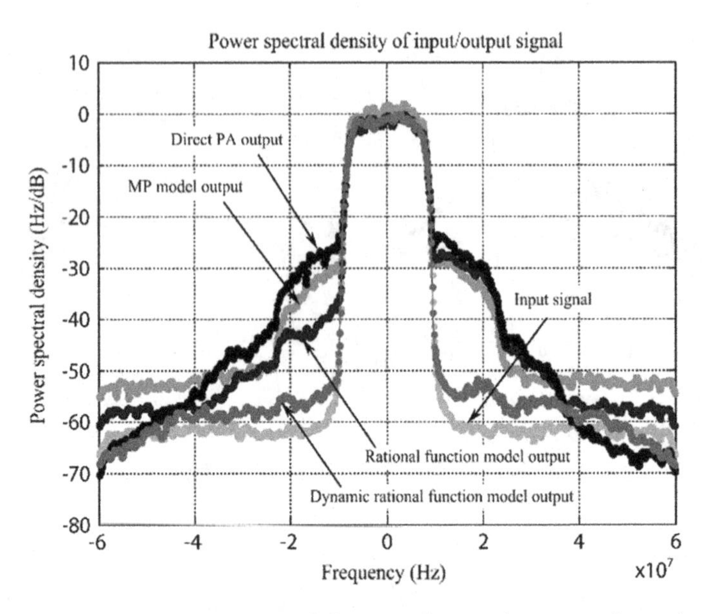

FIGURE 8.3 Output power spectral density after predistortion for each model.

is introduced, compared with direct PA outputs. The MP predistortion model improves the ACPR by about 5 dB compared with direct PA outputs. Based on the MP model, the rational function model improves the ACPR by about 6 dB on the left half spectrum while it does not significantly improve the ACPR on the right half spectrum. Compared with these two models, the dynamic rational function model performs much better, which improves the ACPR by about 20 dB based on the MP model. Comparison results prove the superiority of the model proposed in this section.

Figure 8.4 shows a comparison of AM/AM characteristic curves produced by the MP model, rational function model and dynamic rational function model. As indicated in the figure, the dynamic rational function model produces an AM/AM characteristic curve appearing more like a straight line, compared to the other two models. This superiority means a better linearization effect of predistortion.

The inherent nonlinearities of PA have proved to impact severely on modern communication systems. Investigations have developed PA behavioral models to better analyze the nonlinearities of PA and compensate the nonlinearities by extending the application of behavioral

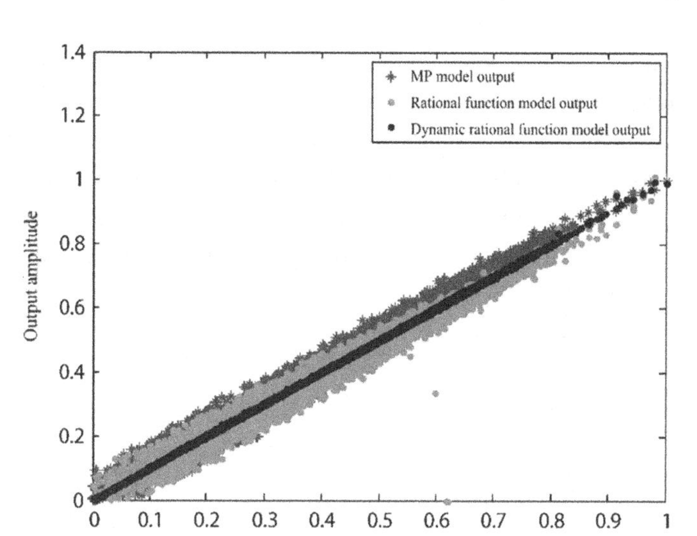

FIGURE 8.4 Comparison of AM/AM characteristics for different models.

models to predistortion. This section proposes a new dynamic rational function model, which considers memory effects with reduced complexity of the model. Simulation results demonstrate that the proposed model has higher accuracy with fewer model coefficients, and effectively compensates the nonlinearities of PA in a predistortion compensation context.

8.2 RF POWER AMPLIFIER MODEL BASED ON PARTICLE SWARM OPTIMIZATION (PSO)_SUPPORT VECTOR MACHINE (SVM)

Neural network and support vector machine (SVM) have demonstrated their respective strengths and weaknesses in RF PA modeling. This section proposes a particle swarm optimization (PSO)_SVM algorithm to combine their strengths and minimize their weaknesses in RF PA modeling. An optimized algorithm is designed for SVM parameter optimization since the selection of SVM parameters is of great significance in the model accuracy and the selection itself is an optimized process of parameter search. The proposed PSO_SVM algorithm uses PSO for SVM parameter optimization, owing to the superiorities of PSO over traditional methods in the parameter optimization process, such as less parameters, easy implementation, high efficiency and fast global optimum finding ability. This section applies the PSO_SVM algorithm to RF PA modeling and compares the established model with the traditional SVM and BP neural network model[7]. Simulation results indicate that the RF PA model based on PSO_SVM is superior to the traditional SVM model and BP neural network (BPNN) model in model accuracy, few-shot learning and approximation capability.

8.2.1 SVM and PSO

1. **Rationale for PSO**[8]

   The PSO algorithm is an emerging intelligent algorithm in recent years, which originates in the study of bird predation behavior and artificial life. In solving an optimization problem, the PSO algorithm regards each particle, that is, a solution to the optimization problem, as a point in the search space. Each particle has a fitness value to be determined by the optimization function, and the movement distance and direction of each particle are determined by the velocity. These particles will move following the current best particle and search generation by generation in the solution space until the optimal solution is finally obtained.

   The PSO algorithm initializes a swarm of particles randomly (random solutions) in the available solution space, and finds the optimal solution by iterations in the solution space. In each iteration, the particle makes an update by tracing the individual extremum p_{best} and the global extremum g_{best}.

   When these two optimal solutions are found, each particle updates its velocity and position based on the following formula:

$$v_{k+1} = c_0 v_k + c_1 \times r_1 (p_{\text{best}_k} - x_k) + c_2 \times r_2 (g_{\text{best}_k} - x_k) \tag{8.14}$$

$$x_{k+1} = x_k + v_{k+1} \tag{8.15}$$

2. **SVM-Related Parameter Optimization**

   The RBF kernel function has only one parameter σ to be determined, and it delivers the best performance and has the most extensive application scope compared to other kernel functions. Consequently, the RBF kernel function is selected as the kernel function for SVM. In SVM parameter optimization using PSO, the unknown SVM parameters, C (penalty coefficient) and σ (width coefficient of RBF kernel function), are firstly specified as the position vector $\boldsymbol{X}$ for the corresponding particle. The specific optimization steps are as follows:

1. Initialize the particle swarm. Specify parameters C and σ as the position vectors for the particle. Initialize the particle's velocity and position.
2. Calculate the fitness value of each particle in the swarm.
3. In every iteration, compare each particle's fitness with its previous best position fitness obtained. If the current value is better, then set p_{best} equal to the current fitness value.
4. Compare each particle's with the swarm's previous best position fitness obtained. Update the swarm global best position (g_{best}) with the greatest fitness value.
5. Adjust the velocity and position of each particle according to Equations (8.14) and (8.15).
6. The procedure ends if predefined criteria are satisfied (good enough position or maximum number of iterations); otherwise, go to step (3) to continue the iterations.

8.2.2 Simulation Experiment and Result Analysis

Based on previous theoretical investigations, a WCDMA signal is used as the input to simulate the PA designed with the MRF6S21140 transistor made by Freescale. The equivalent circuit model is shown in Figure 8.5. ADS software is used for the circuit simulation and extraction of IO data. This subsection selects 450, 350, 180, 100 and 50 sets of data as the training samples respectively, and another 100 sets of data are selected as the test samples. The PSO_SVM, SVM and BP neural networks are used for PA modeling with different training samples.

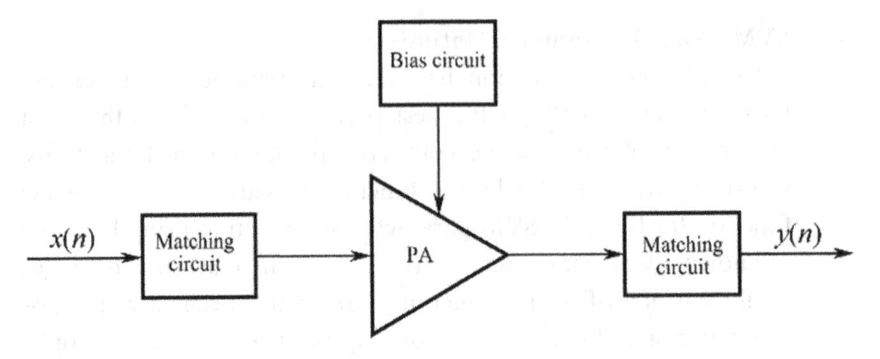

FIGURE 8.5 Equivalent circuit model of PA.

The RMSE of test points serves as the test standard, which is defined as follows:

$$\text{RMSE} = \sqrt{\frac{1}{n}\sum_{i=1}^{n}[\hat{y}_i - y_i]^2} \quad (8.16)$$

where $\hat{y}_i$ is a predicted value and y_i is the real value.

The simulation results are compared in Figures 8.6 and 8.7.

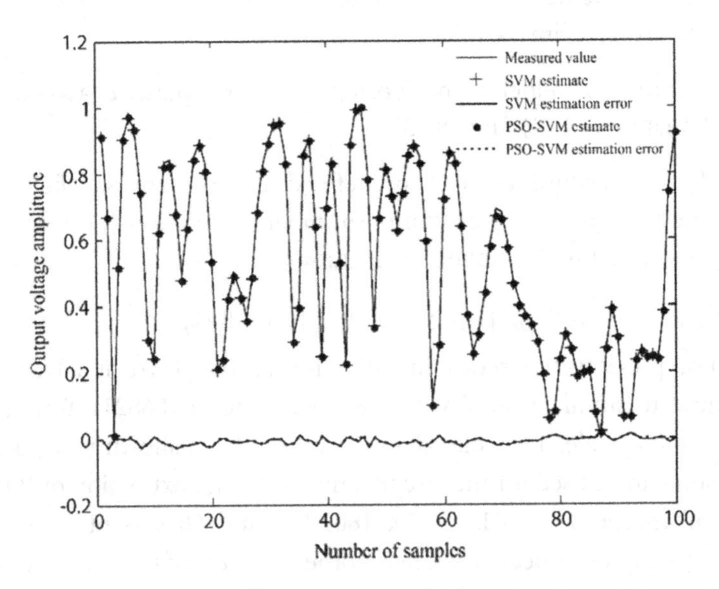

FIGURE 8.6 Comparison of simulation results between PSO_SVM model and SVM model.

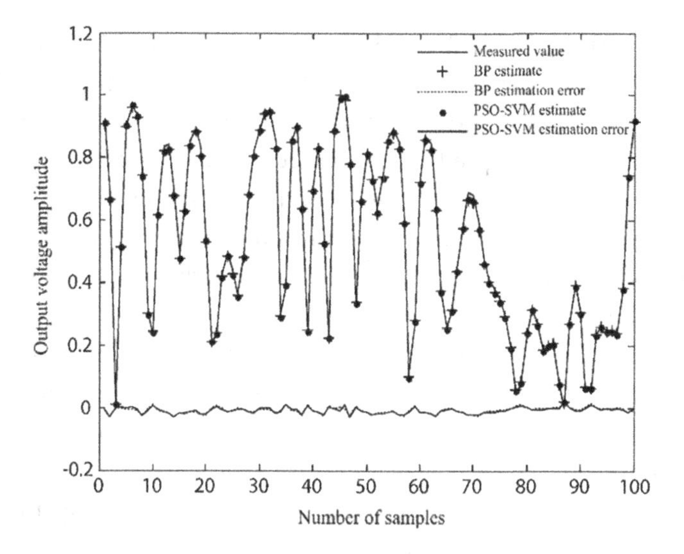

FIGURE 8.7 Comparison of simulation results between PSO_SVM model and BPNN model.

Figure 8.6 shows the simulation results of the PSO_SVM model and the traditional SVM model established with 350 samples. Herein, the traditional SVM model adopts ε-SVM regression, its kernel function uses the RBF kernel function, and parameter settings are $\gamma = 10$, $C = 10$, $\varepsilon = 0.01$.

Figure 8.7 shows the simulation results of the PSO_SVM model and BPNN model established with 350 samples. For the BPNN, the maximum number of iterations is set to 100 and the target RMSE is 0.0001.

Table 8.2 gives a performance comparison of amplifier modeling based on SVM and BPNN models, with R serving as the correlation coefficient.

TABLE 8.2 Performance Comparison of Amplifier Modeling Based on SVM and BPNN Models

Number of Samples	PSO_SVM		SVM		BP Neural Network	
	RMSE	*R*	RMSE	*R*	RMSE	*R*
50	0.0351	99.84%	0.0361	99.83%	0.0485	99.59%
100	0.0339	99.75%	0.0350	99.73%	0.0369	99.70%
180	0.0345	99.87%	0.0347	99.72%	0.0361	99.73%
350	0.0126	99.92%	0.0133	99.91%	0.0163	99.88%
450	0.0128	99.91%	0.0136	99.88%	0.0173	99.90%

As indicated by Table 8.2 and Figure 8.7, the PSO_SVM model produces a smaller RMSE and a greater correlation coefficient, compared to the SVM and BPNN models. Analysis results indicate that the PSO_SVM model delivers higher prediction accuracy. The PSO_SVM model does not rely much on the quality and quantity of sample data. It is able to obtain high accuracy even by learning limited sample data, while the BPNN model experiences a worsening trend with a reduced number of sample points.

In this section, the PSO algorithm is applied to the SVM parameter optimization, which largely overcomes the aimlessness in the artificial selection of SVM parameters. In addition, the selection of SVM parameters based on PSO has a definite theoretical basis. As indicated by the aforementioned analysis and research, PSO_SVM is superior to the traditional SVM and BPNN across all performance indicators. As conclusion, the PSO_SVM modeling scheme demonstrates significant advantages in RF PA modeling.

REFERENCES

1. 牛伟, 王敏锡, 陈凯亚. 基于有理函数的级联自适应预失真器[J]. 微波学报, 2007, (S1):139–142.
   NIU W, WANG X, CHEN K. Cascaded Adaptive Predistorter Based on Rational Function[J]. Journal of Microwaves, 2007, (S1):139–142.
2. 南敬昌, 李厚儒, 方杨. 面向 TD-LTE 通信系统的功放新模型[J]. 计算机工程与应用, 2015, 51(05):261–265.
   NAN J, LI H, FANG Y. New Power Amplifier Model for TD-LTE Communication System[J]. Computer Engineering and Applications, 2015, 51(05):261–265.
3. 王晖, 菅春晓, 李高升, 等. 基于记忆有理函数的功率放大器行为模型[J]. 国防科技大学学报, 2013, 35(3):149–152.
   WANG H, JIAN C, LI G, et al. Power Amplifier Behavioral Modeling Using Memory Rational Function[J]. Journal of National University of Defense Technology, 2013, 35(3):149–152.
4. CUNHA T M, LAVRADOR P M, LIMA E G, et al. Rational Function-Based Model with Memory for Power Amplifier Behavioral Modeling[J]. Workshop INMMIC, 2011:1–4.
5. HAMMI O, YOUNES M, GHANNOUCHI F M. Metrics and Methods for Benchmarking of RF Transmitter Behavioral Models with Application to the Development of a Hybrid Memory Polynomial Model[J]. IEEE Transactions on Broadcasting, 2010, 56(3):350–357.

6. HUANG D, LEUNG H, HUANG X. Experimental Evaluation of Predistortion Techniques for High Power Amplifier[J]. IEEE Transactions on Instrumentation and Measurement, 2006, 55(6):2155–2164.
7. 南敬昌, 任建伟, 张玉梅. 基于 PSO_BP 神经网络的射频功放行为模型[J]. 微电子学, 2011, 41(05):741–745.
NAN J, REN J, ZHANG Y. Study of Dynamic Behavioral Model for RF Power Amplifier Based on PSO BP Neural Network[J]. Microelectronics, 2011, 41(05):741–745.
8. EBERHART R C, KENNEDY J. A New Optimizer Using Particle Swarm Theory. Proceedings of the Sixth International Symposium On Micro Machine and Human Science. New York, NY, 1995[C]. USA: IEEE, 39–43.

CHAPTER 9

Nonlinear Circuit Analysis Methods

9.1 APPLICATION OF HYBRID GENETIC ALGORITHM WITH VOLTERRA SERIES-BASED IMPROVEMENT IN HARMONIC BALANCE

This section proposes an improved hybrid genetic algorithm (GA) to address the frequent problems in harmonic balance simulation for a GA, such as large randomness, slow iterative process and weak local search ability. The improved algorithm combines the Volterra series, quasi-Newton method and GA. It firstly estimates the frequency-domain initial values by using the memory characteristics of the Volterra series, then performs the global optimization by means of the GA and finally performs the local optimization by means of the quasi-Newton method. Harmonic balance simulation results based on the MRF281 indicate that the improved algorithm reduces the number of iterations by about 40% compared with the GA and produces the simulation data fitting well with the measured data. The improved algorithm combines the characteristics of global optimization and local optimization, which improves the accuracy and convergence speed significantly and overcomes the drawbacks of GA (such as large randomness and weak local search ability).

The harmonic balance method represents a mixed domain analysis that combines time-domain components and frequency domain components by means of fast Fourier transform (FFT) and approximates the circuit state variables with Fourier series expansions. Generally, the expansion terms must be sufficiently large to ensure that the impact of high-order harmonics on simulation results is negligible. In other words, the method establishes a harmonic balance equation and then finds its solution using an appropriate algorithm. At present, the algorithms applied to harmonic balance basically include Newton's method [1], neural network algorithm and GA. Newton's method proposed in Literatures [2] and [3] yields an approximation to the root of the equation by using the first few terms of the Taylor series. Newton's method features fast convergence and complex root finding ability. However, its convergence relies much on the selection of initial values, and repeated derivations and inversions are required resulting in high computational intensities. The neural network algorithm proposed in Literatures [4] and [5] does not require an explicit expression of the nonlinear equation, but it is difficult to guarantee the accuracy and easy to fall into local minima for the neural network algorithm. In addition, the network structure is difficult to determine and over-fitting is easy to occur. The GA proposed in Literature [6] overcomes the drawback of Newton's method in the sensitivity to initial intervals and the defects of neural network algorithm, having the ability to yield the root of harmonic balance equation without any fitting method. The algorithm has favorable global search ability, strong robustness, adaptability and parallelism. However, the algorithm has insufficient local search ability, the selection of the initial population affects the convergence speed and the algorithm is easy to fall into the local optimal solution. From a practical point of view, this section proposes an efficient, high-accuracy and high-reliability algorithm to address the problems of inaccuracy and slow computation in harmonic balance simulation.

9.1.1 Harmonic Balance Theory

The small-signal equivalent model [7,8] of MESFET is shown in Figure 9.1. In the figure, the components outside the dashed box are bias-independent parasitic elements, while the components inside the dashed

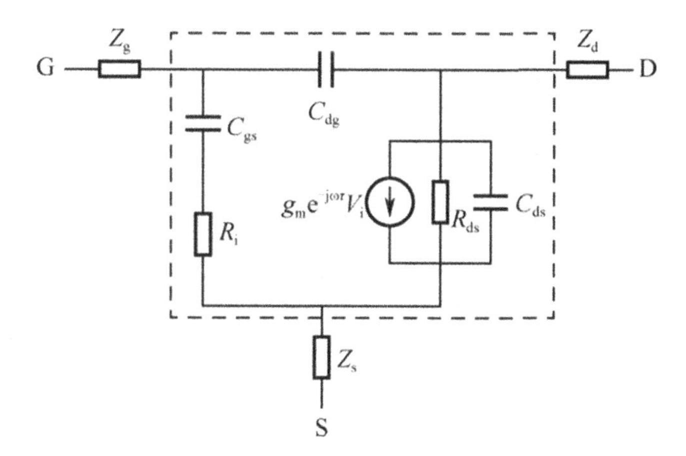

FIGURE 9.1 Small-signal equivalent model of MESFET.

box are bias-dependent intrinsic elements. Small-signal equivalent model and parameter extraction technique are the basis for understanding the physical mechanism of devices and establishing nonlinear equivalent models. It is critical to develop an accurate small-signal equivalent model for solving the large-signal equivalent model. Therefore, the parameters of extrinsic elements are used as independent variables of the parameter function for intrinsic elements to solve the parameters of the small-signal circuit equivalent model. In addition, the extrinsic element of MESFET at zero bias is taken as the initial value, which is optimized to yield the value of the intrinsic element. This method accurately describes the real working conditions of MESFET, features fast convergence and high computational efficiency and therefore meets the requirements of simulation.

The small-signal equivalent model cannot characterize the harmonics in the context of large RF signals, which must be corrected [9]. Based on the small-signal equivalent model, some nonlinear components are used to replace the original linear components, while parasitic elements are less affected by signal transformation and therefore can be retained. In this way, the device can be correctly characterized in the context of large RF signals.

The large-signal equivalent model [10] of MESFET is shown in Figure 9.2, where I_{ds}, I_{gs} and I_{dg} denote the nonlinear components and the other parameters have been solved in the small-signal equivalent circuit.

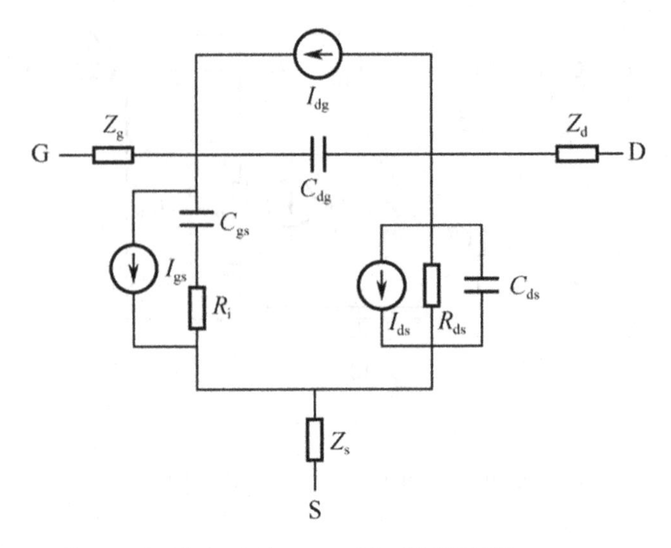

FIGURE 9.2 Large-signal equivalent model of MESFET.

For a nonlinear component, we can measure the volt-ampere characteristic curve in DC state, and then use a curve-fitting approach to yield its expression on MATLAB® software according to a selected empirical formula, thereby extracting the parameters of the large-signal equivalent circuit.

The empirical formula of nonlinear current is as follows:

$$I_{gs} = I_{g0}\left(e^{\alpha_f V_g} - 1\right) \tag{9.1}$$

$$I_{dg} = I_{b0}\left(e^{\alpha_r V_{dg}} - 1\right) \tag{9.2}$$

$$I_{ds} = \frac{\beta (V_g - V_T)^2}{1 + b(V_g - V_T)}(1 + \lambda V_d)\tanh(\alpha V_d) \tag{9.3}$$

In this section, the large-signal equivalent model of MESFET is decomposed into a linear network and a nonlinear network, as shown in Figure 9.3. In the figure, the source impedance and load impedance are added to make the equivalent model more accurate. That is, two voltage sources are added to the linear network so that the entire circuit is simplified into 3 + 2 ports.

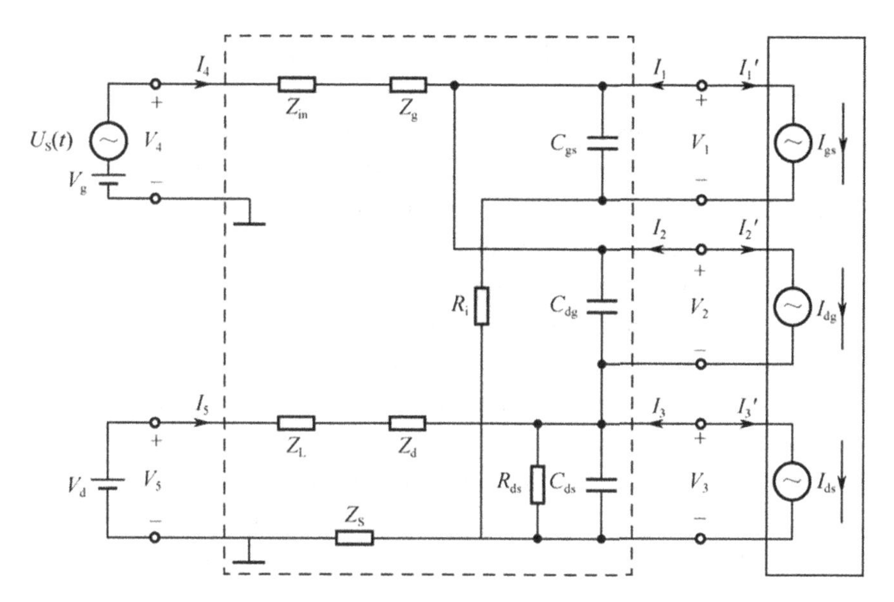

FIGURE 9.3 Decomposed MESFET large-signal equivalent model.

The nonlinear equation set is solved iteratively using an improved hybrid GA. The initial values of iterations V_1, V_2 and V_3 are computed by the Volterra series method. The g-d pole current can be ignored in the computation. The initial values are substituted into Equation (3.54) to solve the current of the linear network. The admittance matrix in Equation (3.55) is derived from the small-signal equivalent parameters. By means of Fourier transform, the linear current and nonlinear current vectors are substituted into Equation (3.56) to yield the final exact solutions by iterative operations. Based on this set of solutions, the current and voltage values for any component can be obtained. The results contain multiple harmonics, favorably characterizing the nonlinearities of the circuit.

9.1.2 Improved Hybrid Genetic Algorithm

A GA is a highly parallel, random and adaptive optimization algorithm developed on the basis of mimicking the natural genetic mechanism and evolution mechanism of natural organisms. It features implicit parallelism and global space search ability without restriction conditions. However, a GA has certain weaknesses, such as premature convergence, low efficiency and absence of effective quantitative analysis approaches regarding the accuracy, feasibility and computational complexity of the algorithm.

To overcome these drawbacks of GA, this subsection proposes an improved hybrid GA. The proposed algorithm combines the Volterra series method, GA and quasi-Newton method, and is divided into three parts, which are initialization, global optimization and local optimization. The algorithm can yield the solution of a harmonic balance equation with a high probability. It requires less iterations and demonstrates high computational efficiency. In contrast, a GA has low computational efficiency and is easy to fall into the local optimal solution, and Newton's method requires derivations and inversions resulting in high computational intensities.

Initialization is to encode the initial population in way of a certain encoding approach. The initial population, in a traditional GA, is produced by a random generation approach [11] and then encoded as per certain encoding rules. However, the initial population produced by this approach is a random type, which makes the initial average fitness quite different from the optimal fitness. Consequently, a large number of iterations are required to yield the exact solutions, which significantly reduces the computational efficiency of GA [12]. The proposed algorithm uses the Volterra series method to estimate the initial population of the equation, replaces the original harmonic input method with the nonlinear current input method and yields the specific expression of high-order harmonic components based on low-order harmonic components. This processing is more suitable for numerical analysis. In addition, real-number coding is used instead of the original binary coding, allowing searches in a larger space. The algorithm also improves the complexity and increases the computational efficiency, enabling the initial population to show a better state in the initial stage.

Volterra series method overcomes the weakness of power series [13]. It introduces memory effects and can describe the nonlinear components of each order in the system separately, thereby accurately characterizing the nonlinearities of the system only with the first three terms. The idea of solving the initial population by the Volterra series method is as follows. Based on the parameters of excitation and other parts in Figure 9.3, the voltage at both ends of the nonlinear conductance is calculated with $k = 3$ (indicating a third-order calculation as the sum of the first three terms is accurate enough for characterization). The yielded voltage components are summed up as an individual in the initial population and encoded as per the real-number coding principle. This procedure is repeated until the number reaches N in the initial population.

The specific initialization steps are as follows.

Step 1: Calculate the linear response, that is

$$V_1 = \frac{1}{G + g_1} \sum_{k=-3}^{k=+3} I_{S,k} e^{j\omega_k t}, \qquad I_{S,k} = \frac{V_{S,k}}{R} \tag{9.4}$$

where k is the number of harmonics. The result of single-tone input is found in (9.4). The second-order nonlinear current component is yielded by using the first-order voltage component.

Step 2: Calculate the second-order nonlinear current and voltage, namely

$$I_2 = \frac{1}{G + g_1} \sum_{k_1=+1}^{k_1=-1} \sum_{k_2=+1}^{k_2=-1} I_{S,k_1} I_{S,k_2} e^{j(\omega_{k1}+\omega_{k2})t} \tag{9.5}$$

Since the input current source has only a first-order response, the second-order voltage is calculated by using a zero-input response.

$$V_2 = \frac{-g_2}{4(G + g_1)^3} \sum_{k_1=+1}^{k_1=-1} \sum_{k_2=+1}^{k_2=-1} I_{S,k_1} I_{S,k_2} e^{j(\omega_{k1}+\omega_{k2})t} \tag{9.6}$$

Step 3: Calculate the third-order nonlinear current and voltage. Similarly, they are also calculated by using a zero-input response.

$$\begin{aligned} I_3 = {} & 2g_2 \frac{-g_2}{4(G+g_1)^3} \sum_{k_1=+1}^{k_1=-1} \sum_{k_2=+1}^{k_2=-1} I_{S,k_1} I_{S,k_2} e^{j(\omega_{k1}+\omega_{k2})t} \\ & + \frac{1}{2(G+g_1)} \sum_{k_1=+1}^{k_1=-1} I_{S,k} e^{j\omega_k t} \end{aligned} \tag{9.7}$$

$$+ g_3 \frac{1}{8(G + g_1)^3} \sum_{k_1=+1}^{k_1=-1} \sum_{k_2=+1}^{k_2=-1} \sum_{k_3=+1}^{k_3=-1} I_{S,k_1} I_{S,k_2} I_{S,k_3} e^{j(\omega_{k1}+\omega_{k2}+\omega_{k3})t}$$

$$V_3 = \frac{-I_3}{G + g_1} \tag{9.8}$$

Step 4: Sum up the yielded first-order, second-order and third-order components to obtain an individual in the initial population, and then perform the coding operation. Repeat this procedure until the number reaches N in the initial population.

Global optimization is to find the optimal solution that satisfies the error in the global scope. A GA is used for global optimization. In a GA, the genetic operators serve as the core of global optimization. A GA reorganizes and mutates the individuals in the selected population by using the genetic operators, which embodies the biological genetic and evolution mechanism ("survival of the fittest") in the nature.

Genetic operators include selection operator, crossover operator and mutation operator.

1. Selection operator

The selection operator is basically used to select individuals with favorable fitness values and discard individuals with unfavorable fitness values. The algorithm combines a random league selection method with an optimal preservation strategy, so that the individuals with favorable fitness values can be copied into the next generation without being destroyed by crossover or mutation operations, thereby ensuring the convergence of the algorithm.

2. Crossover operator

The crossover operator serves as the main approach to generate new individuals, determining the global convergence of GA. It enables two individuals to exchange parts of their genes in a certain way, thus producing two new individuals. The algorithm adopts arithmetic crossing to apply the crossover operation to two individuals X_1 and X_2. The new individuals produced after crossover are

$$\begin{cases} X'_1 = \alpha(X_1 - X_2) + X_2 \\ X'_2 = \alpha(X_2 - X_1) + X_1 \end{cases} \tag{9.9}$$

where the crossover operator α is the parameter. The larger the value of α, the closer X'_1 is to X_1, which means the GA searches within a small scope near the individual X_1 ensuring the accuracy of solutions.

3. Mutation operator

The mutation operator is to change the gene values on some loci of individual strings in the population, thereby producing new individuals. It serves as an auxiliary method to produce new individuals and determines the local convergence of GA. The combination of crossover operator and mutation operator enables the GA to complete the process of finding the optimal solution. The GA adopts the basic bit mutation, which refers to a mutation operation of a certain bit or a certain number of bits randomly assigned to an individual code string with probability p. The global convergence of the algorithm is guaranteed because the better individuals in each generation of the population are retained and the mutation operator itself is ergodic.

As the GA is easy to fall into the local optimal solution and later iterations are less efficient, the quasi-Newton method is adopted for local optimization after global optimization is complete. Quasi-Newton method features fast convergence and sets strict requirements for initial values. It does not converge if initial values are far from the exact solution and has fast convergence if initial values are close to the exact solution. However, quasi-Newton method avoids computing derivations and inversions compared to Newton's method, therefore simplifying the solution process and notably improving the computational efficiency.

Adopting the most effective approach of the quasi-Newton method, BFGS correction formula, gives

$$\boldsymbol{B}_{k+1} = \boldsymbol{B}_k + \frac{\boldsymbol{y}^k(\boldsymbol{y}^k)^{\mathrm{T}}}{(\boldsymbol{y}^k)^{\mathrm{T}}\boldsymbol{s}^k} - \frac{\boldsymbol{B}_k\boldsymbol{s}^k(\boldsymbol{s}^k)^{\mathrm{T}}\boldsymbol{B}_k}{(\boldsymbol{s}^k)^{\mathrm{T}}\boldsymbol{B}_k\boldsymbol{s}^k} \tag{9.10}$$

The overall flowchart of the algorithm is shown in Figure 9.4.

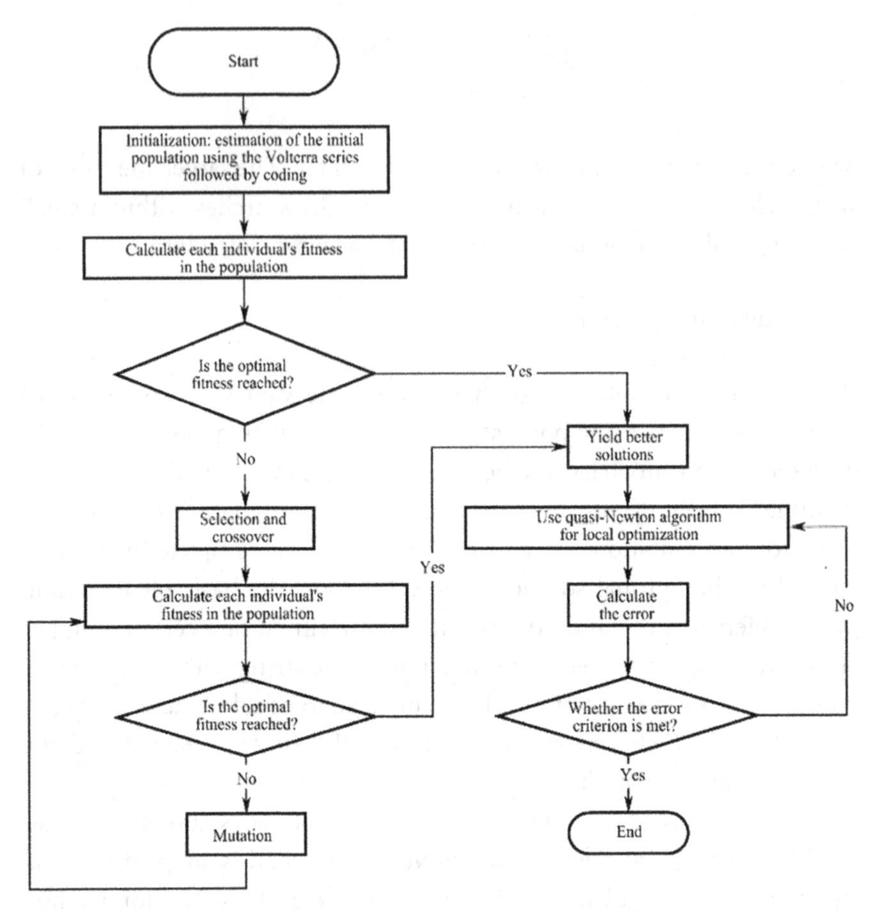

FIGURE 9.4 Overall algorithm flowchart.

Step 1: Perform the initialization. Produce an initial population and encode it as per the real-number encoding rules. Set the population size as N, the GA's maximum number of iterations as D and the genetic operator and target fitness as θ.

Step 2: Evaluate each individual in the population. Define the fitness function as follows:

$$\begin{cases} \text{find: } x = [x_1, x_2, \cdots, x_n] \\ \text{min: fit}(x) = \sum_{i=1}^{n} |f_i(x)| \end{cases} \tag{9.11}$$

Calculate each individual's fitness η_i, $i = 1, 2, \cdots, N$ using the fitness function. If an individual's fitness $\eta_i \leqslant \theta$, $i = 1, 2, \cdots, N$, go to Step 4 for local optimization; otherwise, go to Step 3.

Step 3: Evolve the population. Use the GA to carry out selection, crossover and mutation operations based on the initial population to produce a new population. If the number of iterations reaches D (GA's maximum number of iterations), go to Step 4; otherwise, go to Step 2.

Step 4: Perform local optimization for the better solution yielded by the GA. Continue the iterations with the better solution taken as the initial value for the quasi-Newton method. If the accuracy ε is satisfied, the algorithm ends; otherwise, go to Step 3.

9.1.3 Simulation and Data Analysis

This subsection aims to evaluate the performance of the improved algorithm in harmonic balance. Based on the aforementioned harmonic balance theory, the single-tone and dual-tone harmonic balance simulation analysis on the MRF281 is performed. The harmonic balance simulation platform is set up by means of MATLAB® programming with the use of small-signal equivalent model parameters and large-signal equivalent model parameters (listed in Tables 9.1 and 9.2). The transistor's static operating points are V_{DS} = 24 V and V_{GS} = 4.5 V The power amplifier works in class A, the operating frequency is 2 GHz and the input power is 25 dBm.

Newton's method, quasi-Newton method, GA and improved hybrid GA (improved algorithm) are applied respectively, with their resultant iterations and allowable errors listed in Table 9.3. As indicated in the table, under the conditions with the same comprehensive allowable errors, Newton's method requires an average of 69 iterations and 65.48s to

TABLE 9.1 Small-Signal Equivalent Model Parameters

Parameter	Value	Parameter	Value	Parameter	Value
L_g(nH)	0.303	$R_d(\Omega)$	1.663	C_{ds}(pF)	0.104
L_d(nH)	0.097	$R_s(\Omega)$	0.961	$R_i(\Omega)$	0.097
L_s(nH)	0.128	C_{gs}(pF)	6.420	$R_{ds}(\Omega)$	243
$R_g(\Omega)$	0.319	C_{gd}(pF)	0.107	g_m(mS)	73.9

TABLE 9.2 Large-Signal Equivalent Model Parameters

Parameter	Value	Parameter	Value
I_{g0}(A)	1.009e(−10)	a_f(1/V)	20.259
a_r(1/V)	0.100	β	0.247
b	0.099	λ	0.007
I_{b0}(A)	4.099e(−6)	a	2.52
V_T(V)	-3.979		

TABLE 9.3 Resultant Iterations and Allowable Errors

Iterations/ Second Error	Newton's Method	Quasi-Newton Method	Genetic Algorithm	Improved Algorithm
10^{-3}	37/34.7	16/10.3	35/21.5	21/14.6
10^{-4}	51/50.3	23/15.6	37/22.3	22/14.7
10^{-5}	68/59.4	29/21.2	37/22.5	27/16.3
10^{-6}	83/84.1	40/31.3	41/23.4	30/16.9
10^{-7}	106/98.9	47/33.8	44/23.9	34/18.1

yield the exact solution; the quasi-Newton method requires an average of 31 iterations and 22.28s; the GA requires an average of 38.8 iterations and 22.72s; the improved algorithm requires an average of 26.8 iterations and 16.12s. Therefore, the improved algorithm notably reduces the number of iterations and the time required for computation, delivering higher computational efficiency.

Figures 9.5 and 9.6 depict the iteration curves yielded by the GA proposed in Literature [9] and the improved algorithm, respectively. As indicated by the figure, the initial average fitness of GA is far from the optimal fitness; when the average fitness gets close to the optimal fitness, the convergence rate greatly declines. The average fitness reaches optimal fitness after about 80 iterations. In contrast, the initial average fitness of the improved algorithm is close to the optimal fitness and reaches the optimal fitness after about 47 iterations. Compared with the GA in Literature [9], the improved algorithm significantly reduces the number of iterations and shortens the computation time, demonstrating favorable convergence performance and high computational efficiency.

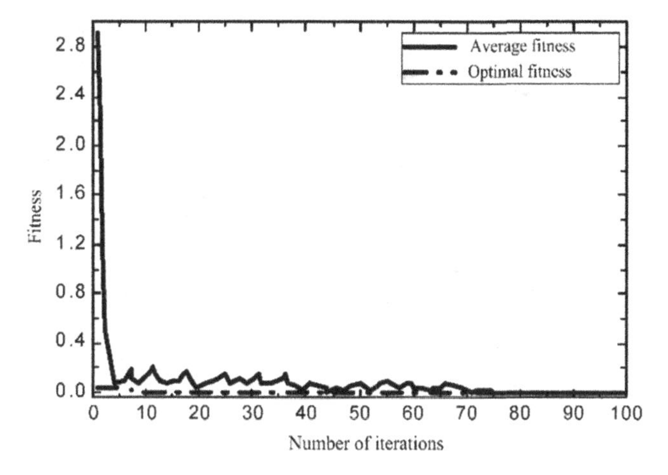

FIGURE 9.5 Iteration curve of GA.

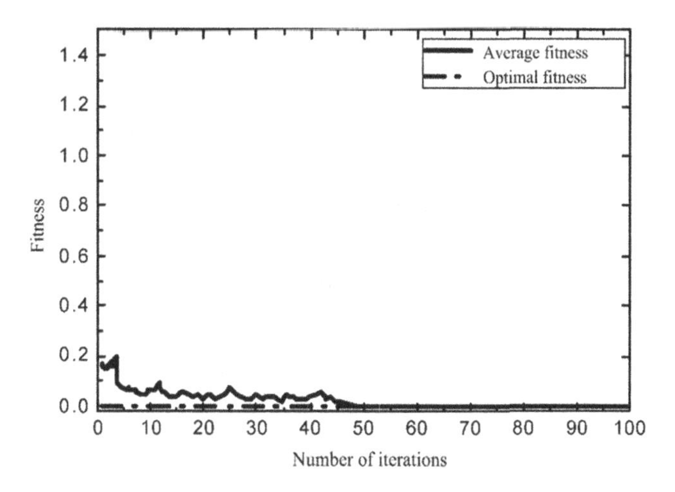

FIGURE 9.6 Iteration curve of improved algorithm.

Figure 9.7 shows the measured values of harmonic characteristics and the simulation curve yielded by the improved algorithm when the input is a 2 GHz single-tone signal. Figure 9.8 shows the measured values of intermodulation characteristics and the simulation curve yielded by the improved algorithm when the input is a 1900/2100 MHz dual-tone signal. In the figures, straight lines represent simulated data and points represent measured data.

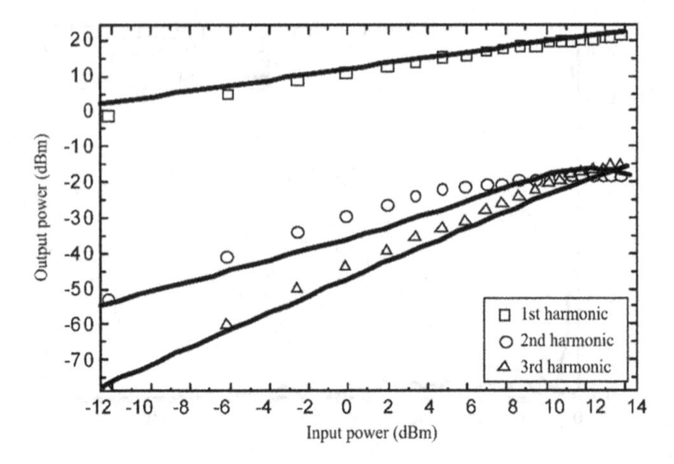

FIGURE 9.7 Harmonic characteristics with a single-tone input signal.

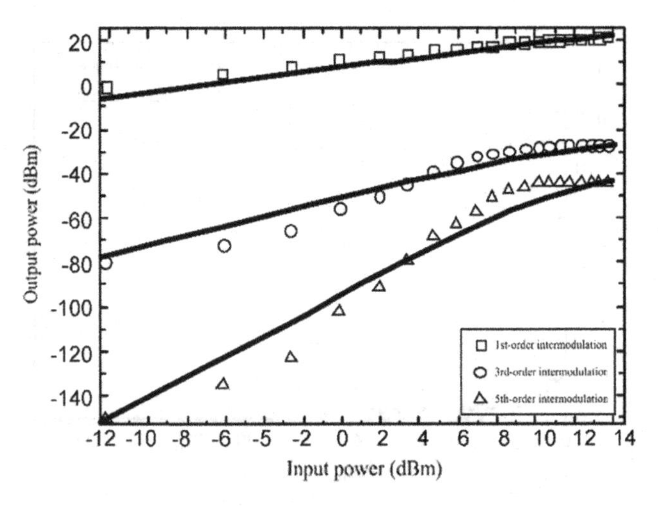

FIGURE 9.8 Intermodulation characteristics of waves of different orders with a dual-tone input signal.

As indicated by Figures 9.7 and 9.8, simulated data and measured data exhibit the same trend but occasionally large errors are observed. This is because the extraction of small-signal equivalent model parameters and large-signal equivalent model parameters cannot take each parameter into account at the same time. As a result, the simulation curve fits well with measurements for a certain segment and not so well for other segments. Overall, the simulation curve is in good agreement with measurements, proving a successful application of the improved algorithm in harmonic balance.

As indicated in Figure 9.7, the fundamental output power grows with the increase of input power until the saturation effect sets in. This is because the network is nonlinear, resulting in the harmonic power. The output power of harmonics increases synchronously with the increase of input power. Due to the internal characteristics of the power transistor, the second harmonic rises and reaches saturation, and then gradually declines. The third harmonic gradually tends to saturation with the continuous increase of input power. As indicated in Figure 9.8, the fundamental power, third-order intermodulation power and fifth-order intermodulation power increase continuously with the increase of input power until the saturation effect sets in.

The improved algorithm uses the Volterra series to improve the initial state of the GA population and enable it to behave in an optimal state. Furthermore, the improved algorithm reduces the complexity and improves the computational efficiency of the GA by use of the strong local convergence ability of the quasi-Newton method. Also, it solves the problems of inaccurate and slow harmonic balance simulation. As indicated by simulation results, the improved algorithm features less iterations, fast convergence, high convergence efficiency and high solution accuracy, greatly improving the computational efficiency of solving harmonic balance equations. The simulation curve fits well with measurements. As a conclusion, the improved algorithm has a favorable application prospect.

9.2 APPLICATION OF QUASI-NEWTONIAN PARTICLE SWARM OPTIMIZATION ALGORITHM IN HARMONIC BALANCE EQUATIONS FOR NONLINEAR CIRCUITS

Solid state microwave devices are all nonlinear to some extent. In a communication PA, any nonlinearity of phase and amplitude in a voltage transmission waveform must be minimized to protect the waveform and spectral content of the signal. However, limiting amplifiers, oscillators, frequency multipliers and mixers rely on the nonlinearities of devices to obtain appropriate operations. In all cases, a complete circuit analysis on these devices requires nonlinear device models and analytical methods to extract device effects, that is, to analyze and obtain the interactions of model circuits. Harmonic balance method has a wide range of applications and can be used in microwave nonlinear circuits such as mixers, frequency multipliers and PAs.

Nonlinear distortion of PA results in intermodulation distortion and adjacent channel interference, severely affecting the system performance. Therefore, the quantitative description and analysis of the amplifier model is very important. Based on the existing harmonic balance analysis methods for nonlinear circuits, this section combines the particle swarm optimization (PSO) algorithm and the quasi-Newton method (referred to as the quasi-Newtonian PSO algorithm) to solve the harmonic balance equation, and also provides the simulation results. As indicated by simulation results, the quasi-Newtonian PSO algorithm improves the convergence speed of the system and fits well with all indicators of the real amplifier. In conclusion, the quasi-Newtonian PSO algorithm has proved valuable referential significance for nonlinear circuit analysis.

9.2.1 Harmonic Balance Theory

The MESFET PA equivalent circuit [14] is outlined in Figure 9.9, including a large-signal model of a nonlinear MESFET circuit, input and output matching circuit, source impedance and load impedance.

Assume that the current flowing through R_g is x. Applying Kirchhoff's current law to the gate points in Figure 9.9 we have

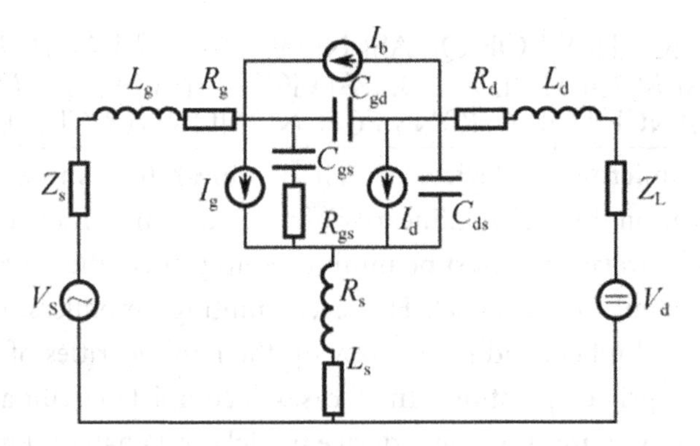

FIGURE 9.9 MESFET PA equivalent circuit.

$$x + y + I_b - I_g - V_g \cdot jk\omega \cdot C_{gs} = 0 \tag{9.12}$$

The equation for the current y on the capacitor C_{gd} is given by

$$\left\{ V_g \cdot jk\omega \cdot C_{gs} \cdot \left[\frac{1}{jk\omega \cdot C_{gs}} + R_{gs} \right] - V_d \right\} \cdot jk\omega \cdot C_{gd} = y \tag{9.13}$$

According to Kirchhoff's voltage law, the following equation applies:

$$\begin{aligned} -V_S + (Z_1 + Z_S)x + I_g \cdot Z_2 + V_g + V_g \cdot jk\omega \cdot C_{gs} \cdot (R_{gs} + Z_2) \\ + I_d Z_2 + V_d \cdot jk\omega \cdot C_{ds} \cdot Z_2 = 0 \end{aligned} \tag{9.14}$$

Substituting Equations (9.12) and (9.13) into Equation (9.14) yields a set of complex algebraic equations:

$$C = AV_{gk} + BV_{dk} \tag{9.15}$$

$$F = DV_{gk} + EV_{dk} \tag{9.16}$$

The set of equations take V_g and V_d as independent variables at each kth harmonic component, where

$$\begin{aligned} A &= 1 + jk\omega\left[(Z_2 + R_{gs})C_{gs} + (Z_1 + Z_S)(C_{gs} + C_{gd} + jk\omega R_{gs} C_{gs} C_{gd})\right] \\ B &= jk\omega\left[-(Z_1 + Z_S)C_{gd} + Z_2 C_{ds}\right] \\ C &= -(Z_1 + Z_2 + Z_S)I_g + (Z_1 + Z_S)I_b - Z_2 I_d + V_S \\ D &= jk\omega\left[Z_2 C_{gs} - (Z_3 + Z_L)(C_{ds} + jk\omega R_{gs} C_{gs} C_{ds})\right] \\ E &= 1 + jk\omega\left[Z_2 C_{gs} + (Z_3 + Z_L)(C_{gd} + C_{gs})\right] \\ F &= -Z_2 I_g - (Z_3 + Z_L)I_b + (Z_2 + Z_3 + Z_L)I_d + V_{DD} \\ Z_1 &= R_g + jk\omega L_d,\ Z_2 = R_s + jk\omega L_s,\quad Z_3 = R_d + jk\omega L_d,\quad k = 1,\ \cdots, N \end{aligned}$$

In this model, C_{gs}, C_{ds}, C_{dg} represent the linear part, and the voltage-controlled current sources I_g, I_b and I_d represent the nonlinear part. Both of them depend on time-domain voltages $v_d(t)$ and $v_g(t)$.

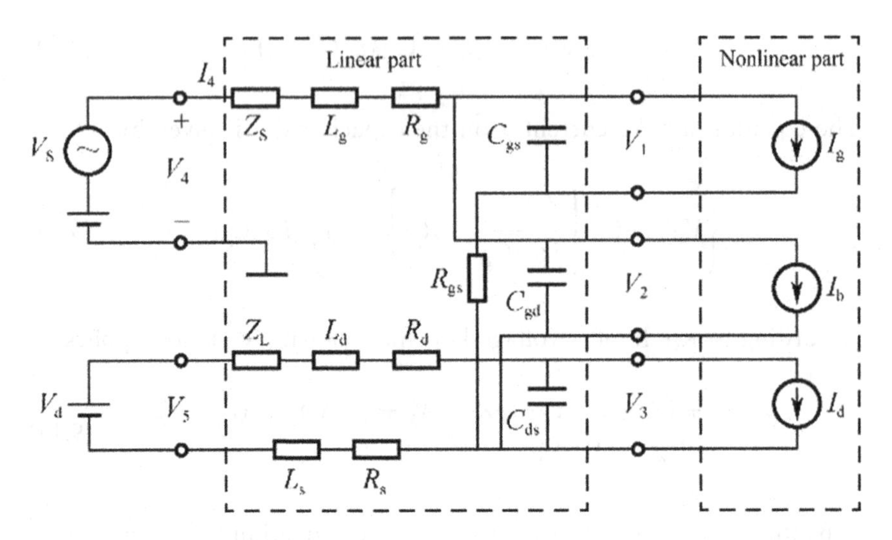

FIGURE 9.10 MESFET circuit divided into linear sub-circuit and nonlinear sub-circuit.

Figure 9.10 depicts an MESFET circuit that is divided into a linear sub-circuit and a nonlinear sub-circuit, with a signal source and DC bias added.

The quasi-Newtonian PSO iterative algorithm is used to solve the equations. The iterative process starts with an initial approximation value. The gate-leakage current I_b can be ignored in the estimation while the output transconductance or conductance of a linear component is assumed. The port initial voltage vector is specified and substituted into Equations (9.15) and (9.16) to calculate the linear network current vector I of the MESFET circuit.

The nonlinear network current $\boldsymbol{I'}$ expressed by a full frequency-domain conversion matrix [15,16] is as follows:

$$\boldsymbol{I'} = \begin{bmatrix} I_g \\ I_d \end{bmatrix} \tag{9.17}$$

Finally, the full frequency-domain harmonic balance equation is expressed as follows:

$$F(\boldsymbol{V}) = \boldsymbol{I} + \boldsymbol{I'} = \boldsymbol{0} \tag{9.18}$$

9.2.2 Quasi-Newtonian PSO Algorithm

Quasi-Newton method is an effective method to solve nonlinear optimization problems. It overcomes the drawbacks of Newton's method (such as derivations and inversions). In the quasi-Newton method, the Jacobian matrix is converted into:

$$\boldsymbol{H}^{p+1} = \boldsymbol{H}^{p} + \Delta\boldsymbol{H}^{p}, \quad p = 1, 2, \cdots \tag{9.19}$$

In this way, the solution process is simplified and the convergence of iterations is guaranteed. The famous BFGS algorithm is given below:

$$\boldsymbol{V}^{p+1} = \boldsymbol{V}^{p} - \boldsymbol{H}^{p}\boldsymbol{F}(\boldsymbol{V}^{p}) \tag{9.20}$$

$$\boldsymbol{H}^{p+1} = \boldsymbol{H}^{p} - \frac{\boldsymbol{H}^{p}\boldsymbol{\gamma}^{p}(\boldsymbol{\gamma}^{p})^{\mathrm{T}}\boldsymbol{H}^{p}}{(\boldsymbol{\gamma}^{p})^{\mathrm{T}}\boldsymbol{H}^{p}\boldsymbol{\gamma}^{p}} + \frac{\boldsymbol{\delta}^{p}(\boldsymbol{\delta}^{p})^{\mathrm{T}}}{(\boldsymbol{\delta}^{p})^{\mathrm{T}}\boldsymbol{\gamma}^{p}} + \boldsymbol{\nu}^{p}(\boldsymbol{\nu}^{p})^{\mathrm{T}} \tag{9.21}$$

$$\boldsymbol{\delta}^{p} = \boldsymbol{V}^{p+1} - \boldsymbol{V}^{p} \tag{9.22}$$

$$\boldsymbol{\gamma}^{p} = \boldsymbol{F}(\boldsymbol{V}^{p+1}) - \boldsymbol{F}(\boldsymbol{V}^{p}) \tag{9.23}$$

$$\boldsymbol{\nu}^{p} = ((\boldsymbol{\gamma}^{p})^{\mathrm{T}}\boldsymbol{H}^{p}\boldsymbol{\gamma}^{p})^{\frac{1}{2}}\left(\frac{\boldsymbol{\delta}^{p}}{(\boldsymbol{\delta}^{p})^{\mathrm{T}}\boldsymbol{\gamma}^{p}} - \frac{\boldsymbol{H}^{p}\boldsymbol{\gamma}^{p}}{(\boldsymbol{\gamma}^{p})^{\mathrm{T}}\boldsymbol{H}^{p}\boldsymbol{\gamma}^{p}}\right) \tag{9.24}$$

The computation steps are as follows:

Step 1: Give the initial estimated voltage vector $\boldsymbol{V}_k^0$. Set the maximum harmonic number as 5 ($K = 0, 1, \cdots, 5$). Give the initial correction matrix $\boldsymbol{H}^0$ and the allowable error ε_0.

Step 2: Find the new estimated voltage vector $\boldsymbol{V}_K^{p+1}$ through Equation (9.20).

Step 3: Yield $\boldsymbol{I'}_K^p$ using the full frequency-domain conversion matrix.

Step 4: Yield $F(\boldsymbol{V}_K^p)$ by means of the equation $F(\boldsymbol{V}) = \boldsymbol{I} + \boldsymbol{I'}$.

Step 5: Determine whether $F(\boldsymbol{V}_K^p) < \varepsilon_0$ applies. If it applies, $\boldsymbol{V}_K^{p+1}$ is yielded and taken as an approximate solution that meets the accuracy requirements, and the algorithm ends. Otherwise, go to Step 6.

Step 6: Compute Equations (9.22), (9.23) and (9.24) to yield $\boldsymbol{\delta}^p$, $\boldsymbol{\gamma}^p$ and $\boldsymbol{\nu}^p$, respectively.

Step 7: Modify the correction matrix and go to Step 2.

In the PSO algorithm, a particle swarm represents a potential solution. Each particle i in the swarm is related to two vectors, velocity vector $\boldsymbol{V}_i = [v_i^1, v_i^2, \cdots, v_i^D]$ and position vector $\boldsymbol{X}_i = [x_i^1, x_i^2, \cdots, x_i^D]$ where D denotes the dimension of the solution space. The velocity and position of each particle are initialized by random vectors in the corresponding range. In the iterative process, the update formulas for the velocity and position of particle i are given by

$$v_i^{d+1} = \omega v_i^d + c_1 \text{rand}_1^d (\text{pBest}_i^d - x_i^d) + c_2 \text{rand}_2^d (\text{nBest}^d - x_i^d) \quad (9.25)$$

$$x_i^{d+1} = x_i^d + v_i^{d+1} \quad (9.26)$$

In Equation (9.25), the inertia weight ω is 0.5; the acceleration coefficients $c_1 = 1$ and $c_2 = 2$; rand_1^d and rand_2^d are two independent random numbers generated in [0, 1]; pBest_i is the best position found by the particle so far; nBest is the swarm's previous best position.

The steps of the quasi-Newtonian PSO algorithm are as follows.

Step 1: Initialize a particle swarm (with the size m).

Step 2: Calculate the fitness value of each particle in the swarm.

Step 3: Compare each particle's fitness with the fitness of its previous best position (individual extremum) experienced. If the current fitness is better, then set the previous best position equal to the current position.

Step 4: Compare each particle's fitness with the fitness of the swarm's previous best position (global extremum) obtained. If the particle's fitness is better, then set the swarm's previous best position equal to the current position.

Step 5: Update the velocity and position of the particle swarm according to Equations (9.25) and (9.26).

Step 6: If the preset maximum number of generations is reached, return the global optimal individual and go to Step 7; otherwise, $d = d + 1$ and go to Step 2.

Step 7: Apply the quasi-Newton method. The global optimal individual returned by Step 6 is taken as the initial point of the quasi-Newton method for iterations.

Step 8: If the given accuracy ε_0 is reached, generate the result and the algorithm ends. Otherwise, go to Step 6.

9.2.3 Experimental Simulation Analysis

Based on the above analysis of harmonic balance and quasi-Newtonian PSO algorithm, this subsection describes a dual-tone simulation analysis on a power transistor with the input power −40 dBm and the operating frequency 1.96 GHz by means of MATLAB® programming.

The circuit diagram of the amplifier is shown in Figure 9.11. The simulation analysis is performed with the Newton's method, quasi-Newton method and quasi-Newtonian PSO algorithm respectively. The Newton's method takes an average of 0.881s per iteration with 42, 56 and 70 iterations, respectively; the quasi-Newton method takes an average of 0.689s per iteration with 13, 18 and 24 iterations, respectively; the quasi-Newtonian PSO algorithm takes an average of 0.471s per iteration with 30 and 31 iterations, respectively. Therefore, the quasi-Newtonian PSO algorithm notably accelerates the iteration speed, demonstrating

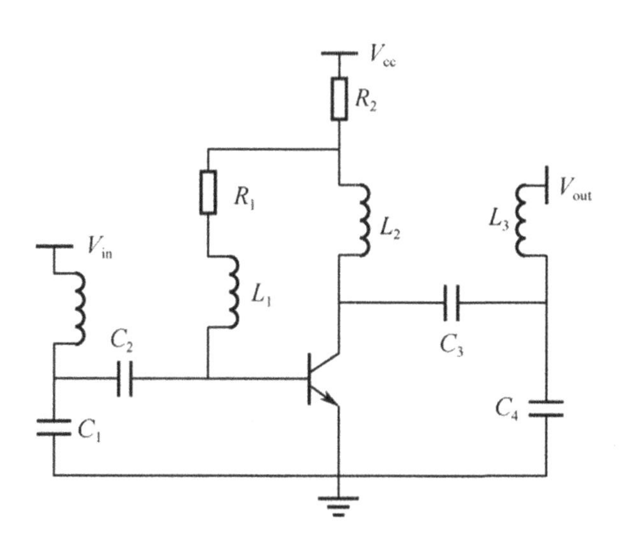

FIGURE 9.11 Circuit diagram of amplifier.

favorable convergence performance and high computational efficiency. The resultant iterations and relative errors are listed in Table 9.4.

Figure 9.12 shows the input-output power relationship, with the input power ranging from −45 dBm to −30 dBm. The fundamental curve, IM2 curve and IM3 curve are depicted in the figure. They all grow with the increase of input power. The fundamental curve grows in the gentlest way, and the slope ratio of IM2 to IM3 is 2:3. Figure 9.13 shows the power components of fundamental wave, third-order intermodulation wave, fifth-order intermodulation wave and seventh-order intermodulation wave when the input is a 1895/1905 MHz dual-tone signal. In general, only the third-order intermodulation wave falls within the

TABLE 9.4 Resultant Iterations and Relative Errors

	Newton's Method	**Quasi-Newton Method**	**Quasi-Newtonian PSO Algorithm**
Allowable Relative Error	**Number of Iterations/ Time (s)**	**Number of Iterations/ Time (s)**	**Number of Iterations/ Time (s)**
10^{-3}	42/38.7	13/9.3	30/14.1
10^{-4}	56/52.3	18/13.1	31/14.6
10^{-5}	70/54.9	24/15.0	31/14.7

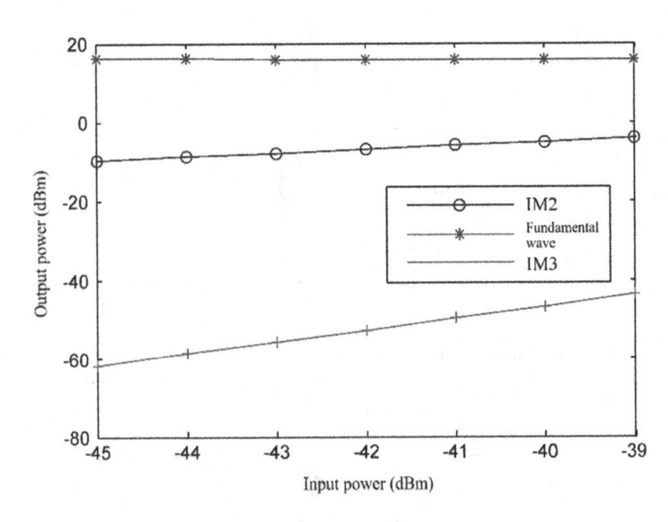

FIGURE 9.12 Input-output power relationship.

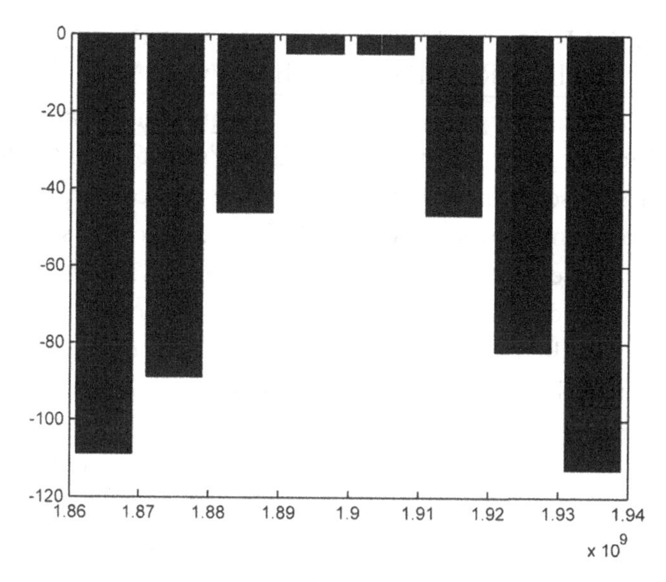

FIGURE 9.13 Power components of intermodulation waves of different orders with a dual-tone input signal.

passband range, which is considered as the major nonlinear product. The simulation analyzes up to the seventh-order intermodulation wave, thereby better fitting the nonlinearities of PA. Harmonic balance simulation using the quasi-Newtonian PSO algorithm can quickly and accurately obtain nonlinear parameters such as intermodulation distortion components, and can obtain steady-state response of RF circuits in a short time.

This section proposes the quasi-Newtonian PSO algorithm designed to solve harmonic balance equations of PA in nonlinear devices, and compares the proposed algorithm with the basic Newton's method and quasi-Newton method for solving harmonic balance equations. As indicated by simulation results, the proposed algorithm is superior to the other two algorithms in convergence speed and can accurately characterize the input and output of PA. The proposed algorithm is a feasible method and has proved valuable significance for the research on harmonic balance analysis methods for nonlinear circuits.

9.3 APPLICATION OF HYBRID ANT COLONY ALGORITHM IN NONLINEAR HARMONIC BALANCE ANALYSIS

In the process of solving a harmonic balance equation, Newton's method proposed in Literature [17] finds the equation's root using the first few terms of Taylor series in the function. It features fast convergence but the computational intensities are extremely high due to the second-order partial derivative matrix and its inverse matrix. The quasi-Newton method proposed in Literatures [18] and [19] overcomes the drawback of Newton's method and reduces the computational intensities, but it is difficult to find the initial point. The neural network algorithm proposed in Literatures [4] and [5] does not require an explicit expression of the nonlinear equation, but it is difficult to guarantee the accuracy and easy to fall into local minima for the algorithm. In addition, network structure is difficult to determine and over-fitting is easy to occur. The ant colony algorithm proposed in Literatures [20] and [21] overcomes the drawback of Newton's method in the sensitivity to initial intervals and the defects of neural network algorithm, having the ability to yield the root of harmonic balance equation without any fitting method. The ant colony algorithm features strong robustness and powerful search ability for the global optimal solution. However, the algorithm has insufficient local search ability and may experience problems such as stagnation, local convergence and slow convergence. Therefore, it is necessary to explore a new method for solving harmonic balance equations in a better way, which increases the convergence speed of solutions and improves the convergence reliability of solutions.

This section proposes a hybrid ant colony algorithm, which is designed to address the problems of the ant colony algorithm, such as insufficient local search ability, frequent stagnation and local convergence, slow convergence and poor efficiency in a harmonic balance context. The proposed algorithm uses the global search ability of ant colony algorithm to search for the initial optimal solution in the global, then uses the strong local search ability of quasi-Newton method for gradual iterations and finally yields the optimal solution. Simulation results demonstrate that the proposed algorithm reduces the number of iterations by 45 and increases the convergence reliability of solutions by 16.23% compared with the ant colony algorithm, and the proposed algorithm produces the simulation data fitting well with the measured data. The hybrid ant colony algorithm combines the advantages of ant colony algorithm and quasi-Newton method. It significantly improves the convergence speed and convergence reliability of solutions, and overcomes the drawbacks of ant colony algorithm such as insufficient local search ability and slow convergence.

9.3.1 Fundamentals of Harmonic Balance

In this subsection, the full frequency-domain harmonic balance method divides a circuit into a linear sub-circuit in the left dashed box and a nonlinear sub-circuit in the right dashed box. See Figure 9.14. To avoid time-frequency domain conversions, the nonlinear sub-circuits described in time domain are analyzed in frequency domain. The relationship between Fourier transform and frequency domain calculation is identified by means of an arithmetic operation method, thereby determining the frequency-domain relationships between excitations and responses of various nonlinear components. In this way, the impact of Fourier transform on the dynamic range and efficiency of solutions can be eliminated.

In Figure 9.14, Z_s and Z_l are the source resistance and load resistance in the circuit, respectively. For convenience of analysis, they are both put into the linear subnetwork. V_s is the input signal voltage. V_g and V_d are the gate and drain bias voltages, respectively. The linear network is a port network, where port 1 is connected to the nonlinear network and ports 2

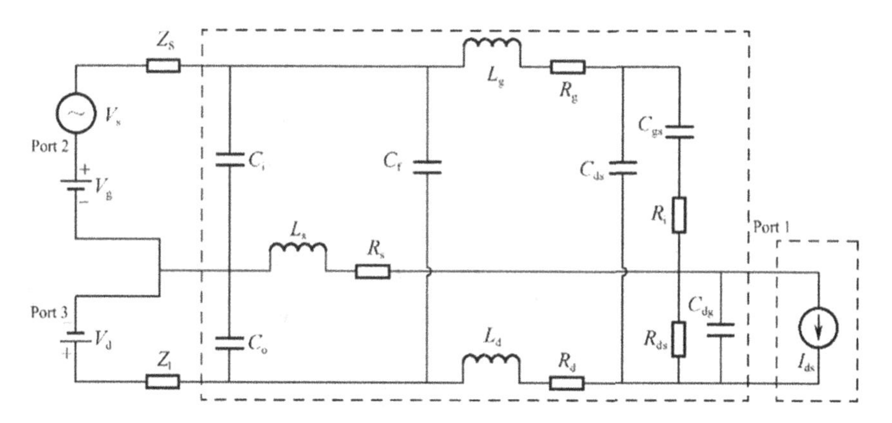

FIGURE 9.14 Decomposed GaAs MESFET large-signal model.

and 3 are the input and output ports of the signal, respectively. For details on the harmonic balance method, see Section 3.3.1.

9.3.2 Hybrid Ant Colony Algorithm

Ant colony algorithm is a novel heuristic algorithm based on population evolution, featuring strong robustness. It is easy to combine with other heuristic algorithms and has a strong solution ability. However, the inherent defects of ant colony algorithm have been observed, such as insufficient local search ability, frequent stagnation and local convergence, slow convergence, etc.

This subsection proposes a novel hybrid ant colony algorithm designed to overcome these drawbacks of ant colony algorithm. The proposed algorithm combines quasi-Newton method with ant colony algorithm, which means a combination of local searches and global searches. The algorithm can solve harmonic balance equations with a large probability, requiring less iterations and delivering high convergence reliability of solutions. It overcomes the drawbacks of ant colony algorithm (e.g., stagnation and insufficient local search ability) and quasi-Newton method (e.g., sensitivity to initial points). Ant colony algorithm is proposed in Literature [22]. For details, see Section 3.3.3.

Quasi-Newton method overcomes the drawbacks of Newton's method while maintaining fast convergence. As a direct extension of Newton's method, quasi-Newton method introduces Newton's condition to determine the line search direction by means of a quadratic approximation near the test point. The best known quasi-Newton methods are DFP and BFGS methods.

An improved BFGS method adopts a new quasi-Newton equation, namely $\boldsymbol{B}_{k+1}\boldsymbol{s}_k = \boldsymbol{y}_k^*$, where $\boldsymbol{y}_k^* = \boldsymbol{y}_k + \boldsymbol{A}_k\boldsymbol{s}_k$ and $\boldsymbol{A}_k$ is a symmetric positive definite matrix. The iterative formula of the improved BFGS method is given by

$$\boldsymbol{B}_{k+1} = \boldsymbol{B}_k - \frac{\boldsymbol{B}_k\boldsymbol{s}_k\boldsymbol{s}_k^{\mathrm{T}}\boldsymbol{B}_k}{\boldsymbol{s}_k^{\mathrm{T}}\boldsymbol{B}_k\boldsymbol{s}_k} + \frac{\boldsymbol{y}_k^*(\boldsymbol{y}_k^*)^{\mathrm{T}}}{\boldsymbol{s}_k^{\mathrm{T}}\boldsymbol{y}_k^*} \tag{9.27}$$

$$\boldsymbol{A}_k = \|\boldsymbol{g}_{k+1} - \boldsymbol{g}_k\|\boldsymbol{I} \tag{9.28}$$

The search criterion (Wolfe's criterion) involved is

$$f(\boldsymbol{x}_k + \alpha_k \boldsymbol{d}_k) \leq f_k + \sigma_1 \alpha_k \boldsymbol{g}_k^{\mathrm{T}} \boldsymbol{d}_k \tag{9.29}$$

$$\nabla f(\boldsymbol{x}_k + \alpha_k \boldsymbol{d}_k)^{\mathrm{T}} \boldsymbol{d}_k \geq \sigma_2 \boldsymbol{g}_k^{\mathrm{T}} \boldsymbol{d}_k \tag{9.30}$$

where $0 < \sigma_1 < \sigma_2 < 1$.

The steps of the improved BFGS method are as follows.

Step 1: Give the initial point $x_0 \in R^n$ and the initial symmetric positive definite matrix $\boldsymbol{B}_0 \in R^{n\times n}$. Assume $\varepsilon > 0$, $k = 0$.

Step 2: If $\|\boldsymbol{g}_k\| \leq \varepsilon$ applies at the iteration point $\boldsymbol{X}_k$ for the gradient function, the iteration ends. Otherwise, solve $\boldsymbol{B}_k \boldsymbol{d}_k + \boldsymbol{g}_k = \boldsymbol{0}$ to determine the search direction $\boldsymbol{d}_k$.

Step 3: Determine the step size α_k based on Wolfe's criterion. Assume $\boldsymbol{X}_{k+1} = \boldsymbol{X}_k + \alpha_k \boldsymbol{d}_k$.

Step 4: Calculate $\boldsymbol{A}_k = \|\boldsymbol{g}_{k+1} - \boldsymbol{g}_k\|\boldsymbol{I}$ and substitute it into Equation (9.27), so that $\boldsymbol{B}_k$ is corrected to $\boldsymbol{B}_{k+1}$.

Step 5: Set $k = k + 1$ and go to Step 2.

The flowchart of the hybrid ant colony algorithm is shown in Figure 9.15.

Step 1: Convert the harmonic balance problem into a minimization problem.

Step 2: Perform parameter initialization. Suppose there are m ants. Each ant starts from a randomly selected position within a given variable interval $[\mu_i, l_i]$. Set the maximum number of cycles $\mathrm{NC}_{\max}$ and the initial number of cycles NC = 0. Solve the initialized pheromone $\tau(k)$, $\tau_{\min}$ and $\tau_{\max}$ for ant $k\,(k = 1, 2, \cdots, n)$. Assume $\Delta\tau(k) = 0$ at the initial moment.

Step 3: Perform the iterative process. Calculate the pheromone concentration of each ant. Select the solution corresponding to the ant when the pheromone concentration $\tau(k)$ is the maximum as the current optimal solution. If it is the current optimal solution, a local update is performed. Otherwise, a global update is performed.

Step 4: Perform a pheromone update for each ant when local and global updates are complete, with pheromones restricted to a given interval $[\tau_{\min}, \tau_{\max}]$.

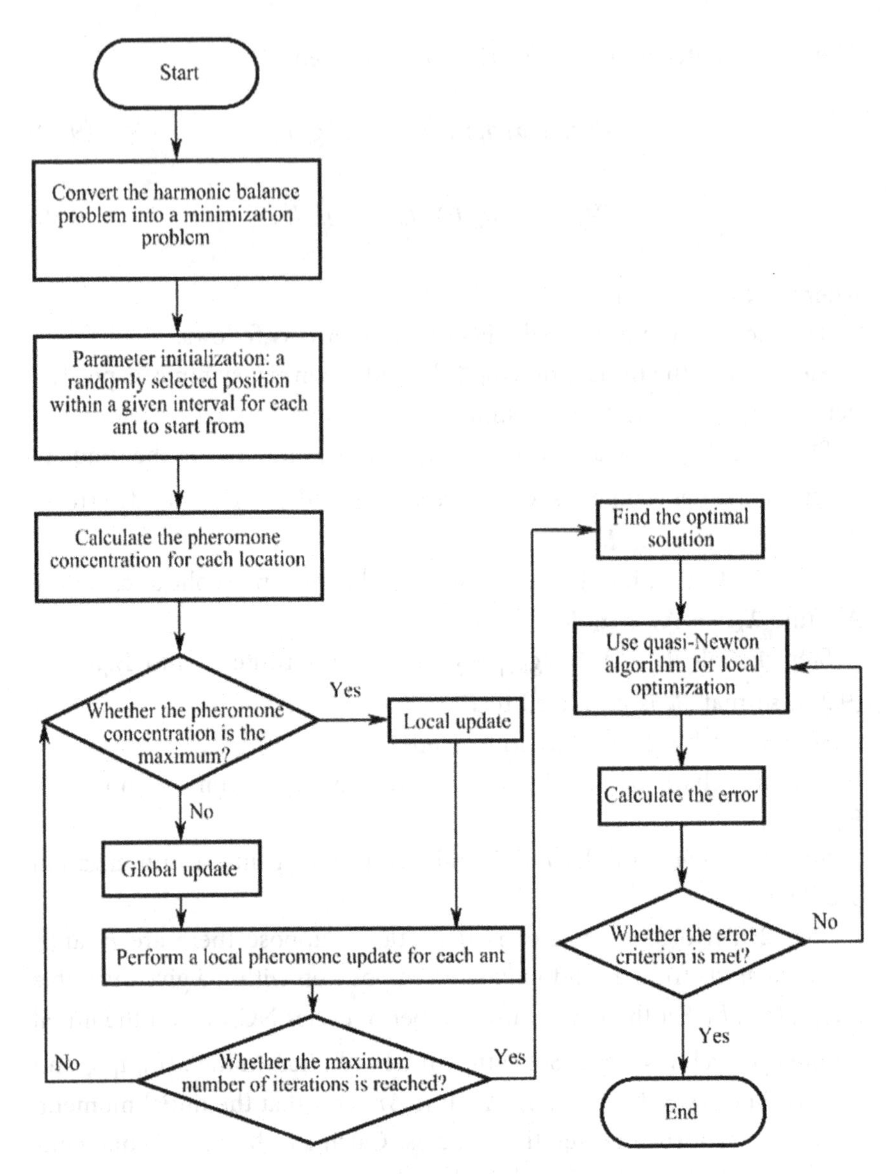

FIGURE 9.15 Flowchart of hybrid ant colony algorithm.

Step 5: Check whether the maximum number of iterations NC_{max} is reached. If so, find the current optimal solution; if not, go to Step 4.

Step 6: Take the current optimal solution as the initial point, and use the improved quasi-Newton method mentioned above to search locally. If the error of the generated optimal solution substituted into the original equation is less than or equal to a preset value (<1.0E-6), the result will be output.

9.3.3 Experimental Simulation Analysis

The simulation uses the FLK057WG power transistor for small-signal and large-signal modeling, and performs the analysis by means of the full frequency-domain harmonic balance method. The single-frequency characteristics of the power transistor are simulated in a MATLAB® context. The power transistor's static operating points are set to $V_{GS} = -0.84$ V and $V_{DS} = 10$ V, the operating frequency is $f = 12.5$ GHz and the source impedance and load impedance are both set to 50 Ω.

Figures 9.16 and 9.17 show the simulation diagrams of convergence speed obtained by using ant colony algorithm and hybrid ant colony algorithm to solve the full frequency-domain harmonic balance equation through the MATLAB® platform, respectively. As indicated by Figure 9.16, the average pheromone concentration of ant colony algorithm is far from the optimal pheromone concentration, and the algorithm requires about 80 iterations making the average pheromone concentration remain constant with the optimal pheromone concentration. As indicated by Figure 9.17, the average pheromone concentration of the hybrid ant colony algorithm gets pretty close to the optimal pheromone concentration, and the algorithm requires about 35 iterations making the average pheromone concentration remain constant with the optimal pheromone concentration. Therefore, the hybrid ant colony algorithm requires 45 less iterations compared with the ant colony algorithm, improving the convergence speed.

In this subsection, the convergence reliability in the MATLAB® simulation and solution process is investigated based on the quasi-Newton method, ant colony algorithm and hybrid ant colony algorithm, respectively. The comparison of convergence reliabilities is listed in Table 9.5. As indicated by the table, the convergence reliabilities of quasi-Newton method, ant colony algorithm and hybrid ant colony algorithm are 49.54%, 82.33% and 98.56%, respectively. That is, the hybrid ant colony algorithm produces a convergence reliability 16.23% higher compared with the ant colony algorithm. It is noted that the quasi-Newton method and ant colony algorithm demonstrate poor reliability

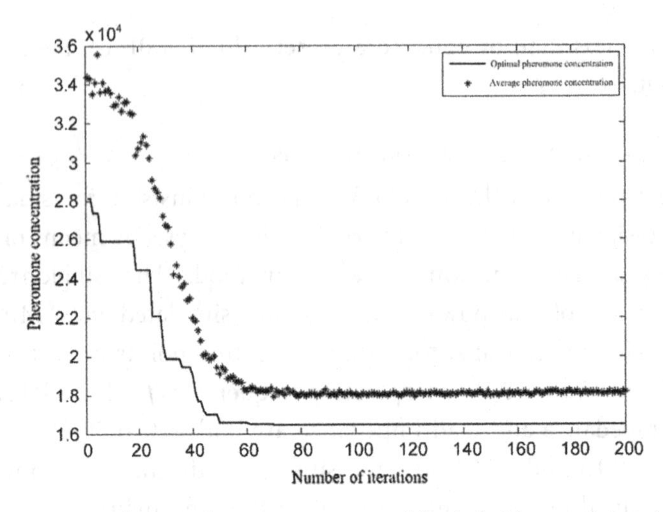

FIGURE 9.16 Simulation of convergence speed for hybrid ant colony algorithm.

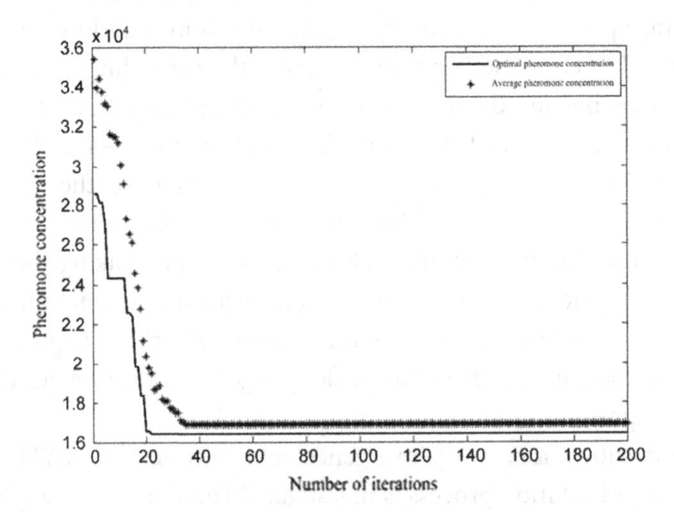

FIGURE 9.17 Simulation of convergence speed for hybrid ant colony algorithm.

TABLE 9.5 Comparison of Convergence Reliabilities Based on Different Algorithms

Algorithm	Quasi-Newton Method	Ant Colony Algorithm	Hybrid Ant Colony Algorithm
Reliability	49.54%	82.33%	98.56%

in solving harmonic balance equations, while the hybrid ant colony algorithm has improved convergence reliability in these cases.

Figure 9.18 shows the measured values of harmonic characteristics and the simulation curve yielded by the hybrid ant colony algorithm when the input is a 12.5 GHz single-tone signal. In the figure, straight lines represent measured data and points represent simulated data. As indicated by the figure, simulated data and measured data exhibit the same trend but occasional errors are observed. This is because the extraction of small-signal equivalent model parameters and large-signal equivalent model parameters cannot take each parameter into account. As a result, the simulation curve fits well with measurements for a certain segment and not so well for other segments. Overall, the simulation curve is in good agreement with measurements, proving a successful application of the proposed algorithm in harmonic balance.

This section describes the nonlinear analysis of PA. A large-signal equivalent circuit model is yielded on the basis of a small-signal circuit model, and the single-frequency characteristics of the model are analyzed. The full frequency-domain harmonic balance equation established by the model is solved by the quasi-Newton ant colony algorithm. Simulation results indicate that the proposed algorithm significantly

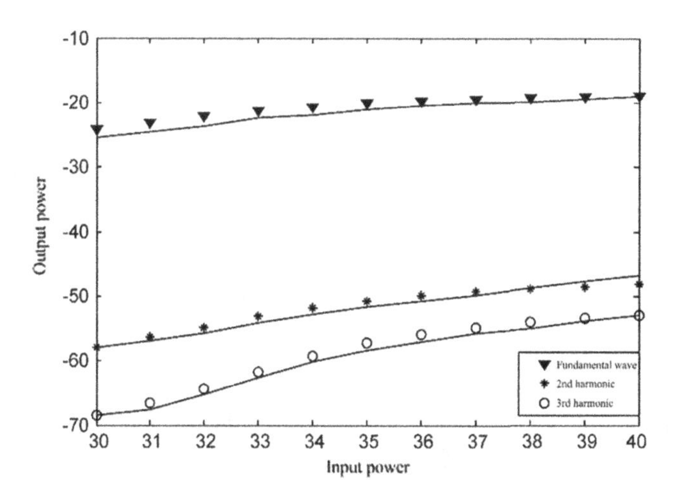

FIGURE 9.18 Simulation curve of single-frequency power harmonic characteristics based on FLK057WG.

improves the convergence reliability and reduces the number of iterations. The hybrid ant colony algorithm proves to be applicable in harmonic balance and has a good prospect.

REFERENCES

1. 郑博仁．射频/微波放大器非线性特性分析方法比较[J]．信息与电子工程, 2005(03):201–206.
   ZHENG B. Comparison of Analysis Methods for Radio & Microwave Amplifier Nonlinearity[J]. Information and Electronic Engineering, 2005(03):201–206.
2. 郑敏毅，胡辉，郭源君．一个强非线性振子的牛顿谐波平衡解[J]．湖南科技大学学报：自然科学版, 2010, 25(3):121–123.
   ZHENG M, HU H, GUO Y. Newton-Harmonic Balancing Approach for a Strongly Nonlinear Oscillator[J]. Journal of Hunan University of Science & Technology (Natural Science Edition), 2010, 25(3):121–123.
3. GHADIMI M, KALIJI HD. Application of the Harmonic Balance Method on Nonlinear Equations[J]. World Applied Sciences Journal, 2013, 22(4):532–537.
4. 孙银慧，白振兴，王兵，等．求解非线性方程组的迭代神经网络算法[J]．计算机工程与应用, 2009, 45(6):55–59.
   SUN Y, BAI Z, WANG B, et al. Iterative Algorithms of Neural Network for Nonlinear Equation Groups[J]. Computer Engineering and Applications, 2009, 45(6):55–59.
5. FAROUK MKADEM, MORSI B AYED, SLIM BOUMAIZA. Behavioral Modeling and Digital Predistortion of Power Amplifiers with Memory using Two Hidden Layers Artificial Neural Networks. Microwave Symposium Digest (MTT), Isfahan, Iran, 2010[C]. Piscataway: IEEE, 656–659.
6. 严刚峰，黄显核．基于遗传算法的改进谐波平衡算法[J]．电子测量与仪器学报, 2009, 23(10):96–100.
   YAN G, HUANG X. Improved Harmonic-Balance Algorithms Based on the Genetic Algorithms[J]. Journal of Electronic Measurement and Instrument, 2009, 23(10):96–100.
7. GAO J. RF and Microwave Modeling and Measurement Techniques for Field Effect Transistor[M]. SciTech Publishing, 2009.
8. 张玉兴，赵宏飞．射频与微波功率放大器设计[M]．北京：电子工业出版社, 2006.
   ZHANG Y, ZHAO H. RF and Microwave Power Amplifier Design[M]. Beijing: Publishing House of Electronics Industry, 2006.
9. 胡辉勇，张鹤鸣，吕懿．SiGe HBT 大信号等效电路模型[J]．物理学报, 2006, 55(1):403–408.
   HU H, ZHANG H, LV Y. SiGe HBT Large Signal Equivalent Circuit Model[J]. Acta Physica Sinica, 2006, 55(1):403–408.

10. PETER AAEN, JAIME A PLA, JOHN WOOD. Modeling and Characterization of RF and Microwave Power FETs[M]. Cambridge University Press, 2007.
11. 边霞，米良．遗传算法理论及其应用研究进展[J]．计算机应用研究, 2010, 27(7):2425–2429.
BIAN X, MI L. Development on Genetic Algorithm Theory and Its Applications[J]. Application Research of Computers, 2010, 27(7): 2425–2429.
12. 葛继科，邱玉辉．遗传算法研究综述[J]．计算机应用研究, 2008, 25(10): 2911–2916.
GE J, QIU Y. Summary of Genetic Algorithms Research[J]. Application Research of Computers, 2008, 25(10):2911–2916.
13. ANDING ZHU, JOSÉ C PEDRO, THOMAS J, et al. Dynamic Deviation Reduction-Based Volterra Behavioral Modeling of RF Power Amplifiers[J]. IEEE Transactions on Microwave Theory and Techniques, 2006, 54(12): 4323–4332.
14. PETERSON D L, PAVIO A M, KIM B. A GaAs FET Model for Large-Signal Applications[J]. IEEE Transactions on Microwave Theory Technology, vol. MTT-32, 276–281, March 1984.
15. 张祖舜，沈灿．微波非线性电路全频域谐波平衡分析[J]．电子学报, 1995, 23(23):62–67.
ZHANG Z, SHEN C. Frequency-Domain Harmonic Balance Simulation of Nonlinear Microwave Circuits[J]. Acta Electronica Sinica, 1995, 23(23):62–67.
16. 赵世杰．基于谐波平衡法非线性散射函数仿真技术的研究[D]．西安: 西安电子科技大学, 2010.
ZHAO S. The Research of Simulation Technology for Nonlinear Scattering Function Based on Harmonic Balance[D]. Xi'an: Xidian University, 2010.
17. 谭振江，肖春英．非线性方程数值解法的研究[J]．吉林师范大学学报: 自然科学版, 2014. 8(3):102–105.
TAN Z, XIAO C. The Research of the Solution to the Nonlinear Equation [J]. Jilin Normal University Journal (Natural Science Edition), 2014. 8(3):102–105.
18. 张安玲，刘雪英．求解非线性方程组的拟牛顿-粒子群混合算法[J]．计算机应用与研究, 2008, 44(33):41–42.
ZHANG A, LIU X. Hybrid Quasi-Newton/Particle Swarm Optimization Algorithm for Nonlinear Equations[J]. Computer Engineering and Applications, 2008, 44(33):41–42.
19. 丁知平．拟牛顿粒子群优化算法求解调度问题[J]．计算机应用研究, 2012, 29(1):140–142.
DIGN Z. Quasi-Newton Method Particle Swarm Optimization Algorithm for Solving Scheduling Problem[J]. Application Research of Computers, 2012, 29(1):140–142.

20. 张冰冰．求解非线性方程组的蚁群算法[J]．工业控制计算机，2013, 26(1):63–64.
ZHANG B. Ant Colony Algorithm for Solving Nonlinear Equations[J]. Industrial Control Computer, 2013, 26(1):63–64.
21. SHIGANG CUI, SHAOLONG HAN. Ant Colony Algorithm and Its Application in Solving the Traveling Salesman Problem. Instrumentatio, Measurement, Computer, Communication and Control (IMCCC), Shenyang, China, 2013[C]. Piscataway： IEEE, 21–23.
22. 吴晓维．求解旅行商问题和非线性方程组的蚁群算法[D]．西安：陕西师范大学, 2008.
WU X. Ant Colony Algorithm for Solving the Traveling Salesman Problem and Nonlinear Equations[D]. Xi'an: Shaanxi Normal University, 2008.

CHAPTER 10

Predistortion Algorithms and Applications

RF PA is the most critical and expensive device in a wireless communication system and is also the principal nonlinear device. It comes with AM/AM and AM/PM distortion characteristics. Such nonlinear distortion effects severely degrade the quality of communication. In addition, modern wireless communication systems, such as CDMA, WCDMA and EDGE, have adopted complex digital modulation formats with high spectrum utilization (such as π/4-DQPSK, M-QAM and MPSK) to increase the data transmission rate and channel capacity. These digital modulation formats result in memory effects as a response to signal envelope changes. Memory effects introduce greater out-of-band diffusion, increase adjacent channel interference and also produce in-band distortion. Consequently, the bit error rate becomes more serious.

Generally, efficiency and linearity are two mandatory factors to be considered in the RF PA design, in addition to meeting the requirements of PA's output power. It is well-known that efficiency and linearity of PA are opposite to each other. To make the distortion acceptable, Class A amplifiers with power back-off are usually used at the expense of reduced PA efficiency and increased heat loss and cost. Similarly, a PA is designed to work in the highly nonlinear region for a high efficiency at the expense of nonlinearities, which lead to spectral regrowth. In this case, adjacent channel interference and in-band distortion are generated, deteriorating the bit error rate. To solve this problem, a high-efficiency PA is generally designed, with the linearity improved by means of linearization techniques. Various linearization techniques for improving the linearity of RF PA have been applied to modern wireless communication systems. As one of these linearization techniques, feed-forward linearization can provide both wide bandwidth and favorable intermodulation distortion compression, but it has poor efficiency. Overall, feed-forward linearization is a complex and expensive solution [1,2]. The feedback method suffers from serious problems of instability and bandwidth limitation [3]. Predistortion techniques come with the superiorities in stability, high efficiency, wide bandwidth and adaptability. Predistortion can be enforced in the RF or baseband part. The analog RF predistortion technique features a simple structure, low cost and moderate linearization [4]. However, like other linearization techniques, the analog RF predistortion technique involves considerable additional analog hardware or requires a number of nonlinear devices, which make it difficult for the technique to produce the required accuracy. The digital predistortion technique appears more complicated due to the use of DSP technology, but this technique provides better intermodulation distortion compression. Digital baseband predistortion is the most cost-effective linearization technique, which uses a few analog devices at a low implementation cost as opposed to the feed-forward predistortion technique. Digital baseband predistortion implements signal processing in the baseband, which is independent of the system's operating frequency and therefore especially suitable for Software-defined Radio [5]. Moreover, the adaptive technique can be easily applied to a digital predistortion structure. Predistortion PA inherently produces memory effects, and its linearity varies with the input signals.

10.1 THEORETICAL ANALYSIS AND SIMULATION IMPLEMENTATION OF DIGITAL BASEBAND PREDISTORTION FOR POWER AMPLIFIER

The adaptive digital baseband predistortion technique has become a preferred solution that eliminates the distortion caused by nonlinearities of PA and meets the formulated requirements of power spectral density, owing to its superiorities such as favorable linearity, wide bandwidth, high efficiency and full adaptability. This section describes the principle of digital baseband predistortion, constructs a digital baseband predistortion circuit and realizes system-level simulation in an ADS context. Simulation results indicate that the constructed digital baseband predistortion system attains a favorable predistortion effect and an improvement in the power spectral density up to 30 dBm. The developed system-level simulation circuit is of important guiding significance to the real digital predistortion system design.

10.1.1 Digital Baseband Predistortion Structure

Figure 10.1 shows the hybrid digital RF predistortion structure [6]. The main path signal of this structure is an analog RF signal, and the control signal given to the vector modulator is a digital signal requiring DSP operations and the data stored in the LUT. Therefore, such a system has a hybrid predistortion structure. The hybrid RF digital predistortion structure provides a compromise between analog RF predistortion and digital baseband predistortion. The compromise scheme is adaptable to wideband operation because the bandwidth is not limited by the computational speed of digital signal processing. The main weakness of this hybrid predistortion system lies in an insufficient ability to correct the memory effects occurring in the RF bandwidth.

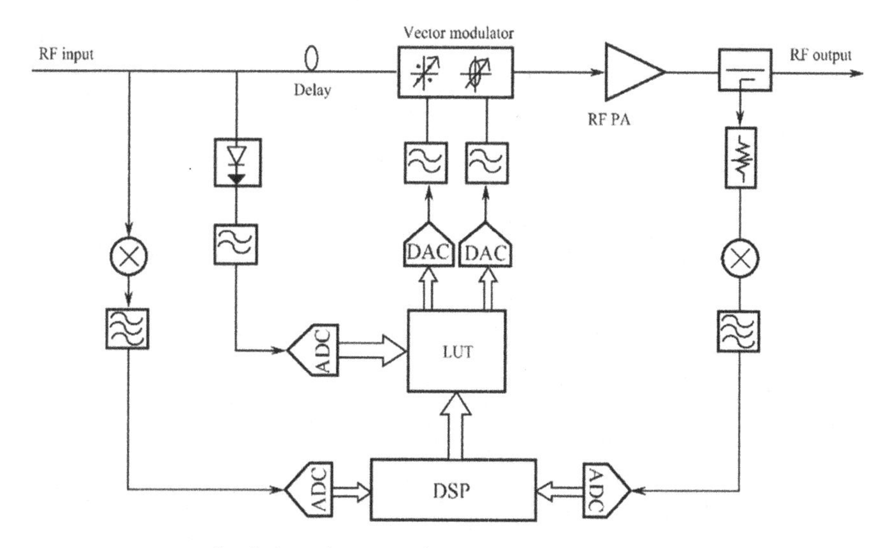

FIGURE 10.1 Hybrid digital RF predistortion structure.

The most common types of digital baseband predistortion are mapping predistortion and constant gain predistortion [7]. The mapping predistorter uses two LUTs, each of which is a function of in-phase and quadrature components. This type of predistorter provides favorable performance. However, in terms of the LUTs and update speed, the predistorter has a large storage and processing burden as well as slow convergence. The constant gain predistorter only requires a one-dimensional LUT, which is indexed by signal envelope values. Consequently, it is easy to implement while requiring less memory for a given performance and adaptive time. Constant gain predistortion uses the LUT to force the predistorter and the subsequent PA to obtain a constant gain and phase at all envelope levels. In some literatures, the two methods are sometimes referred to as complex vector mapping LUT technique and complex gain LUT method.

Figure 10.2 shows the block diagram of an adaptive digital baseband predistortion system [5], which is in a digital baseband domain. In the digital predistorter, the digital complex baseband input signal samples are multiplied by complex coefficients drawn from the LUT entries to yield a predistortion function that is the inverse of the amplifier's characteristics, thereby achieving linearization. Here, the coefficients drawn from the LUT implement the predistortion function. The adaptive algorithm determines the coefficient values by comparing the feedback signal with the delayed input signal. Some Literatures [5,8,9] have reported the digital baseband predistortion structure, describing a consistent principle and structure. At present,

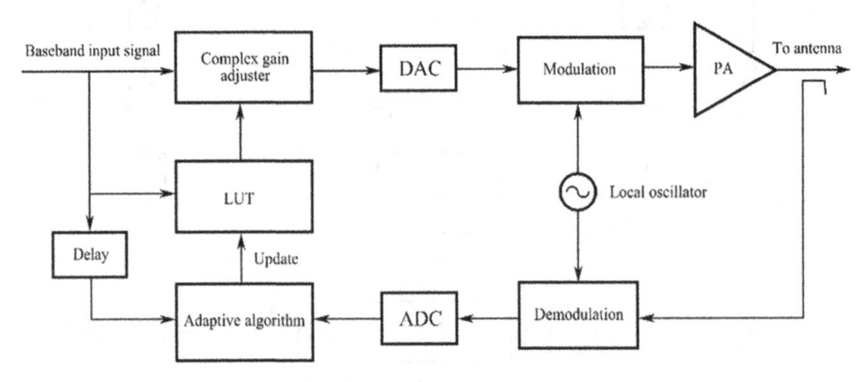

FIGURE 10.2 Block diagram of adaptive digital baseband predistortion system.

major research highlights include DSP adaptive algorithm, nonuniform LUT indexing [9], adaptive predistortion based on neural network and PA predistortion with memory effects.

10.1.2 Theoretical Derivation of Transfer Function for Digital Predistorter

The gain function of digital baseband predistorter can be derived given the characteristics of PA (i.e., AM/AM and AM/PM functions) and linearization requirements. The theoretical derivation in this subsection assumes that the PA has no memory effects, meaning that its behavior is frequency-independent. The theoretical derivation of the digital predistortion function is given below.

When the PA has no memory effects, its gain can be essentially expressed by its AM/AM and AM/PM functions. In polar and rectangular coordinates, the general complex gain model of PA is expressed as follows:

$$\begin{aligned} G_{PA}(|r|) &= G(|r|)\{\cos[\varphi(|r|) + j\sin[\varphi(|r|)]\} \\ &= G_i(|r|) + jG_q(|r|) \end{aligned} \tag{10.1}$$

To yield a favorable predistortion effect, the gain function of predistorter should be the exact inverse of the PA's gain. In polar and rectangular coordinates, the predistortion gain is defined as follows:

$$\begin{aligned} G_{pre}(|r|) &= P(|r|)\{\cos[\theta(|r|)] + j\sin[\theta(|r|)]\} \\ &= P_i(|r|) + jP_q(|r|) \end{aligned} \tag{10.2}$$

where $P(\ )$, $\theta(\ )$, $P_i(\ )$和$P_q(\ )$ are all nonlinear functions of the input signal amplitude $|r|$. For the complete removal of nonlinearities, the following conditions must be met:

$$\begin{aligned} P(|r_s|)G(|r_p|) &= 1 \\ \theta(|r_s|) &= -\varphi(|r_p|) \end{aligned} \tag{10.3}$$

where $|r_s|$ is the amplitude of the source signal and $|r_p|$ is the amplitude of the predistorted signal.

Based on Equations (10.1), (10.2) and (10.3), the in-phase and quadrature components of the predistortion gain are given by

$$P_i(|r_s|) = \frac{G_{in}(|r_p|)}{G_i^2(|r_p|) + G_q^2(|r_p|)} = \frac{G_i(|r_p|)}{G^2(|r_p|)} \tag{10.4}$$

$$P_q(|r_s|) = -\frac{G_q(|r_p|)}{G_i^2(|r_p|) + G_q^2(|r_p|)} = -\frac{G_q(|r_p|)}{G^2(|r_p|)} \tag{10.5}$$

As indicated by these formulas, the in-phase component and quadrature component of the predistortion complex gain can be expressed by the two complex gain components of PA, and the predistortion gain function can be yielded based on these computations. These formulas are theoretical representation of the digital baseband predistortion principle. In a real system, the DSP algorithm is required to calculate the coefficients of the complex gain adjuster based on the input-output error signals, which is a complicated computation and adjustment process.

10.1.3 Simulation Implementation of Digital Baseband Predistortion

The simulation circuit for the digital baseband predistortion system is created in an ADS context where a constant gain digital predistorter structure is adopted. In this structure, the input signal is a complex baseband signal and the whole predistortion process is implemented in the digital baseband signal domain. After a digital-to-analog conversion, the predistorted baseband signal is modulated to RF for transmission. This is how the above system differs from the digital RF predistortion system.

The constructed simulation circuit consists of baseband modulated signal generator, complex gain adjuster, LUT address generator, LUT, symbol synchronization (feedback delay determination), coefficient error calculation and convergence control, DAC, input signal selector, etc. See Figure 10.3. The later parts of this diagram for the simulation structure circuit, such as modulator and PA, are included in Figure 10.2. The simulation circuit works as follows. Various functional modules are integrated into a digital baseband predistortion system, where DSP algorithm and adaptive technique (implemented by using a constructed

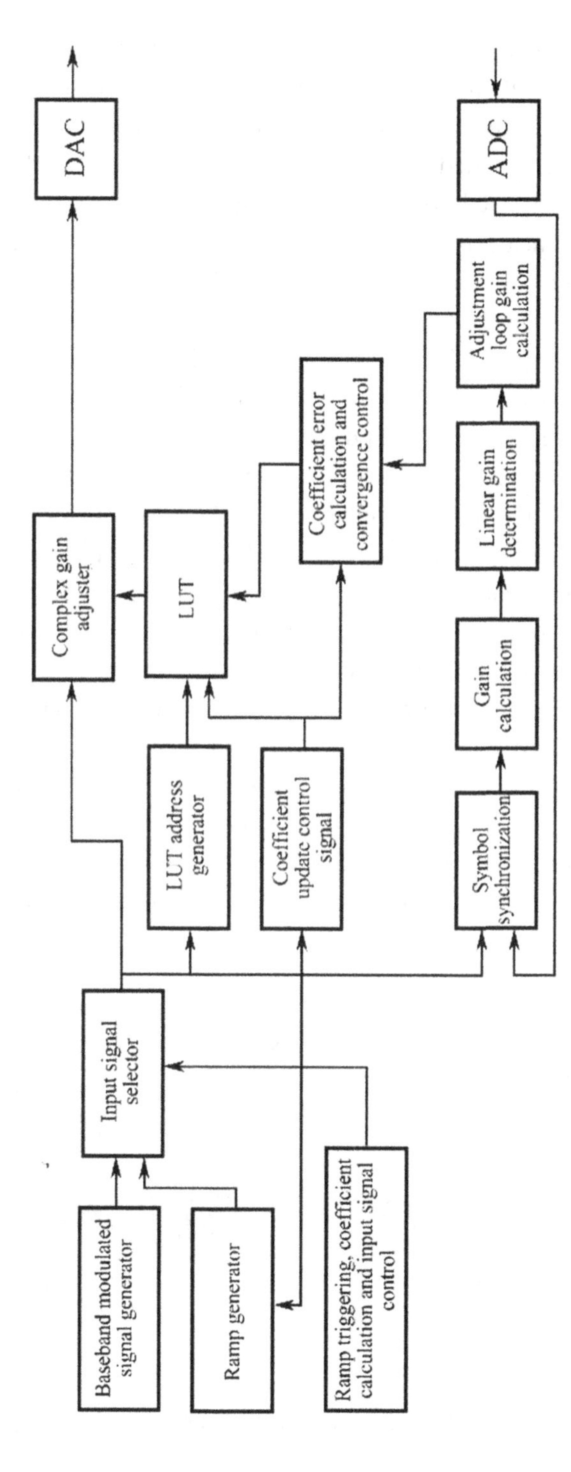

FIGURE 10.3 Block diagram of digital baseband predistortion simulation structure.

circuit) are applied. The coefficients of the complex gain adjuster are computed based on the input-output error signals. Symbol synchronization requires careful consideration. In this way, the transfer function of the gain adjuster yielded by multiplying the complex baseband input signal by the generated complex coefficients is the exact inverse of the amplifier's nonlinearities. Consequently, the final transfer function of the predistorter and PA remains linear, achieving the linearization effect. The major steps in calculation of control coefficients of the complex gain adjuster include gain calculation, linear gain determination, adjustment loop gain calculation, coefficient error calculation and convergence control, LUT and coefficient update, etc. This series of calculation and control are implemented by the DSP chip in a real system.

The signal source in the simulation is a WCDMA forward link baseband signal with a spectral width of 5 MHz. The complex gain adjuster performs the multiplication of two complex signals (input baseband signal and generated complex control coefficients). The PA uses an equation-based behavioral model amplifier, of which the model is a memoryless nonlinear amplifier behavioral model realizing the amplifier's characteristics expressed by the given AM/AM and AM/PM functions. ADC and DAC perform analog-to-digital conversion and digital-to-analog conversion, respectively. The LUT size is set to 256. After the system simulation circuit is constructed and parameter settings are configured, the simulation is started and simulation results are yielded. See Figure 10.4.

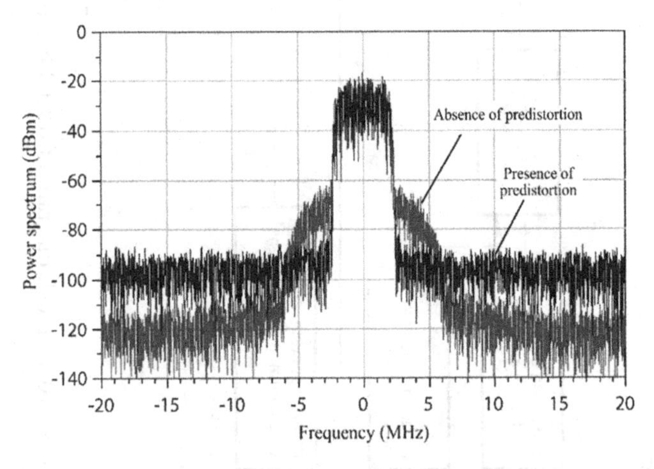

FIGURE 10.4 Power spectrum of output signal before and after predistortion.

As indicated by simulation results, the constructed digital baseband predistortion structure can produce a favorable predistortion effect and meet the power spectral density requirements formulated by standardization organizations. The energy of the input power is mainly concentrated in a range of 5 MHz around the center frequency, with low spectral diffusion and small distortion. The improvement in power spectral density reaches up to 30 dBm.

This simulation system allows you to alter the input baseband signal, PA model, LUT size, coefficient calculation and convergence method, so that simulation effects varying with models, sizes and methods can be observed. In a real system, coefficient calculation and convergence control are obtained by means of the DSP algorithm.

As indicated in simulation results, digital baseband predistortion delivers better performance compared with RF/digital hybrid predistortion, although the former has relatively complex implementation.

High output power requirements and especially the application of complex digital modulation formats with high spectrum utilization have highlighted the nonlinearities of RF PA and the amplitude and phase distortion caused in signals, which severely affect the system performance and have become a crucial problem. The digital predistortion technique is an effective method to improve the nonlinear distortion of PA. This digital predistortion structure is implemented by means of LUT and DSP or FPGA algorithm, thereby achieving linearization. Based on the digital baseband predistortion structure proposed in this section, an adaptive digital baseband predistortion system is developed in an ADS context. This section also describes the selection of major modules and the establishment of coefficient calculation and convergence control module, and finally provides the simulation results. Simulation results indicate that the constructed digital baseband predistortion system yields a favorable predistortion effect. The simulation circuit in this subsection is an indispensable key step in designing the real digital predistortion circuit. You can understand the overall system performance by modifying models and their parameters. This kind of simulation circuit is of important guiding significance to the design of real digital baseband predistortion system.

10.2 RESEARCH ON DIGITAL PREDISTORTION METHOD OF DOUBLE-LOOP STRUCTURE

This section proposes a double-loop predistortion method, aiming to address the weaknesses of the traditional predistortion structure. The proposed method combines direct learning structure and indirect learning structure to overcome their respective weaknesses, and uses the least square method twice to yield the parameters of predistorter model with better linearization performance. In addition, a pre-equalizer is added after the predistorter and a post-equalizer is added to the adaptive control module, thus eliminating the linear distortion inherently produced by the predistorter itself. Then, the proposed method is applied to the nonlinear PA with memory effects for simulation validation. Simulation results indicate that the predistortion method proposed in this section can favorably suppress in-band distortion and out-of-band spectrum spreading, reduce the bit error rate of the system and improve the linearization performance.

10.2.1 Predistortion Structure of Double-Loop Structure

Figure 10.5 shows the predistortion system with the presence of equalizers. The pre-equalizer processes the output signal of the digital predistortion module to pre-compensate the linear distortion in the forward branch. The post-equalizer processes the feedback signal to eliminate the impact of linear distortion in the feedback branch on the system.

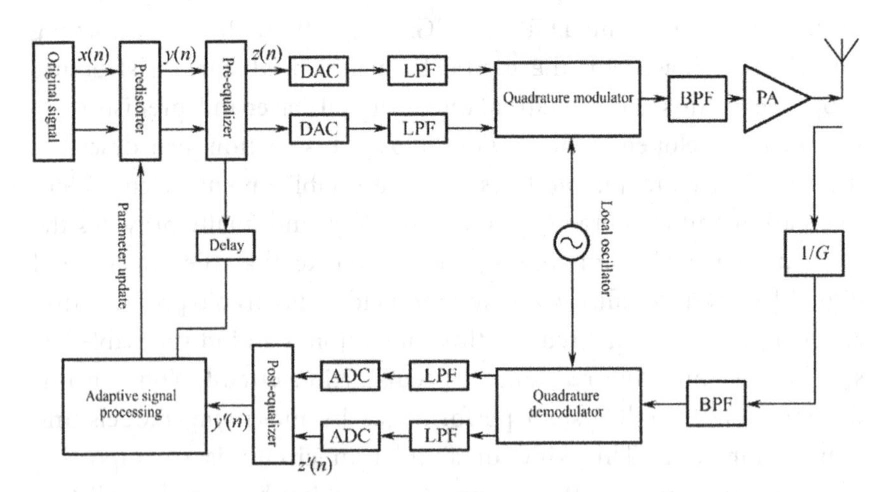

FIGURE 10.5 Block diagram of predistortion system with the presence of equalizers.

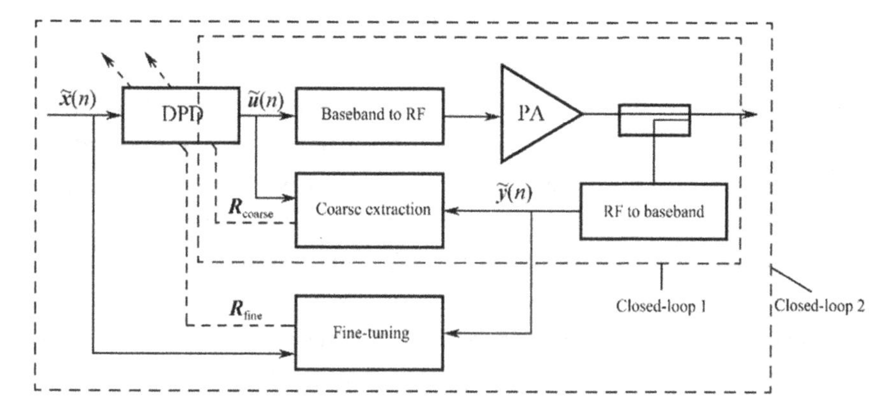

FIGURE 10.6 Double-loop predistortion structure.

Figure 10.6 shows the double-loop predistortion structure proposed in this section. Closed-loop 1 operates at the beginning. That is, the indirect learning structure performs a coarse extraction of model parameters. Although the inverse model parameters obtained by this structure deviate from the ideal values to some extent, they are still close to the best model parameters. When the additive noise is not too large, the obtained predistorter model parameters have jumped out of the local optimal range. Then, closed-loop 2 starts to work. That is, the direct learning structure fine-tunes the parameters obtained by coarse extraction, and the converged weight vector in closed-loop 1 is used as the initial value of the weight vector in closed-loop 2. Consequently, the double-loop structure significantly improves the system performance but does not increase the complexity of the digital predistortion system too much.

In the coarse extraction loop, the model parameters are extracted using the least square method, which can be written as

$$\boldsymbol{R}_{\text{coarse}(i)} = (\boldsymbol{Y}_{(i-1)}^{\text{H}}\, \boldsymbol{Y}_{(i-1)})^{-1} \boldsymbol{Y}_{(i-1)}^{\text{H}}\, \boldsymbol{U}_{(i-1)} \tag{10.6}$$

where $\boldsymbol{R}_{\text{coarse}(i)}$ represents the predistorter's parameter vector, which contains all unknown coefficients in the DPD model. The first iteration does not pass through the DPD module, so the initial DPD output is the same as the original input. From the second iteration, the parameters are calculated according to Equation (10.6) and the new DPD output comes from Equation (10.7).

$$U_{(i)} = X_{(i)} R_{\text{coarse}(i)} \tag{10.7}$$

The behavioral model of PA simulates the characteristics of a real amplifier, playing an important role in simulations of communication subsystems and PA linearization systems. In the first iteration, the output of non-linearized PA differs a lot from the original input due to nonlinear amplification processing of the PA, thus resulting in a difference between postinverse and preinverse. This mistake affects the system performance in the iterative process. The signal is predistorted by the "inaccurate" DPD model and then is output as an ideal signal in the next iteration. After several iterations, although the output of PA is close to the original input, there is still a certain error between the two signals due to the inherent defect in the model parameter extraction process.

To compensate for the residual error after coarse extraction of model parameters, this subsection proposes another parallel branch model, which fine-tunes the coefficients of DPD. After coarse extraction, the residual error between the original input $\tilde{x}(n)$ and the linearized output $\tilde{y}(n)$ becomes very small, and the output error may be approximately equal to the expected error of PA input. That is,

$$\tilde{e}(n) = G^{-1}(\tilde{y}(n) - \tilde{x}(n)) \approx \tilde{y}(n) - \tilde{x}(n) \tag{10.8}$$

where $G^{-1}(\cdot)$ is the inverse of the PA's transfer function. The output error of PA can be eliminated by subtracting the error from the output of DPD. The deviation coefficient is given by

$$R_{\text{fine}(i)} = (X_{(i-1)}^{\text{H}} X_{(i-1)})^{-1} X_{(i-1)}^{\text{H}} E_{(i-1)} \tag{10.9}$$

Matrix $X_{(i-1)}$ consists of the original input signal $\tilde{x}(n)$ in the same form as Matrix $Y_{(i-1)}$. $E_{(i-1)}$ is the resulting error vector from $\tilde{e}(n)$. Subtracting the deviation coefficient from the existing coefficient, we have:

$$R_{(i)} = R_{(i-1)} - \lambda R_{\text{fine}(i-1)} \quad (0 < \lambda \leq 1) \tag{10.10}$$

where λ denotes the sensitivity factor controlling the convergence speed. In the simulation, $\lambda \approx 0.707$ and the initial value of $R_{(i-1)}$ is the value yielded by the coarse extraction branch for model parameters. Finally, the output of DPD is expressed as

$$\boldsymbol{U}_{(i)} = \boldsymbol{X}_{(i)}\boldsymbol{R}_{(i)} \tag{10.11}$$

10.2.2 Experimental Validation and Result Analysis

This subsection describes a simulation validation for the nonlinear PA with memory effects by building the predistortion system on the MATLAB® platform, thereby validating the correctness of the proposed predistortion structure.

A fifth-order memory polynomial containing only odd terms is used as the PA distortion model, with the memory depth equal to 3. A 16QAM signal is used as the test signal.

Figure 10.7 shows the power spectral density of the output signal when additive noise is present in the nonlinear system of PA. Curves ① and ④ depict the power spectral densities in absence of predistortion and distortion, respectively. Curves ② and ③ depict the power spectral densities in presence of the indirect learning structure and the structure proposed in this section, respectively. As indicated in the figure, out-of-band spectrum spreading is suppressed in presence of predistortion, and the proposed predistortion structure outperforms the indirect learning structure. Specifically, the proposed structure has a stronger anti-noise ability and produces a power

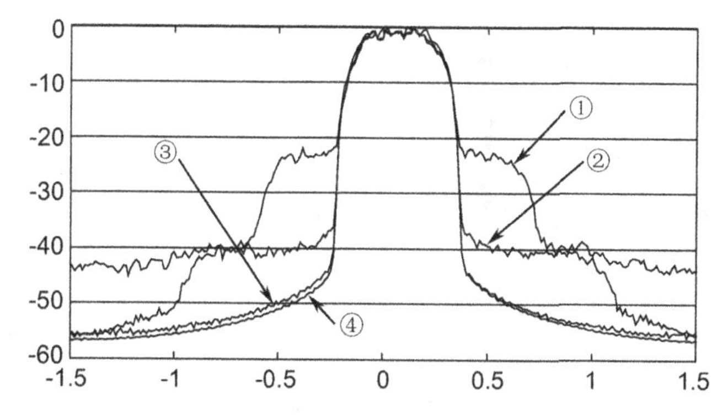

FIGURE 10.7 Power spectral density of output signal.

spectral density curve of the output signal almost coinciding with that of the original signal.

The constellation diagrams with and without predistortion are shown in Figure 10.8. Absence of predistortion causes a blurred constellation diagram with a large deflection angle, while presence of predistortion produces a clearer constellation diagram. An indirect learning structure yields a limited predistortion compensation effect due to the additive noise. In contrast, the predistortion method proposed in this section effectively overcomes the constellation deflection and diffusion problem

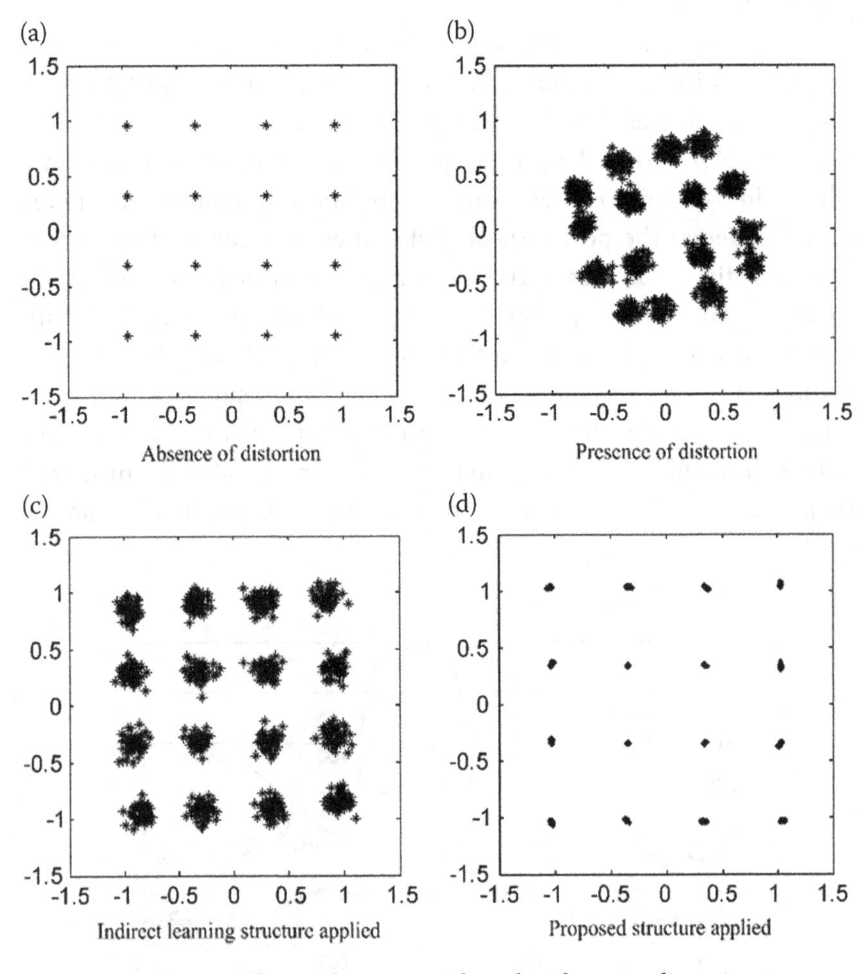

FIGURE 10.8 Constellation diagrams with and without predistortion.

due to nonlinearities and memory effects of PA, yielding a better predistortion effect.

In addition, the bit error rate (BER) of the PA nonlinear system is simulated to validate the reliability of the proposed predistortion method that is applied to a PA with memory effects. The simulation is shown in Figure 10.9. Curve ① is the BER in absence of predistortion. Curves ② and ③ are BERs in presence of the indirect learning structure and the proposed predistortion structure, respectively. As indicated in the figure, presence of predistortion notably reduces the BER, and the proposed predistortion structure produces a lower BER compared with the indirect learning structure.

This section proposes a digital predistortion method with a double-loop structure, which combines the traditional direct learning structure with the indirect learning structure. The proposed method introduces pre-equalizer and post-equalizer in the predistortion system to eliminate the linear distortion in the forward branch and feedback branch. As indicated by simulation results, this predistortion method favorably suppresses in-band distortion and out-of-band spectrum spreading, notably outperforming the direct learning structure and indirect learning structure. As a conclusion, the predistortion method proposed in this section significantly improves the performance of the digital baseband predistortion system without largely increasing the complexity and cost of implementation.

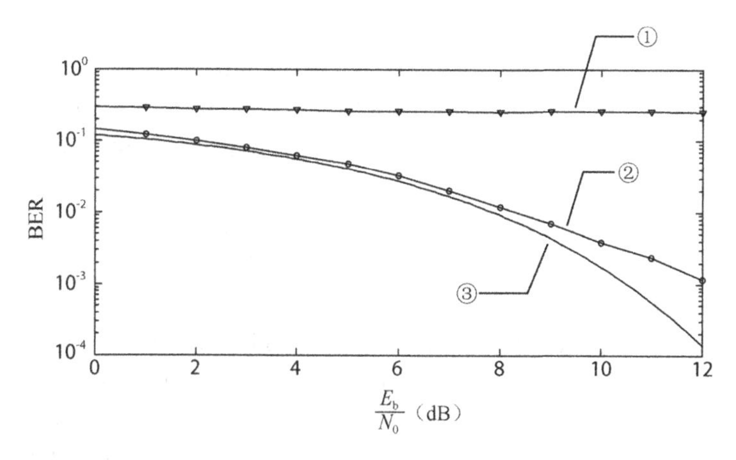

FIGURE 10.9 BER simulation diagram.

10.3 APPLICATION OF PEAK-TO-AVERAGE RATIO SUPPRESSION AND PREDISTORTION IN OFDM-ROF SYSTEM

Orthogonal Frequency Division Multiplexing-Radio over Fiber (OFDM-ROF) system will represent the future development direction in the communications industry, as this system suits the requirements for wireless applications, broadband transmission and high transmission quality. To address the inherent problems of this system such as high PAPR and nonlinearities in the transmission link, this section proposes a co-simulation system combining PAPR suppression and novel double-loop digital predistortion. Simulation results indicate that the improved OFDM-ROF system produces a significant improvement in all indicators.

10.3.1 OFDM-ROF System Analysis

OFDM is a multi-carrier modulation technique that adopts the cyclic prefix to preserve orthogonality between carriers after the signal passes through dispersive channels. OFDM exponentially increases the channel capacity and spectrum utilization without requiring additional spectrum resources [10]. The OFDM technique also enables the ROF system to have increased coverage and further higher data rates. Nowadays, the combination of OFDM and ROF has become an effective approach to extend the bandwidth of ROF system and increase the transmission distance of wireless signals. Compared with traditional wireless communication techniques, OFDM has demonstrated its superiorities in effective inter-carrier interference (ICI) and inter-symbol interference (ISI) suppression, high-speed data transmission in the presence of multipath and fading, sufficient anti-interference capability and high spectrum utilization.

ROF is a novel technique that enables radio signals to transmit over optical fiber links in the form of light. ROF uses optical fibers as the transmission medium, optical waves as the carrier, high frequency microwaves and millimeter waves as the modulation wave for wireless transmission. It consists of central station, optical fiber link, base station and subscriber unit [11]. A central station is connected to a plurality of base stations via optical fibers, and each base station is wirelessly connected to users within its cellular coverage area. The central station completes most of the system's functions (such as signal processing, modulation and demodulation, up-conversion to generate millimeter wave signals, etc.), while the base station basically completes the photoelectric conversion of downlink signals and electro-optical conversion and amplification of uplink signals, which leads to a simpler structure.

The ROF communication network has demonstrated the benefits of low loss, high bandwidth, simplified structure, signal format transparency, strong flexibility, easy management, strong resistance to electromagnetic interference and chemical corrosion.

The OFDM-ROF system, in which the baseband OFDM signal is transmitted, is a novel optical fiber wireless communication system. Figure 10.10 depicts a full-duplex OFDM-ROF system.

The central station completes the generation of downlink optical OFDM wireless signal and the detection and reception of uplink data. The base station basically performs the following two tasks.

1. Yielding an electrical OFDM wireless signal by photoelectric detection of the downlink optical OFDM wireless signal, and transmitting the electrical OFDM wireless signal through the antenna.

2. Receiving the uplink signal from the antenna and modulating it onto an optical carrier to generate an uplink optical signal. The mobile terminal (also known as subscriber unit) receives the wireless OFDM signal through the antenna, obtains the baseband signal through down-conversion, receives and demodulates the baseband signal and also transmits the uplink signal. Usually, complex processes and expensive equipment are deployed in the central station to reduce the cost of the entire communication network and facilitate management.

OFDM-ROF fully demonstrates the superiorities as a combination of the optical communication network featuring a huge transmission capacity and the wireless network featuring sufficient flexibility, promising a broad prospect in the development of ultra-wideband wireless communication in the future. However, a relatively high PAPR of the OFDM

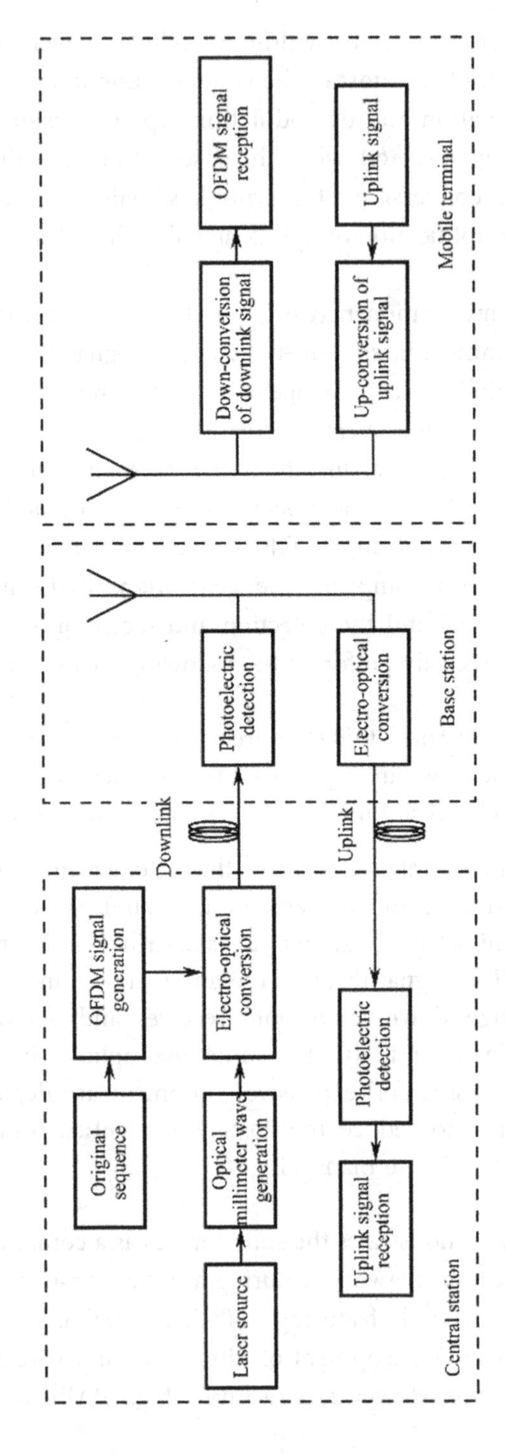

FIGURE 10.10 Block diagram of full-duplex OFDM-ROF system.

signal makes it prone to be affected by nonlinear devices in the transmission link, resulting in nonlinearities. Consequently, the transmission signal waveform tends to be distorted, which leads to a sharp deterioration in the performance of the entire OFDM-ROF system. To address these weaknesses, the OFDM-ROF system needs a further optimization and improvement.

10.3.2 Nonlinear Distortion Analysis of OFDM-ROF System

The peak-to-average power ratio (PAPR) is the ratio of peak power to average power of a signal. It is often used to characterize the signal envelope variation. The PAPR of an OFDM signal is defined by

$$\mathrm{PAPR(dB)} = 10\log_{10}\frac{\max\{|x_n|^2\}}{E\{|x_n|^2\}} \tag{10.12}$$

where x_n denotes the output signal after an inverse fast Fourier transform (IFFT) operation. Suppose there are N sub-channels in the OFDM system. When these N sub-channels are summed in the same phase, the yielded peak power of signal is N times of the average power. If the number of sub-channels is large, the OFDM system will have a very high PAPR, meaning a faster and larger signal envelope variation. Consequently, greater nonlinear distortion is easy to occur deteriorating the overall performance of OFDM-ROF system.

The part in an ROF system that produces nonlinearities is defined as a nonlinear link, and the digital predistortion technique can be used to compensate the nonlinearities caused by the nonlinear link. Intensity Modulated-Direct Detection (IM-DD) is one of the frequently used structures in the ROF system. This subsection investigates the application of digital predistortion technique in the OFDM-ROF system employing the IM-DD structure. Nonlinear devices in the IM-DD structure mainly include laser diode, single-mode fiber, photodiode receiver and PA [12]. The following gives a brief analysis on these nonlinear devices.

1. Laser diode

A laser diode produces static nonlinearity and dynamic nonlinearity due to stimulated emission, spontaneous emission and gain suppression.

The output power of laser diode is a nonlinear function of the input signal. Generally, the dynamic nonlinearity is characterized by the laser rate equation. When the resonant frequency of the laser is 5 times higher than the modulation frequency, the dynamic nonlinearity is considered to be frequency-independent[13].

2. Single-mode fiber

Single-mode fiber has attenuation and dispersion effects. Dispersion of an optical fiber can be expressed by the transmission equation as follows

$$H(f) = e^{-j\alpha l}(f - f_0)^2 \tag{10.13}$$

where α, l, f and f_0 denote dispersion coefficient, fiber length, modulated wireless frequency and optical carrier frequency, respectively.

3. Photodiode receiver

In the traditional ROF link, the nonlinearity of photodiode is not the main factor affecting the nonlinearity of the whole link, while the rated power of photodiode has the same important impact as the nonlinearity of other devices, which jointly affects the dynamic range of the whole ROF system.

4. PA

To meet the transmission requirements of antenna, a PA is required after the photodiode completes the photoelectric conversion of signal. The PA is an important nonlinear device in the ROF system, and its nonlinearity has an important impact on the communication quality of the whole system.

10.3.3 Co-Simulation System Establishment

It is unrealistic to accurately compensate for nonlinearities caused by all nonlinear devices since there are so many devices resulting in nonlinearities in the ROF system. Consequently, this subsection uses an adaptive digital predistortion technique with feedback loop to linearize the system. The ROF system is designed with the classic IM-DD structure, and the block diagram of ROF system in the presence of predistortion is shown in Figure 10.11. The nonlinear link in the ROF system is regarded as a black box model, and we only consider its input and output characteristics. The adaptive algorithm automatically updates the predistorter parameters by comparing the error between the negative feedback signal and the reference signal, thereby achieving linearization.

PAPR suppression by use of the peak cancellation method is an improved PAPR suppression technique based on direct clipping [14]. This algorithm is simple and easy to implement, compared with direct clipping. In addition, this algorithm effectively improves in-band distortion and out-of-band spectrum spreading caused by direct clipping. The specific steps of the algorithm are as follows.

1. Implement peak clipping in polar coordinates for the complex signal input, with the signal's phase component preserved.
2. Use a low-pass filter with the same frequency band as the input signal to shape and filter the error pulse signal generated after the first peak clipping, thereby filtering out the out-of-band spectrum spreading caused by direct peak clipping.
3. Enable the filtered signal to be reversely superimposed on the corresponding input signal to achieve the effect of peak cancellation.

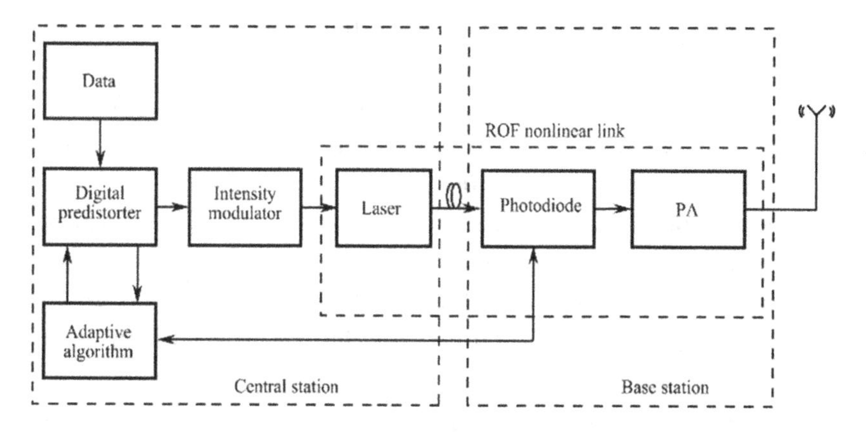

FIGURE 10.11 Block diagram of OFDM-ROF adaptive predistortion system.

Introduce a second peak clipping in polar coordinates to avoid PAPR recovery caused by the low-pass filter [15].

Figure 10.12 depicts the block diagram of a co-simulation system combining PAPR suppression and digital predistortion. A seventh-order memory polynomial containing only odd terms is used as the PA distortion model, with the memory depth equal to 3. A 64QAM signal is used as the test signal. The input signal experiences a power back-off first and then a PAPR suppression by means of the peak cancellation (clipping) method, and finally is linearized by an adaptive predistortion system.

10.3.4 Co-Simulation Result

The AM/AM and AM/PM characteristics of PA are shown in Figures 10.13 and 10.14, respectively. The input-output relationship of PA is not a one-to-one mapping function due to memory effects. In the absence of predistortion, the system's AM/AM and AM/PM characteristics are quite poor with points widely scattered. In the presence of adaptive predistortion, the system's AM/AM and AM/PM characteristics are significantly improved and basically appear as straight lines, which meet the system requirements.

Spectrum simulation results of the OFDM-ROF system are shown in Figure 10.15. Out-of-band spectrum spreading is severe in the absence of predistortion, as indicated by curve ①. In the presence of polynomial predistortion, out-of-band spectrum spreading is somewhat improved, as indicated by curve ②. In the presence of adaptive predistortion proposed in this section, a notably greater improvement is observed compared with that achieved in polynomial predistortion, as indicated by curve ③. When the peak cancellation method is applied and the PAPR of signals is suppressed by 1.5 dB, the signal's linearity is pretty good and basically coincides with the original signal, as indicated by curve ④. Curve ⑤ is the original signal without predistortion.

Distortion of in-band signals can be assessed by the constellation diagram and error vector magnitude (EVM). Figure 10.16 shows simulated constellation diagrams of OFDM-ROF system. In the presence of double-loop digital predistortion or peak clipping in polar coordinates, the constellation diagram becomes clear only with a minor difference from the original constellation diagram, and the constellation deflection

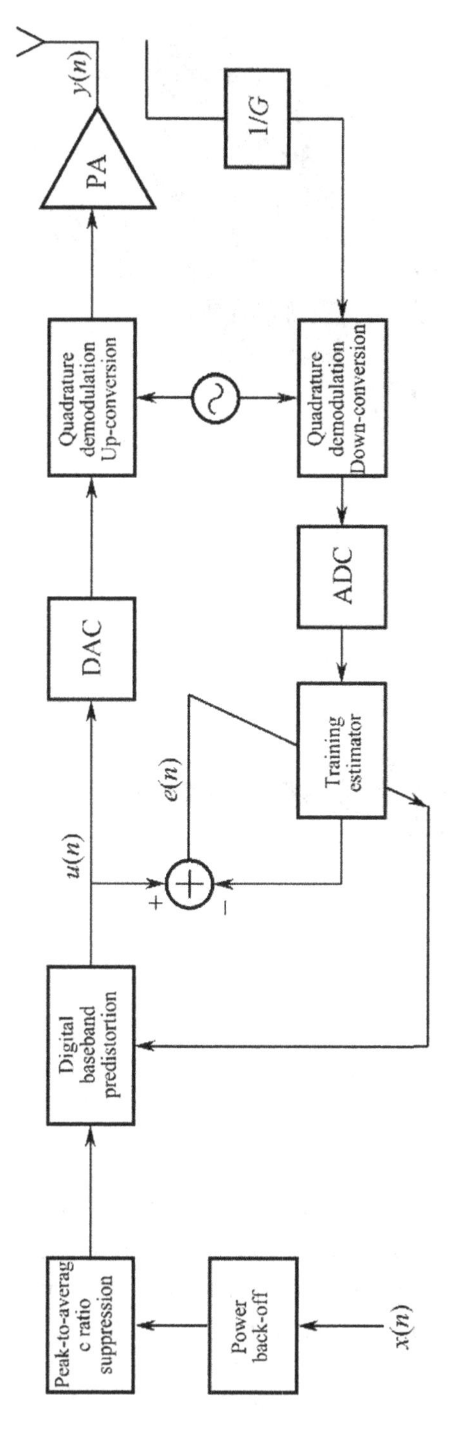

FIGURE 10.12 Block diagram of co-simulation system.

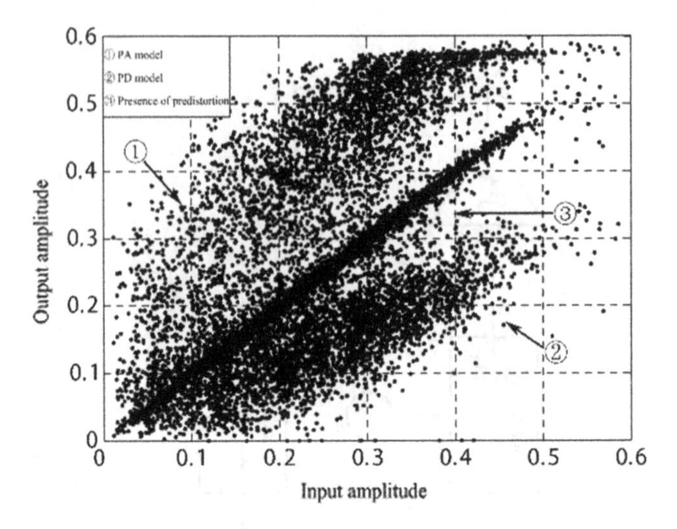

FIGURE 10.13 AM/AM characteristics of PA.

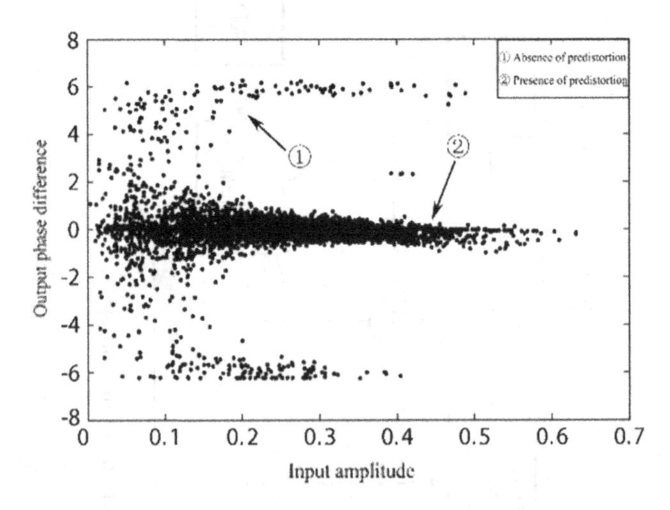

FIGURE 10.14 AM/PM characteristics of PA.

and diffusion problem is notably improved. In addition, simulation results indicate that the EVM is 51.2902% if the signal directly passes through the PA, 1.1874% if peak cancellation is applied, and 1.6406% in the presence of adaptive digital predistortion. The latter two EVMs are completely capable of allowing correct signal demodulation.

As part of the global microwave interconnection and wireless local area network, OFDM combined with ROF improves the signal quality and

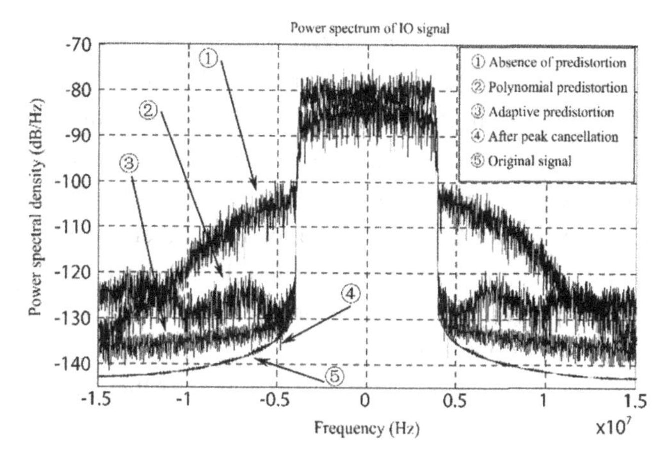

FIGURE 10.15 Spectrum simulation of OFDM-ROF system.

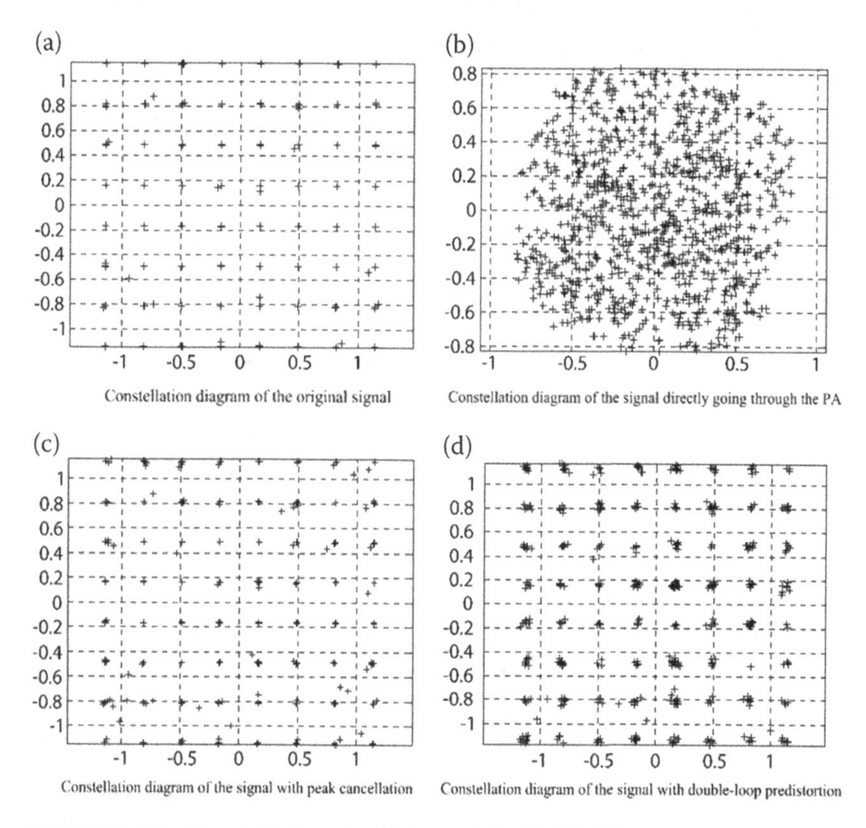

FIGURE 10.16 Constellation simulation of OFDM-ROF system.

simplifies the complexity of digital signal processing while effectively reducing multipath fading of wireless signals. The OFDM-ROF system is of great significance to high-speed and high-fidelity communications nowadays. This section proposes a co-simulation system combining PAPR suppression and digital predistortion to improve the overall linearity of the system, aiming to address the problems of high PAPR and nonlinearities in the OFDM-ROF system. As indicated by simulation results, the improved system achieves a significant improvement in all performance indicators compared with the system in the absence of peak cancellation and predistortion. In conclusion, the scheme combining PAPR suppression and digital predistortion in the OFDM-ROF system is applicable and effective.

10.4 COMBINED SCHEME OF PEAK-TO-AVERAGE RATIO SUPPRESSION AND PREDISTORTION TECHNOLOGY WITH IMPROVED ALGORITHM

In this section, the adaptive predistortion method applied to the high power amplifier (HPA) in the OFDM transmission system is investigated.

Literatures [16] to [18] investigate various PAPR suppression algorithms, such as partial transmit sequence, selective mapping and clipping. Although these algorithms reduce the PAPR favorably, the signal that enters the PA after experiencing PAPR suppression still suffers from distortion in the nonlinear dynamic range. If signal peaks are suppressed only within the linear dynamic range of PA, the efficient operation of PA cannot be guaranteed. In Literatures [19] and [20], only the adaptive predistortion technique is used to compensate the nonlinearities of PA. However, the nonlinear range of PA is limited. If the linear output value corresponding to the HPA input signal's amplitude is greater than the maximum output signal's amplitude, its nonlinear distortion cannot be compensated by the predistorter. Therefore, Literature [21] proposes a scheme combining PAPR suppression and predistortion. The scheme utilizes the peak clipping method to reduce the PAPR of an OFDM signal and the least mean square (LMS) error method to implement adaptive predistortion, thereby attaining 20 dB out-of-band spectrum suppression gain.

To mitigate the impact of HPA nonlinearities on the OFDM transmission system, an improved clipping method and the adaptive predistortion technique are combined in this section to reduce the PAPR of an OFDM signal and extend the linearization range of HPA, thereby compensating the nonlinearities of PA. This section proposes a variable step size LMS error algorithm, which is applied to the digital predistortion system based on a memory polynomial model. Simulation results indicate that the combination of PAPR suppression and adaptive predistortion outperforms adaptive predistortion alone, and the new algorithm effectively improves the nonlinearities of PA, attaining up to 28 dB out-of-band spectrum suppression gain.

10.4.1 System Model

Figure 10.17 outlines an adaptive predistortion transmitter model for the OFDM system. The TX end converts the binary digital signal to be transmitted into a mapping of subcarrier amplitude and phase. Following a serial-to-parallel conversion, it then transforms the spectral representation of digital data into the time domain using an IFFT, and performs a parallel-to-serial conversion to complete OFDM modulation. In addition, the predistorter performs nonlinear predistortion to overcome the nonlinear distortion of amplifier, and the adaptive algorithm updates the predistorter parameters.

If the input signal bandwidth is small enough that the memory time constants of a nonlinear system are less than the inverse value of the maximum envelope frequency of the input signal, memory effects in the system can be ignored. However, the OFDM system has a wide input signal bandwidth, and therefore the HPA's memory effects cannot be ignored in the system. A PA model with memory effects is applicable to the characterization of the HPA's memory nonlinearity in the OFDM system. This subsection uses a Wiener model, which consists of a linear filter cascaded with a memoryless nonlinear system. See Figure 10.18.

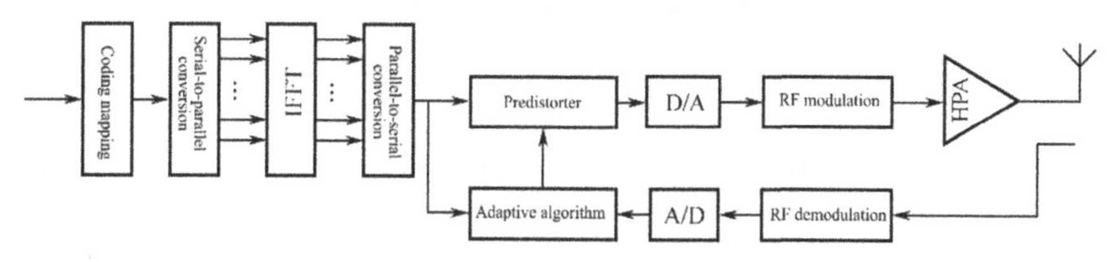

FIGURE 10.17 Simple block diagram of adaptive predistortion transmitter model for OFDM system.

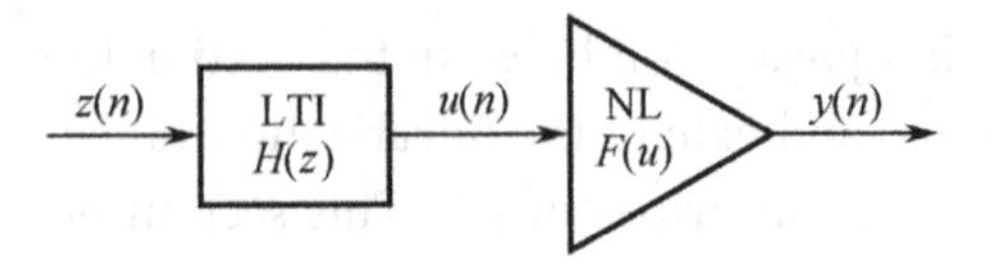

FIGURE 10.18 Wiener model.

In the Wiener model, $H(z)$ and $F(u)$ are the transfer functions of linear filter system and memoryless nonlinear system, respectively; $z(n)$ and $y(n)$ are the input and output signals, respectively.

$$H(z) = \frac{1 + 0.5z^{-2}}{1 - 0.2z^{-1}} \tag{10.14}$$

$$y(n) = \sum_{\substack{k=1 \\ \text{odd}}}^{5} b_k u(n) |u(n)|^{k-1} \tag{10.15}$$

For the coefficients in Equation (10.15), $b_1 = 1.0108 + j0.0858$, $b_3 = 0.0879 - j0.1583$, $b_5 = -1.0992 - j0.8891$.

10.4.2 Digital Predistortion System

The basic principle of digital predistortion linearization is to utilize an auxiliary distortion unit to generate the auxiliary distortion signal containing an intermodulation interference component, thereby suppressing the nonlinear distortion of PA. A nonlinear module is set up in front of the PA. The predistorter cascaded with the PA enables the OFDM signal to be compensated before entering the PA. In this way, the characteristic curve of digital predistortion system is complementary to the characteristic curve of PA, achieving linear amplification.

The predistorter uses a memory polynomial structure, which is expressed as

$$z(n) = \sum_{k=1}^{K} \sum_{q=0}^{Q} a_{kq} x(n-q) |x(n-q)|^{k-1} \tag{10.16}$$

where $x(n)$ and $z(n)$ denote the input and output of the predistorter, respectively; K denotes the nonlinear order; Q denotes the memory depth; a_{kq} denotes the predistorter coefficient. The predistorter is based on a memory polynomial model with 5th-order nonlinear 10th-order memory effects.

Adaptive predistortion algorithm is the key technique for a predistortion system. LMS algorithm often serves as a predistortion algorithm owing to its superiorities in low computational intensity, favorable stability, simple structure and easy implementation. However, its convergence speed is inversely proportional to the step size factor and its steady-state error is directly proportional to the step size factor. Consequently, a fixed step size LMS algorithm cannot have desired convergence speed and steady-state error at the same time. To solve the contradiction between convergence speed and steady-state error, a large number of literatures have proposed variable step size LMS algorithms.

Literature [22] proposes a variable step size VSS_LMS algorithm. Assume $\boldsymbol{d}(n)$ denotes the ideal output signal, $\boldsymbol{W}(n)$ denotes the tap coefficient of the filter, $\boldsymbol{e}(n)$ denotes the error signal and $\boldsymbol{x}(n)$ denotes the input signal. Then we have

$$\boldsymbol{e}(n) = \boldsymbol{d}(n) - \boldsymbol{x}^{\mathrm{T}}(n)\boldsymbol{W}(n) \tag{10.17}$$

$$\boldsymbol{W}(n+1) = \boldsymbol{W}(n) + \mu(n)\boldsymbol{e}(n)\boldsymbol{x}(n) \tag{10.18}$$

where $\mu(n)$ is a step size variable. The update for step size is expressed by

$$\mu(n+1) = \begin{cases} \mu_{\min}, & \mu(n) < \mu_{\min} \\ \alpha\mu(n) + \beta l, & otherwise \\ \mu_{\max}, & \mu(n) > \mu_{\max} \end{cases} \tag{10.19}$$

where α determines the step size value when the algorithm converges and β controls the steady-state error and convergence time of the algorithm. $0 < \beta < 1$. β is set to a small value and α is approximately set to 1. When the algorithm converges and the steady-state error l is close to zero, the step size variation is insignificant, thus ensuring a very small steady-state error after algorithm convergence. In the earlier stage of the adaptive algorithm, the error signal is large and the step size is also large, so the algorithm converges fast. When the algorithm converges to a steady state, the error signal becomes small and the step size also turns small, resulting in a small steady-state error. The value range of step size is specified by μ_{min} and μ_{max}, ensuring a small steady-state error and convergence of the algorithm. The value of μ_{min} is determined by the steady-state error and convergence speed. Generally, it is greater than and approximate to zero, so that the step size is positive. The value of μ_{max} is less than and approximately equal to $1/\lambda_{max}$. λ_{max} represents the maximum eigenvalue of the autocorrelation matrix of the input signal.

If the desired signal $\boldsymbol{d}(n)$ contains a noise signal n_0, the error signal $\boldsymbol{e}(n)$ cannot indicate the real error. Consequently, the step size update is affected by the noise signal, and the iterations of filter weight coefficient cannot approach the optimal value. As demonstrated in Literature [23], the variable step size_least mean square (VSS_LMS) algorithm is prone to be disturbed by the noise signal. Therefore, VSS_LMS algorithm is improved into VFSS_LMS algorithm, in which the update for step size is expressed by

$$\mu(n+1) = \begin{cases} \mu_{min}, & \mu(n) < \mu_{min} \\ \alpha\mu(n) + \beta p^2(n), & otherwise \\ \mu_{max}, & \mu(n) > \mu_{max} \end{cases} \tag{10.20}$$

$$p(n) = \gamma p(n-1) + (1-\gamma)e(n)e(n-1) \tag{10.21}$$

where $0 < \gamma < 1$ for controlling the convergence time. The estimation error $p(n)$ replaces $\boldsymbol{e}(n)$ in the step size update formula of the VSS_LMS algorithm. The iterative update of $p(n)$ represents the error signal variation. The autocorrelation estimation of the current error signal $\boldsymbol{e}(n)$ and the previous error signal $\boldsymbol{e}(n-1)$ is used to control the step size update.

The input signal $\boldsymbol{x}(n)$ is not correlated with the noise signal. Consequently, the input signal $\boldsymbol{x}(n)$ will never be affected by the error signal $\boldsymbol{e}(n)$, no matter whether the ideal signal $\boldsymbol{d}(n)$ contains the noise signal (that is, whether the error signal $\boldsymbol{e}(n)$ indicates the real error). Also, the correlation between $\boldsymbol{x}(n)$ and $\boldsymbol{e}(n)$ is stronger than the autocorrelation of $\boldsymbol{e}(n)$. Therefore, the step size update is adjusted by use of

$\boldsymbol{e}(n)x(n)$ instead of $\boldsymbol{e}(n)\boldsymbol{e}(n-1)$ in the improved algorithm to eliminate the impact of noises contained in the expected signal, allowing the iterations of filter weight coefficient to approach the optimal value even at low signal-to-noise ratio. Based on the above analysis, the VFSS_LMS algorithm is improved into the VFFSS_LMS algorithm, in which the update for step size is expressed as

$$\mu(n+1) = \begin{cases} \mu_{\min}, & \mu(n) < \mu_{\min} \\ \alpha\mu(n) + \beta p^2(n), & otherwise \\ \mu_{\max}, & \mu(n) > \mu_{\max} \end{cases} \tag{10.22}$$

$$p(n) = \gamma p(n-1) + (1-\gamma)e(n)x(n) \tag{10.23}$$

where $\mu(0) = \mu_{\max}$ and $p(0) = \mu(0)$ enabling the adaptive algorithm to have a fast convergence speed in the initial stage. Parameters α, β, $\mu_{\min}$ and $\mu_{\max}$ take the same values as those in the VSS_LMS algorithm. The VFFSS_LMS algorithm uses $\mu_{\min}$ to limit the minimum step size guaranteeing a small misadjustment near the optimal weight coefficient, uses $\mu_{\max}$ to limit the maximum step size guaranteeing the algorithm convergence. In the initial stage of the adaptive algorithm, the error is large and the estimation error $p(n)$ is also large (indicating a large step size), which accelerates the algorithm convergence. As the adaptive process proceeds, the error gradually decreases and the estimation error $p(n)$ is also reduced. In this situation, even if there is noise, the step size becomes small, resulting in a small steady-state error and small weight coefficient misadjustment, which improves the anti-noise interference performance of the algorithm.

10.4.3 Combined Scheme of Predistortion and Peak-to-Average Ratio Suppression

For a system with a high-PAPR input signal, the overall performance will be notably improved if the PAPR suppression technique is used to appropriately reduce the PAPR on the premise of a certain level of performance indicators and then the nonlinearities of PA are corrected by means of predistortion processing.

The simplest way to reduce the PAPR of an OFDM signal is clipping. The clipping method is to limit the peak amplitude to a preset maximum level. That is

$$\bar{u}_n = \begin{cases} u_n, & |u_n| \leq \text{Th} \\ \text{Th}e^{j\angle u_n}, & |u_n| > \text{Th} \end{cases}, \quad 0 \leq n \leq N-1 \tag{10.24}$$

where Th is the maximum signal amplitude allowed by the system. The part exceeding Th is clipped.

Direct clipping applied to distort the amplitude of an OFDM signal will result in interference to the system itself, thus deteriorating the bit error rate of the system. The nonlinear distortion of OFDM signal will lead to an increased out-of-band emission, because the clipping operation can be considered as a multiplication of the sampled OFDM symbols with a rectangular window function and the waves above a certain threshold are clipped. Therefore, the bandwidth after clipping is determined by both the sampled OFDM symbols and the rectangular window function. Clipping followed by filtering can effectively mitigate out-of-band emission but still cannot eliminate in-band distortion. In a word, direct clipping reduces the bit error rate and spectral efficiency of the entire system.

Literature [24] proposes an improved clipping method, which consists of J-fold oversampling and clipping filtering. The improved clipping method effectively improves the out-of-band spectrum spreading and in-band distortion resulted from direct clipping. Figure 10.19 depicts the schematic diagram of improved clipping.

The improved clipping method firstly performs J-fold oversampling of the OFDM signal (i.e., inserting $N(J-1)$ zeros into the OFDM data vector), and then converts the OFDM data vector from frequency domain to time domain using an inverse Fourier transform. The oversampling technique effectively prevents spectrum aliasing. After that, clipping and filtering are performed. Specifically, the signal is converted from time domain to frequency domain using a Fourier transform, the out-of-band signal is set to zero artificially to suppress out-of-band spectral regrowth, and then the signal is converted back to the time domain. Clipping and filtering completely remove out-of-band spectrum

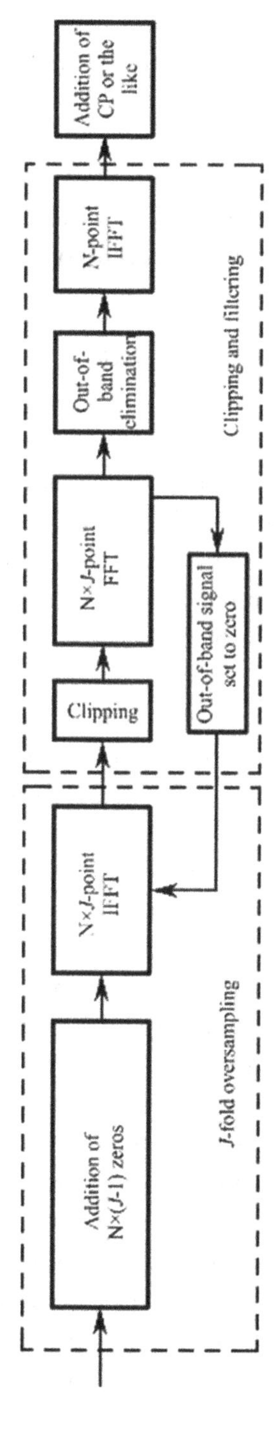

FIGURE 10.19 Schematic diagram of improved clipping.

spreading of an OFDM signal but may cause the signal's PAPR to recover. We utilize repeated clipping and filtering to reduce signal peaks until a desired PAPR is achieved.

The adaptive structure of predistorter is an indirect learning structure, which does not require the assumption of PA model or parameter estimation. Its superiorities also lie in simple structure and easy implementation. The combination of adaptive predistortion and PAPR suppression is depicted in Figure 10.20.

10.4.4 Experimental Result and Analysis

The simulation is performed in an additive white Gauss noise (AWGN) channel. N (number of effective OFDM subcarriers) = 3780, CP (cyclic prefix length) = 420, IBO (input power back-off) = 4 dB. A 64QAM modulation scheme is used. The simulation uses the improved clipping method to make the input signal amplitude of PA not exceed the saturation voltage. The oversampling factor J is equal to 4 and the clipping is repeated three times. The signal is subject to 1.5 dB PAPR suppression, followed by an adaptive predistortion amplification system.

Nonlinearities of PA are basically manifested in out-of-band spectrum spreading and in-band signal distortion, which are investigated in this subsection as the major indicators for system performance improvement. Figure 10.21 shows different signal power spectrums measured in the system, where the LMS predistortion algorithm is applied. As indicated in the figure, the spectrum of the signal processed by improved clipping basically coincides with the original signal spectrum, demonstrating that improved clipping effectively suppresses the out-of-band spectrum. Both traditional polynomial predistortion and PAPR suppression combined with predistortion improve the out-of-band spectrum spreading by approximately 20 dB, compared with the spectrum of signal directly entering the PA. PAPR suppression combined with predistortion slightly outperforms traditional polynomial predistortion.

Figures 10.22 shows the power spectrum of the adaptive predistortion algorithm. As indicated in the figure, clipping combined with VSS_LMS predistortion and clipping combined with VFFSS_LMS predistortion improve the out-of-band spectrum suppression by approximately 23 dB and 28 dB, respectively, compared with the spectrum of signal directly entering the PA.

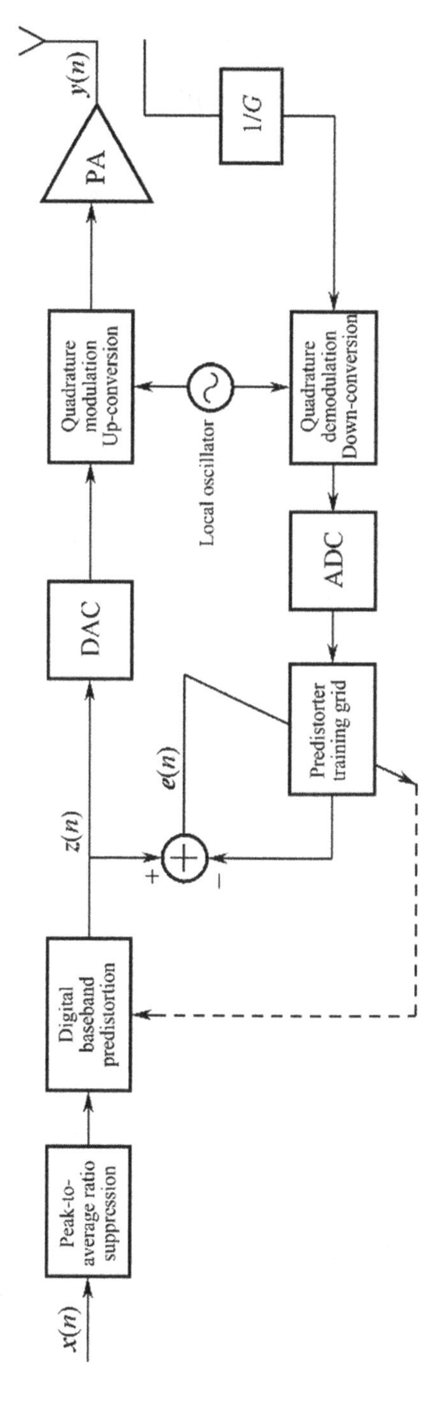

FIGURE 10.20 Combined scheme of adaptive predistortion and PAPR suppression.

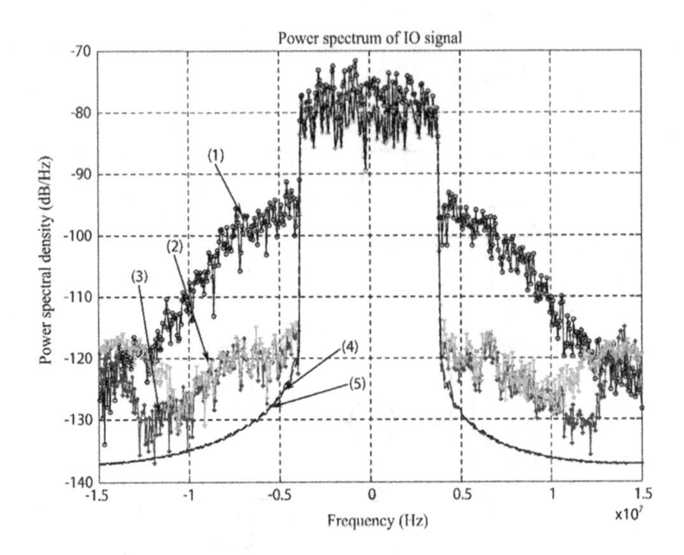

FIGURE 10.21 Signal power spectrums in predistortion. (1) Absence of predistortion (2) Polynomial predistortion (3) PAPR suppression combined with predistortion (4) Presence of clipping (5) Original signal.

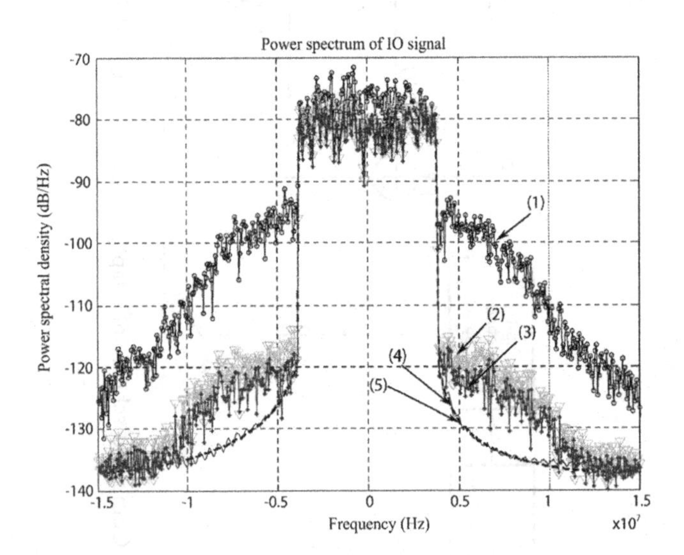

FIGURE 10.22 Signal power spectrums in adaptive predistortion. (1) Absence of predistortion (2) VFSS_LMS predistortion (3) VFFSS_LMS predistortion (4) Presence of clipping (5) Original signal.

Distortion of in-band signals can be assessed by the constellation diagram and error vector magnitude (EVM). Figure 10.23 shows different constellation diagrams measured in the system. Figure 10.23(a) shows the original OFDM signal's constellation diagram. Figure 10.23(b) shows the constellation diagram of signal processed by improved clipping, which appears clearer with an EVM 0.6370%. Figure 10.23(c) shows the constellation diagram of signal directly entering the PA, which indicates a severe constellation diffusion and deflection problem with an EVM 66.2130%. Figure 10.23(d) shows the constellation diagram of signal processed by VFFSS_LMS predistortion, which appears clearer with an EVM 1.2912%. VSS_LMS predistortion produces an EVM 1.4604%, which is greater than that of VFFSS_LMS predistortion.

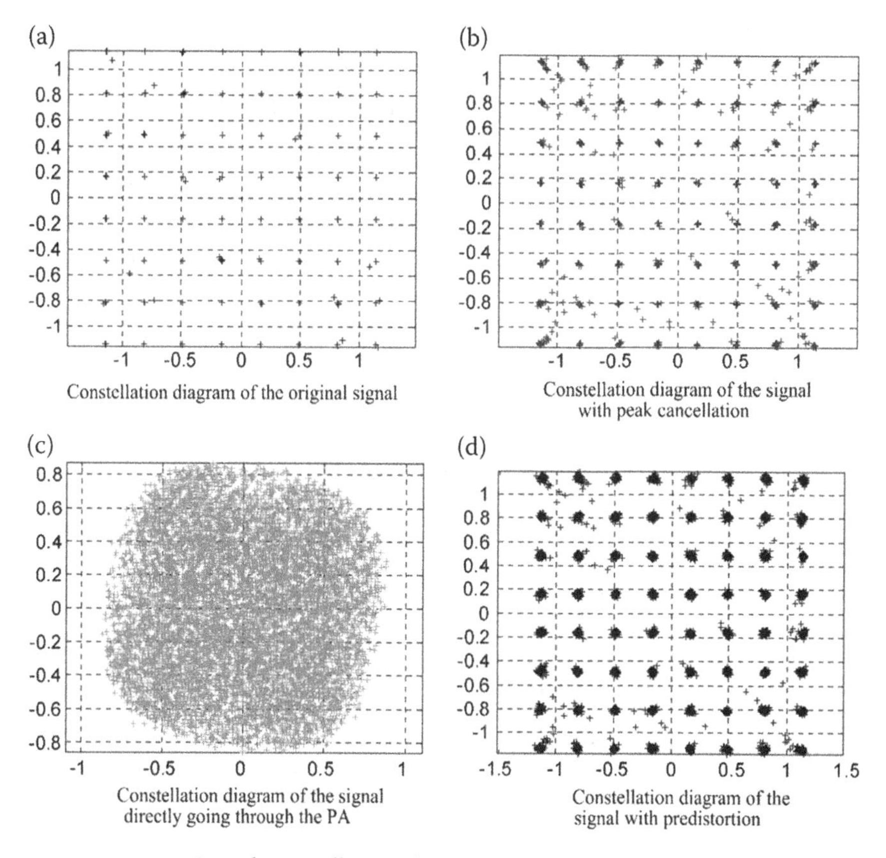

FIGURE 10.23 Signal constellation diagrams in predistortion.

Based on ACPR and EVM measurements, VFFSS_LMS predistortion outperforms VSS_LMS predistortion, and PAPR suppression combined with predistortion outperforms predistortion alone.

As shown in Figure 10.24, the system's AM/AM characteristics are quite poor with points widely scattered in the absence of predistortion. However, in the presence of PAPR suppression combined with VFFSS_LMS predistortion, AM/AM characteristics are significantly improved and basically appear as a straight line, which meets the system requirements.

Figure 10.25 shows the convergence curve of VFFSS_LMS algorithm. As indicated in the figure, the algorithm has converged within 500 iteration points and the error is in the order of 10^{-3} after the convergence stabilizes. The yielded convergence speed and stability meet the design requirements.

In this section, PAPR suppression combined with predistortion is applied, and VFFSS_LMS algorithm is proposed on the basis of VSS_LMS algorithm to demonstrate fast convergence and strong anti-noise capability. In addition, these two adaptive algorithms are applied to predistortion. Simulation results indicate that PAPR suppression combined with predistortion can favorably suppress in-band distortion and out-of-band spectrum spreading, and the combination scheme proves to outperform

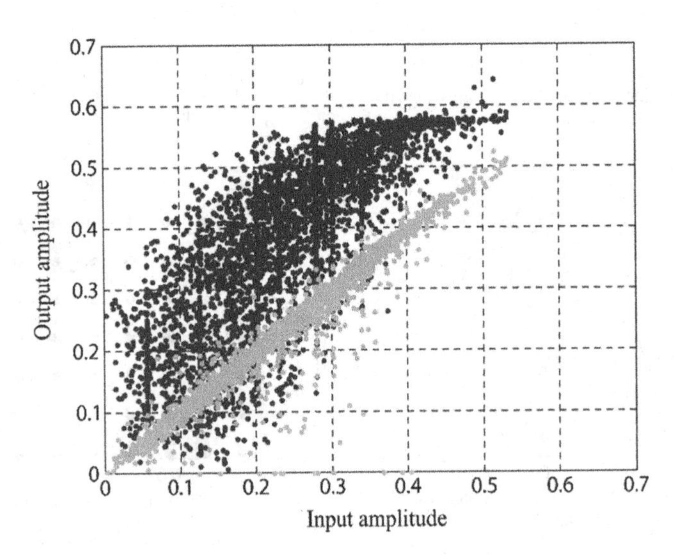

FIGURE 10.24 Comparison of AM/AM characteristics with and without VFFSS_LMS predistortion.

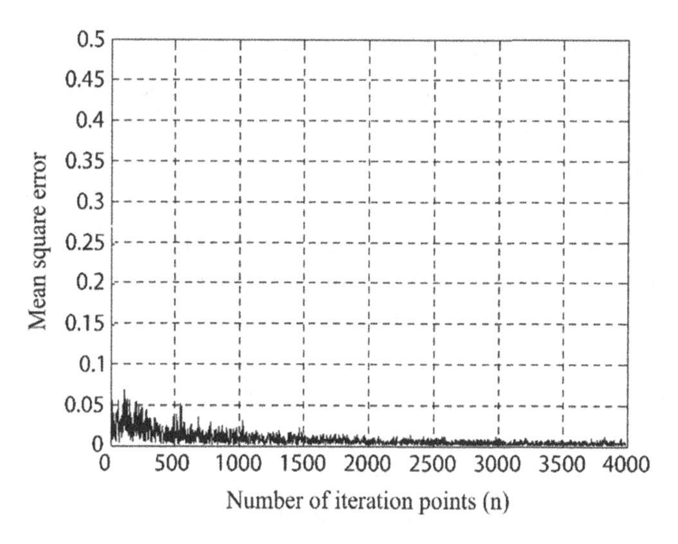

FIGURE 10.25 Convergence curve of VFFSS_LMS algorithm.

predistortion that is used alone. This section also demonstrates that VFFSS_LMS algorithm proposed outperforms VSS_LMS algorithm and performs much better than a commonly used LMS algorithm.

10.5 SPARSE NORMALIZED POWER AMPLIFIER MODEL AND PREDISTORTION APPLICATION

Frequently-used memory nonlinear PA models include Volterra series behavioral model and neural network behavioral model. The neural network model can vividly characterize nonlinear functions but the model structure has an extremely complicated implementation. The pruned model has a relatively simpler expression at the expense of reduced accuracy. To accurately characterize the nonlinearities of PA, we need to increase the model's nonlinear order and memory depth, resulting in higher complexity of the model. Therefore, this approach cannot be widely applied. Literatures [25] to [27] apply compressed sensing (CS) to PA models, although there are few literatures concerning the PA behavioral model research by means of CS algorithm until now. Literature [28] proposes a simplified second-order dynamic deviation reduction Volterra model. Literature [25] proposes a Volterra filter polynomial model using CS algorithm to reduce the sparsity, applies batch estimations to the polynomial model and validates the proposed model using an adaptive RLS algorithm. Literature [26] proposes a predistortion analysis based on the CS method, which is applied to weakly nonlinear systems. Literature [27] proposes a two-part Volterra series model using unstructured pruning.

This section investigates the nonlinearities of RF PA and proposes a novel sparse Volterra series model. Based on the CS algorithm, the proposed model equates sparse system identification to signal reconstruction, attains sparsity in the kernel coefficients by using the regularized orthogonal matching pursuit (ROMP) algorithm and selects the active kernel coefficients. The model proposed in this section improves the modeling accuracy by 10.7 dB with 25% less coefficients, compared with the memory polynomial (MP) model; improves the modeling accuracy by 3.9 dB with 84.58% less coefficients compared with the general memory polynomial (GMP) model. As indicated by simulation results, the proposed method delivers favorable predistortion linearization performance, notably reduces the model coefficients and is superior to traditional PA behavioral models.

10.5.1 Model Description

The MP model is the simplest of all simplified Volterra series models. It ignores the kernel functions of non-diagonal terms in a Volterra series model and simplifies the expression of the model. Moreover, the MP model has a relatively high accuracy and reduced complexity while taking memory effects into account. In a weak memory effect model, the MP model is expressed as

$$y_{\mathrm{MP}}(n) = \sum_{m=0}^{M} \sum_{n=1}^{N} a_{mn} x(n-m) |x(n-m)|^{n-1} \tag{10.25}$$

where x and y_{MP} are the input and output signals of the MP model, respectively; M, N and a_{mn} are the memory depth, nonlinear order and coefficient of the MP model, respectively; $|x(n-m)|$ is the modulus of the input signal $x(n-m)$.

The expression of MP model is so simple that it cannot be applied to the PA model having strong memory effects. To accurately characterize the strong nonlinearities of PA, the MP model introduces two additional memory cross terms (i.e., leading and lagging envelope terms) to enhance the characterization of memory effects for the model. The GMP model is expressed as

$$y_{\mathrm{GMP}}(n) = \sum_{m=0}^{M_a-1} \sum_{k=0}^{N_a-1} a_{mk} x(n-k) |x(n-k)|^m + \sum_{m=1}^{M_b} \sum_{k=0}^{N_b-1} \sum_{l=1}^{L_b} b_{mkl} x(n-k) |x(n-k-l)|^m + \sum_{m=1}^{M_c} \sum_{k=0}^{N_c-1} \sum_{l=1}^{L_c} c_{mkl} x(n-k) |x(n-k+l)|^m \quad (10.26)$$

where x and y_{GMP} denote the input and output signals of the GMP model, respectively; a_{mk}, b_{mkl} and c_{mkl} denote the GMP model's coefficients, coefficients of the signal and lagging envelope terms and coefficients of the signal and leading envelope terms, respectively; M_a, M_b and M_c denote the memory depths of each term, respectively; N_a, N_b and N_c denote the nonlinear orders of each term, respectively; L_b and L_c denote the lagging cross terms index and leading cross terms index, respectively.

The GMP model is able to characterize the nonlinear memory effects of PA, but the characterization of high-order nonlinearities requires all memory terms, resulting in high complexity of the model. Also, the nonlinear order and memory depth must be increased for an accurate characterization of a high-order nonlinear PA model. A nonlinear system model can be represented by a general dynamic nonlinear Volterra series, and the baseband discrete time-domain Volterra series model can be expressed as

$$y(n) = \sum_{n=1}^{N} x(n) \cdots \sum_{m_1=0}^{M} \sum_{m_2=m_1}^{M} \cdots \sum_{m_n=m_{n-1}}^{M} \sum_{m_{n+1}=0}^{M} \sum_{m_{n+2}=m_{n+1}}^{M} \cdots \cdots \sum_{m_{2n-1}=m_{2n-2}}^{M} h_{2n-1}^{(m_1,\cdots,m_{2n-1})} \Psi_{2n-1}^{(m_1,\cdots,m_{2n-1})}(n) + e(n) \quad (10.27)$$

where $x(n)$ is the complex input envelope signal of the model; $y(n)$ is the complex output envelope signal of the model; M is the pruned memory length; $(2n - 1)$ is the order of the model; $e(n)$ is the measurement error; $h_{2n-1}^{(m_1,\cdots,m_{2n-1})}$ is the normalized coefficient of the $(2n - 1)$th-order Volterra series kernel. Assuming $\Psi_{2n-1}^{(m_1,\cdots m_{2n-1})}(n)$ is the regressor corresponding to the PA and $E[\cdot]$ is the expectation factor, we have

$$\Psi_{2n-1}^{(m_1,\cdots m_{2n-1})}(n) = (E[|x(n - m_1)\cdots x(n - m_{2n-1})|^2])^{-\frac{1}{2}} \times \prod_{i=1}^{n} x(n - m_i) \prod_{i=n+1}^{2n-1} x^*(n - m_i) \tag{10.28}$$

10.5.2 Model Sparsification and Identification

Signal reconstruction has always been a hot research topic in compressed sensing (CS) and is favored by researchers. E. Candes and T. Tao et al. have demonstrated that the fundamental problem of CS reconstruction can be formulated as an l_0 norm optimization problem, where l_0 norm represents the number of non-zero elements in the vector. Typically, solutions to the l_0 norm optimization problem include convex relaxation algorithm, matching pursuit class algorithm and iterative threshold algorithm. CS not only introduces significant changes to signal processing but also presents new opportunities in a variety of scientific research fields, such as sparse channel estimation, data acquisition, medical imaging, etc.

In this section, sparsity of the kernel function $\boldsymbol{h}$ of Volterra series model is utilized to yield the kernel function subset of Volterra series model and to provide an optimal input-output waveform of PA. Given that the coefficients of Volterra series model are linear, Equation (10.27) is written as a matrix:

$$\boldsymbol{y} = \boldsymbol{X}\cdot\boldsymbol{h} + \boldsymbol{e} \tag{10.29}$$

where $\boldsymbol{y}$ is a column vector containing continuous M samples in the envelope of the output signal; $\boldsymbol{e}$ is a noise vector of Gaussian distribution with zero mean and variance σ_e^2; $\boldsymbol{h}$ is a vector in which coefficients of Volterra series model are arranged in sequence; $\boldsymbol{X}$ is a measurement matrix having the same order and rank (both having a full rank) as the regressor in Equation (10.28). The expression of the measurement matrix is given by

$$X = [\Psi_1^{(0)}, \cdots, \Psi_1^{(M)}, \cdots, \Psi_{2N-1}^{(M, \cdots, M)}] \tag{10.30}$$

The RLS algorithm is usually used to estimate the kernel vector since the model coefficients are linear, and the estimated value is

$$\hat{h} = (X^{\mathrm{H}} \cdot X)^{-1} X^{\mathrm{H}} \cdot y \tag{10.31}$$

Model sparsification is represented by Equation (10.31), and sparse system identification is equivalent to signal reconstruction. Therefore, we can achieve system sparsity by solving l_0, use the sub-optimal algorithm to solve the l_0 norm, transform the non-convex optimization algorithm into a convex optimization algorithm and use ROMP in the matching pursuit class algorithm to solve the problem [29]. In Literature [30], the orthogonal matching pursuit (OMP) algorithm is used for model sparse identification of the system. At each iteration, the OMP algorithm only selects the column most related to the residual while the ROMP algorithm can select several more columns. Figure 10.26 shows a simple block diagram of ROMP algorithm, which basically consists of four parts. Support set update is the most important part.

$$\begin{aligned} &\text{minimize} \quad ||h||_0 \\ &\text{subject to} \quad ||y - Xh||_2 \leq \varepsilon \end{aligned} \tag{10.32}$$

In selection of atoms, the ROMP algorithm identifies and groups the most matched K atoms into the candidate set, or selects all nonzero

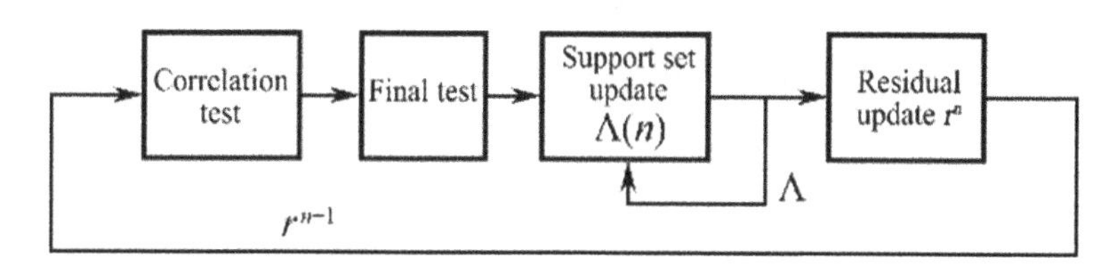

FIGURE 10.26 Simple block diagram of ROMP algorithm.

columns of all inner products if these inner products have fewer than K nonzeros (indicating the sparsity). Next, the ROMP algorithm regularizes the set containing K columns (i.e., the selected column vectors), and selects the maximum absolute value of all column vectors and inner products of the residuals. The maximum absolute value cannot exceed twice the minimum value, ensuring that the selected atoms carry information approximate to a certain value. The atoms carrying smaller values are discarded and constitute a regularized subset of the candidate set. The atomic sequence numbers in the subset are stored in an index set. ROMP yields the approximate solution by means of the least square method, performs a residual update and proceeds with the iterations until the termination criteria applies. Eventually, the estimated values of sparse coefficients are attained. The following lists the steps in the ROMP algorithm to estimate the sparse elements of kernel vectors. $\boldsymbol{\lambda}_s$ denotes the vector inner product of signal $\boldsymbol{X}$ and residual $\boldsymbol{Y}$.

1. *Initialize:* $\boldsymbol{r}^0 = \boldsymbol{y}$, $\Lambda(0) = [\ ]$, $n = 1$;
2. $J = \max_s \{\boldsymbol{\lambda}_s = |\langle \boldsymbol{X}_{\{s\}}^{\mathrm{H}} \boldsymbol{r}^{n-1} \rangle|, K\}$;
3. $J_0 = \max_{J_0} ||\lambda_{J_0}||_2 \quad |\lambda_s| \leq 2|\lambda_q|,\ s,\ q \in J_0\ ,\ J_0 \in J$;
4. $\Lambda(n) = \Lambda(n-1) \cup \{J_0\}$;
5. $\overset{\wedge(n)}{\boldsymbol{h}} = (\boldsymbol{X}_{\Lambda(n)}^{\mathrm{H}} \cdot \boldsymbol{X}_{\Lambda(n)})^{-1} \boldsymbol{X}_{\Lambda(n)}^{\mathrm{H}} \cdot \boldsymbol{y}$;
6. $\overset{\wedge(n)}{\boldsymbol{y}} = \boldsymbol{X}_{\Lambda(n)} \overset{\wedge(n)}{\boldsymbol{h}}$;
7. $\boldsymbol{r}^{(n)} = \boldsymbol{y} - \overset{\wedge(n)}{\boldsymbol{y}}$.

If $n = K$ does not apply, return to (2). If it applies, the iterative process terminates.

After attaining sparsity in the kernel coefficients, the ROMP algorithm yields the estimated values $\hat{\boldsymbol{h}}^{(n)}$ of model coefficients and the noise variance $\hat{\sigma}_e^2$, and completes the extraction of model parameters.

$$\overset{\wedge(n)}{\boldsymbol{h}} = (\boldsymbol{X}_{\Lambda(n)}^{\mathrm{H}} \cdot \boldsymbol{X}_{\Lambda(n)})^{-1} \boldsymbol{X}_{\Lambda(n)}^{\mathrm{H}} \cdot \boldsymbol{y} \tag{10.33}$$

$$\hat{\sigma}_e^2 = \frac{1}{M}\left\| y - X_{\Lambda(n)}\hat{h}^{(n)} \right\|^2 \tag{10.34}$$

The PA behavioral model is a time-domain waveform model and the NMSE is used to directly evaluate the accuracy of the model. The selected model size should lead to a small NMSE value corresponding to a minor number of coefficients, so that a simplified model can be selected.

Model parameters are given by Equation (10.34) and the nonlinear order and memory depth of the model are swept. In a reasonable sweeping range, the nonlinear order and memory depth of the model are determined to be 0–15 and 0–5, respectively. During the experiment, nonlinearities of PA gradually decline as the nonlinear order increases. However, the accuracy of the model must take into account both memory depth and nonlinear order with reasonable parameter settings. Finally, the nonlinear order and memory depth of MP model, GMP model and proposed model are determined to be 11 and 4, 11 and 4, 11 and 4, respectively.

10.5.3 Model Performance Validation

In this subsection, a total of 8000 sets of data are sampled, of which 4000 sets are used to identify model parameters and the other 4000 sets are used to validate the model accuracy. An LDMOS Doherty amplifier with central frequency at 1.96 GHz is used, attaining 50 dB gain. The test signal is a 16QAM signal with a chip rate of 15 Mcps and a signal bandwidth of 15 MHz.

Table 10.1 lists the model parameters, NMSEs and total coefficients of the sparse Volterra series model, MP model and GMP model. When using the same settings of model parameters, the MP, GMP and sparse Volterra series models yield the NMSEs −32.7 dB, −39.5 dB and −43.4 dB

TABLE 10.1 Comparison of Models

Model	Model Parameter	NMSE (dB)	Total Coefficients
MP	$M = 4, \; N = 11$	-32.7	48
GMP	$M_a = 4, \; N_a = 11$ $M_b = 3, \; N_b = 3, \; L_b = 1$ $M_c = 3, \; N_c = 3, \; L_c = 1$	-39.5	85
CS-DPD	$M = 4, \; 2N - 1 = 11$	-43.4	36

respectively, and the three models use a total of 48, 85 and 36 coefficients respectively. Compared with MP model, the sparse Volterra series model improves the NMSE performance by 10.7 dB with 25% less coefficients. Compared with GMP model, the sparse Volterra series model improves the NMSE performance by 3.9 dB with 84.58% less coefficients. In general, the proposed model contributes to significant improvement in the simplification of model coefficients without deteriorating the accuracy. Rationality of the sparse model has been successfully validated.

Figure 10.27 illustrates the output power spectrums of the proposed sparse Volterra series model and the GMP model. As indicated in the figure, both models can simulate the power spectrum characteristics of the real PA, but the proposed model produces a smaller output error against measurements compared with the GMP model. The proposed model has higher modeling accuracy at the center of the horizontal axis of the spectrum diagram.

10.5.4 Predistortion Application

Figure 10.28 shows the AM/AM and AM/PM characteristics exhibited by the MP, GMP and proposed sparse Volterra series models. As indicated by the figure, both AM/AM and AM/PM characteristics indicate the presence of significant nonlinearities for the signal

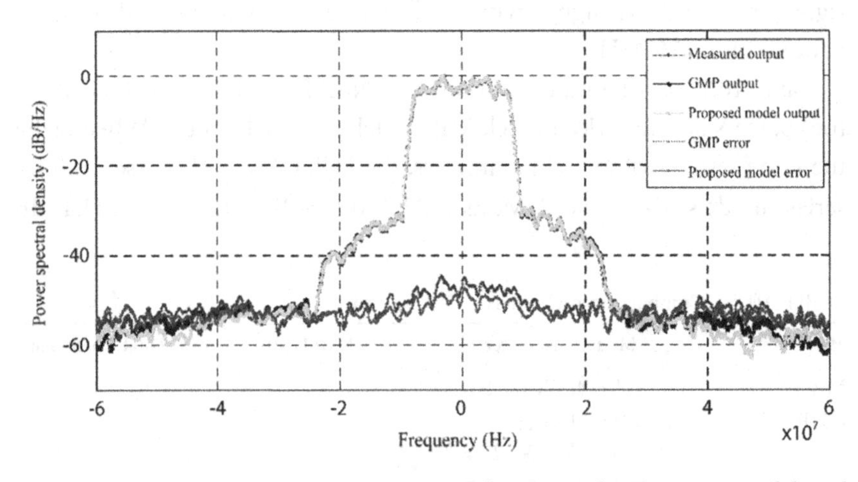

FIGURE 10.27 Output power spectrums of sparse volterra series model and GMP model.

(a)

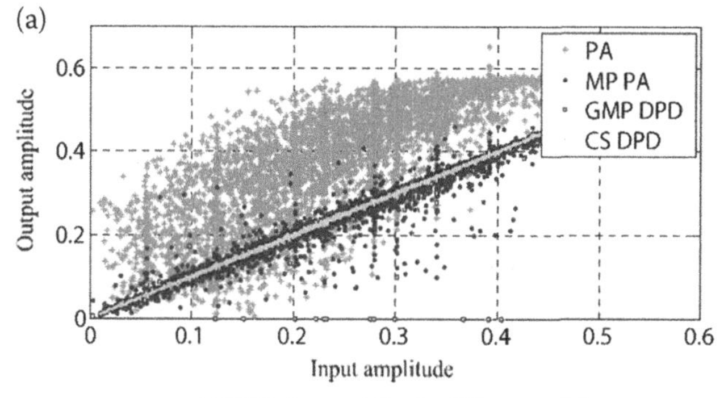

AM/AM characteristics for different models

(b)

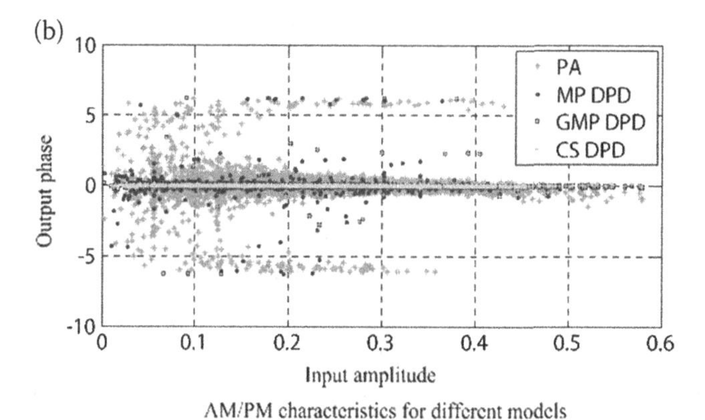

AM/PM characteristics for different models

FIGURE 10.28 AM/AM and AM/PM characteristics of different models.

directly passing through the PA. The three models have improved the nonlinearities of PA by varying degrees using their predistortion techniques, thereby enhancing the PA linearization characteristics. The predistortion effect yielded by the model proposed in this section is significant and better than that yielded by MP and GMP models. Its AM/AM and AM/PM characteristics basically appear as straight lines, sufficiently compensating for the nonlinear distortion of PA. The favorable predistortion effect of the proposed model has been successfully validated.

Figure 10.29 shows the power spectral densities in predistortion outputs of the MP, GMP and proposed sparse Volterra series models, with their parameter settings in strict compliance with those listed in Table 10.1. As indicated in the figure, all three models effectively suppress spectral regrowth, and the proposed model and GMP model

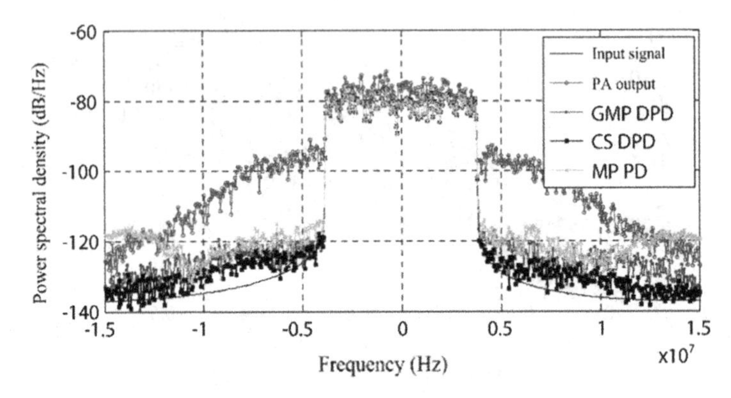

FIGURE 10.29 Output power spectral density after predistortion for each model.

yield much better predistortion performances compared with the MP model. Figure 10.27 have demonstrated that the proposed model produces a better power spectrum characterizing the real PA performance more accurately, compared with the GMP model. In Figure 10.29, the power spectral density in the predistortion output of the proposed model is sufficiently close to the measured spectral density curve of the input signal. Furthermore, the proposed model significantly simplifies the complexity compared with MP and GMP models and also improves the ACPR. The superiority of the proposed model is successfully validated.

Volterra series model plays an important role in the behavioral model of PA. In recent years, linearization of PA is facing severe challenges due to the shortage of spectrum resources. Although Volterra series can be pruned for a simplified model, it still fails to meet the requirements of PA linearization techniques. This section proposes a novel sparse normalized Volterra series model and compares it with MP and GMP models. As indicated by the experimental results, the proposed model simplifies the complexity with 84.58% less coefficients compared with the GMP model; the proposed model also delivers an improved performance. While ensuring the accuracy of PA model, the proposed model lowers the complexity of digital predistortion. In conclusion, the proposed model is of great significance for behavioral modeling of PA and development of digital predistortion.

10.6 COMBINED PREDISTORTION METHOD OF SIMPLIFIED FILTER LOOK-UP TABLE AND NEURAL NETWORK

Nowadays, popular predistortion models are based on polynomial, neural network or look-up table (LUT). Literature [23] proposes a two-dimensional indexing method for memory LUT predistortion, which introduces a second dimension index to distinguish the memory effects of input signal at different moments. This method alleviates signal distortion resulted from memory effects, improves phase distortion of signal and suppresses out-of-band spectrum spreading. However, quantization noise and storage capacity limit the development of this technique. Literature [31] proposes the use of multiple LUTs to compensate the HPA's memory effects and the exponential decay weighted average power to characterize the HPA's memory effects. The LUTs are flexible to use and modify but require larger storage space and longer time for adaptive convergence. Literature [30] proposes a filter look-up table (FLUT) method, which uses a memoryless predistorter connected in series with a filter to simulate the inverse characteristics of the PA's memory nonlinearity. The FLUT method has proven to deliver favorable compensation performance and is easy to implement.

This section proposes an improved FLUT predistortion method to address the distortion problem of communication systems due to the memory nonlinearity of PA. Based on FLUT, the proposed method improves the FLUT predistortion structure and simplifies the adaptive update part. It trains the neural network model utilizing a transmission narrowband sequence to compensate the nonlinearities of PA, and uses a two-dimensional filter code table to compensate memory effects. In other words, the proposed method reduces the complexity of model coefficient identification by processing the nonlinearities and memory effects of PA separately. As indicated by simulation results, the improved FLUT method effectively reduces the system's bit error rate, suppresses out-of-band spectrum spreading and mitigates in-band distortion. It yields better linearization effect on the PA with memory effects, compared with the original FLUT method.

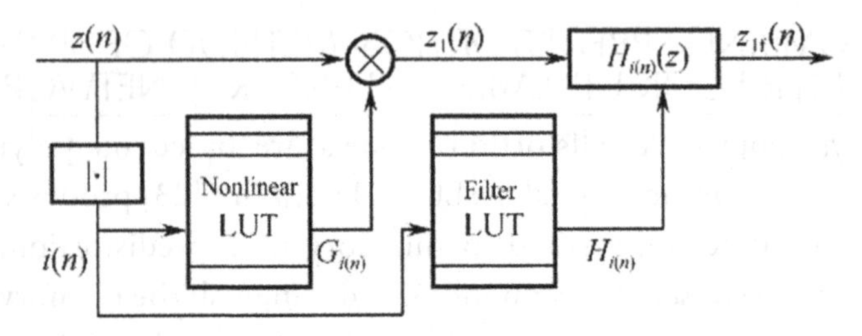

FIGURE 10.30 Block diagram of filter look-up table.

10.6.1 Filter Look-Up Table Predistortion

The FLUT method uses LUT and filter for predistortion simultaneously, taking into account both the nonlinearities and memory effects of PA. Figure 10.30 depicts the FLUT method consisting of an LUT and a filter that are combined for predistortion of the input signal.

The LUT's input and output signals are represented by $z(n)$ and $z_1(n)$, respectively. Their relationship is

$$z_1(n) = G_{i(n)} z(n) \tag{10.35}$$

In Equation (10.35), $i(n)$ determines which entry in the filter will be applied to the LUT-predistorted signal $z_1(n)$. In the filter, the transfer function for the jth entry is given by

$$H_j(z) = \sum_{k=0}^{L-1} h_j(k) z^{-k} \tag{10.36}$$

Therefore, the predistortion signal $z_{1f}(n)$ of the global FLUT system is expressed as

$$z_{1f}(n) = \sum_{k=0}^{L-1} h_{i(n)}(k) z_1(n-k) = \sum_{k=0}^{L-1} h_{i(n)}(k) G_{i(n-k)} z(n-k) \tag{10.37}$$

where $G_{i(n-k)}$ is the gain value indexed in the LUT corresponding to the quantized value of the magnitude $|z(n-k)|$.

Literature [30] proposes a basic structure of FLUT predistortion, as shown in Figure 10.31. FLUT predistortion adopts an indirect learning structure in the feedback branch to enable the amplifier's output signal

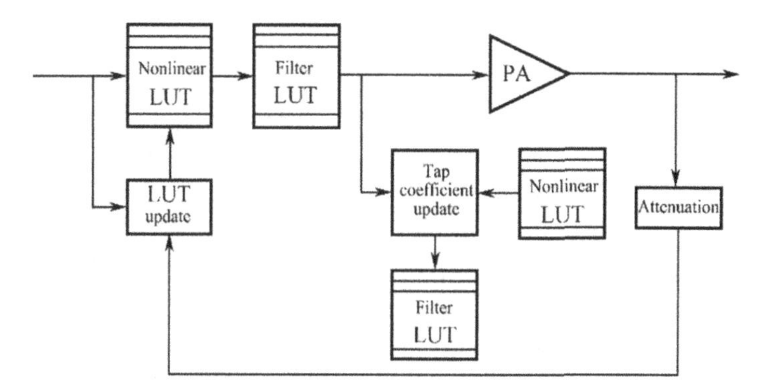

FIGURE 10.31 Basic structure of FLUT predistortion.

to pass through the nonlinear LUT and filter LUT having the same parameter settings; in addition, the nonlinear LUT is updated using a direct learning structure and the filter's tap coefficients are updated using an indirect learning structure. The feedback branch of FLUT predistortion is structurally complicated. The index of the filter LUT is only the magnitude of the current input signal and does not consider the impact of previous inputs.

10.6.2 Predistortion Scheme Combining Improved Filter Look-Up Table and Neural Network

This subsection proposes an improved FLUT predistortion method based on the original FLUT structure. This improved method decomposes the predistorter model containing a memory amplifier into a filter LUT subsystem in series with a neural network subsystem, which compensate the amplifier linear subsystem and memoryless nonlinear subsystem respectively. In addition, the neural network and adaptive filter LUT are both updated by means of a direct learning method. Figure 10.32 depicts the block diagram of the improved method.

The improved FLUT predistortion method applies a swap between the nonlinear LUT and the filter LUT, and replaces the nonlinear LUT with a neural network while Equation (10.37) still applies compensating the memory nonlinearity of PA. At the beginning, switches K1, K2 and K3 connect to the corresponding pin 1 or pin 2. The training sequence generation unit supplies a narrowband ramp training sequence with a bandwidth much smaller than the bandwidth of PA.

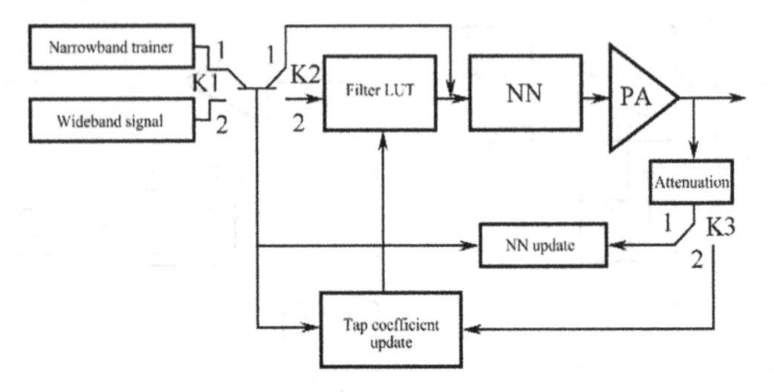

FIGURE 10.32 Block diagram of improved FLUT predistortion.

The neural network compensates the nonlinearities of PA and the neural network's weights are updated by using a direct learning structure. When the error function in the adaptive algorithm gradually converges to a preset threshold, the update of neural network's weights is complete and the nonlinear distortion of PA is compensated. At this time point, switches K1, K2 and K3 connect to pin 2. The filter LUT is enabled to eliminate memory effects of PA and a direct learning structure is used to update the filter LUT, thus compensating the PA distortion caused by memory effects.

Literature [30] holds that the amplifier's nonlinear distortion impacts the accuracy of coefficient update for the adaptive filter, and suggests using an indirect learning structure to update the filter LUT. Compared with Figure 10.31, the improved FLUT predistortion structure uses less one nonlinear LUT and one filter LUT in the feedback branch, notably reducing the system complexity against FLUT predistortion.

The neural network predistortion subsystem adopts a direct learning structure to adaptively update the neural network's weights and thresholds, as shown in Figure 10.33. With the continuous operation of

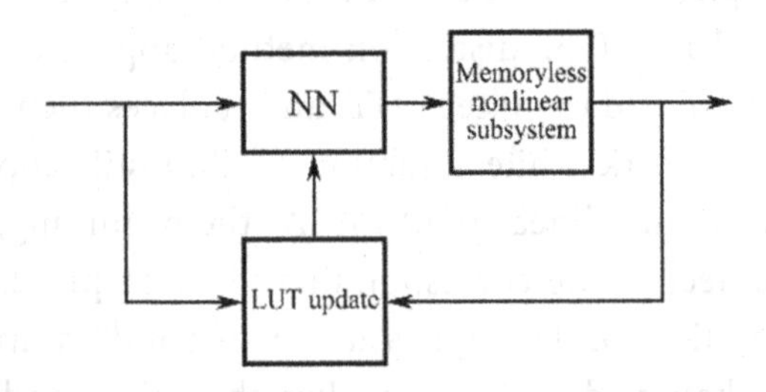

FIGURE 10.33 Neural network prediction subsystem.

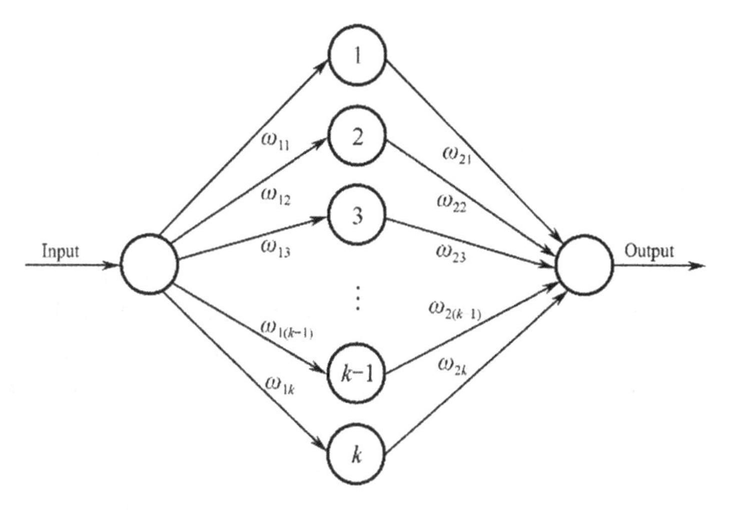

FIGURE 10.34 Single-input single-output three-layer feedforward neural network.

the predistortion system, the error of the adaptive algorithm gradually decreases and the neural network tends to converge. When the neural network eventually converges, the nonlinearities of PA have been compensated. The neural network system only fits the inverse characteristics of the memoryless nonlinear subsystem of PA, so the system can be regarded as an adaptive predistortion system of a memoryless PA. The neural network model adopts a single-input, single-output, three-layer feedforward neural network having a simple structure, as shown in Figure 10.34.

There are multiple adaptive update algorithms available for use in predistortion of a memoryless PA. The neural network predistortion subsystem uses a gradient descent training algorithm containing a momentum factor to update the weights. This algorithm combines a training algorithm containing a momentum factor and the Levenberg-Marquardt (LM) learning algorithm to accelerate the network training process and prevent the network from falling into the local minimum. The calculation formula is given by

$$\boldsymbol{e}(t) = \frac{1}{2}(\boldsymbol{y}(t) - \boldsymbol{d}(t))^2 \tag{10.38}$$

$$\Delta\boldsymbol{\omega}(t) = -\eta\frac{\partial e(t)}{\partial \boldsymbol{\omega}(t)} + \alpha\Delta\boldsymbol{\omega}(t-1) \tag{10.39}$$

where $\boldsymbol{y}$ is the measured network output; $\boldsymbol{d}$ is the desired network output; $\boldsymbol{e}$ is the network output error; η is the network learning rate; α is the momentum factor; η, $\alpha \in [0, 1)$.

FLUT investigations have found that the selection of filter coefficient vectors depends only on the magnitude of the current input signal, and therefore memory effects are not sufficiently characterized. To characterize the inverse memory effects of PA more accurately, the selection of filter coefficient vectors should consider the impact of more input signals. If the memory depth of the predistorter is L, the impact of L input signals will be considered, which makes the indexes of filter coefficient vectors extremely complicated and also introduces an increased quantization error. To solve the contradiction between an accurate characterization of memory effects and a complicated indexing process, this subsection proposes a filter matrix $N \times N_{\text{FIR}}$ to realize the filter LUT predistortion subsystem. Each element of the filter matrix indicates a set of filter coefficient vectors $\boldsymbol{h}_{i,j}$. The filter matrix adopts a two-dimensional indexing technique, which uses two index parameters related to the input sequence to determine the selection of filter coefficient vectors. The first index parameter is the magnitude of the current input signal and the second one is associated with previous inputs. In Literature [23], the second dimensional index in the memory LUT indicates the sum of the magnitudes of previous input signals. Although this method is applicable, it cannot reflect the impact of previous inputs on the current input. Therefore, we use a variable that reflects the impact of previous inputs on the current input to represent the second index parameter. The variable is defined as γ_n, which is quantized as follows to be used as the second index parameter of the filter LUT:

$$\gamma_n = \frac{||\boldsymbol{x}(n) - [\boldsymbol{x}(n-1) + \boldsymbol{x}(n-2) + \cdots + \boldsymbol{x}(n-L)]/L||}{||\boldsymbol{x}(n)||} \tag{10.40}$$

The filter LUT predistortion subsystem adopts a direct learning structure, which is depicted in Figure 10.35. The LMS algorithm is used to adaptively update the filter coefficient $\boldsymbol{h}_{i,j}$, so that the output signal of

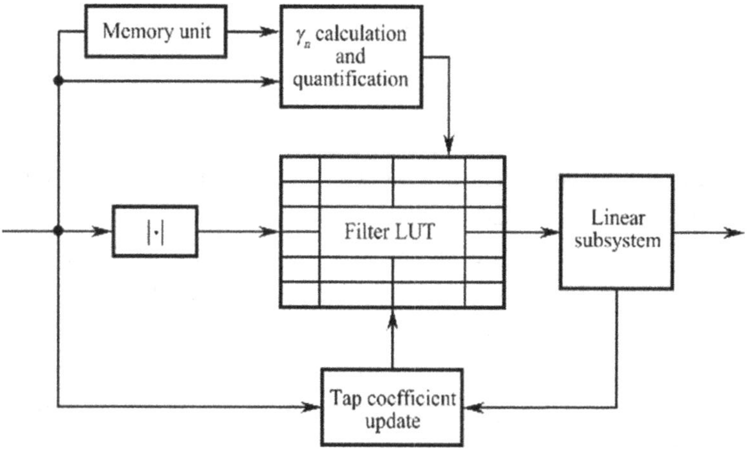

FIGURE 10.35 Block diagram of filter LUT predistortion subsystem.

the filter LUT predistortion subsystem approaches the input signal. When the final output is equal to the input, it indicates that the filter LUT is able to eliminate the distortion caused by memory effects.

The adaptive update formula of LMS algorithm is given by

$$\boldsymbol{h}_{\mathrm{ij}}(n+1) = \boldsymbol{h}_{\mathrm{ij}}(n) + \lambda \boldsymbol{x}(n)\boldsymbol{e}^{*}(n) \tag{10.41}$$

where $\boldsymbol{h}_{i,j} = (\boldsymbol{h}_{i,j}(0), \boldsymbol{h}_{i,j}(1), \cdots, \boldsymbol{h}_{i,j}(L-1))$; $\boldsymbol{x}(n)$ is the input signal of the filter LUT subsystem; $\boldsymbol{e}(n) = \boldsymbol{x}(n) - \boldsymbol{y}(n)$; $\boldsymbol{y}(n)$ is the feedback signal; "$*$" indicates the conjugate; λ is the step size factor.

10.6.3 Experimental Result and Analysis

In the simulation system, the input signal is modulated by 64QAM and then passes through a raised cosine roll-off filter with the roll-off coefficient 0.5 and the up-sampling rate 8. To facilitate simulation, the filtered signal is normalized and experiences 4 dB power back-off. Suppose the hidden layer of the neural network has 5 neurons and the initial weight is 1. For the filter matrix, the number of rows N is set to 16 and the number of columns N_{FIR} is set to 4. The impulse response of the filter is initialized as a unit pulse.

The Saleh model with memory is used as the PA model [32]. A memoryless Saleh model serves as the memoryless nonlinear subsystem of PA with the parameters

$$\begin{cases} f(A) = \dfrac{2A}{1+A^2} \\ g(A) = \dfrac{\pi A^2}{3(1+A^2)} \end{cases} \tag{10.42}$$

The linear subsystem of PA is simulated by a third-order FIR filter with coefficients [0.7932, 0.1824, 0.0793].

Figure 10.36 shows the signal spectrums produced by the FLUT method and the improved method. Both methods effectively suppress out-of-band spectrum spreading. The improved method produces better spectrum sidelobe suppression with an improvement about 10 dB compared to the FLUT method, which means a more effective suppression of adjacent channel interference.

The constellation diagram is a vector diagram showing the co-directional component (horizontal coordinate axis) versus orthogonal component (vertical coordinate axis) of the signal to represent signal characteristics. Figures 10.37 shows the constellation diagrams of output signals when the input signal directly goes through the PA and goes in presence of FLUT predistortion or improved predistortion method, plus the constellation diagram of the original signal.

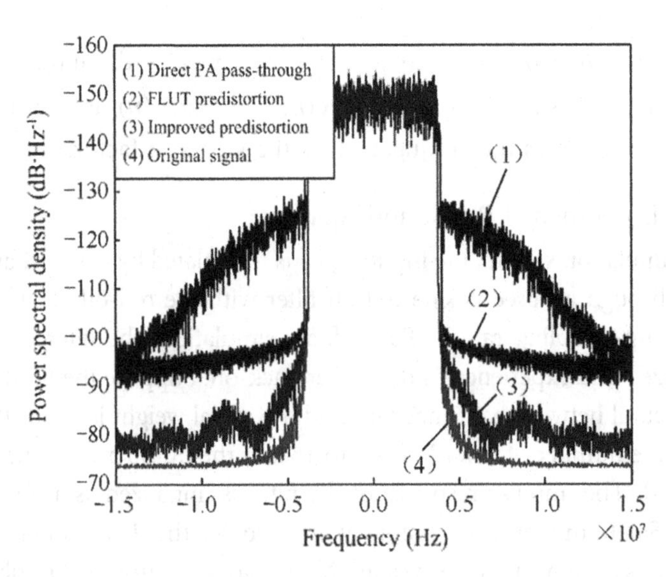

FIGURE 10.36 Signal spectrum comparison.

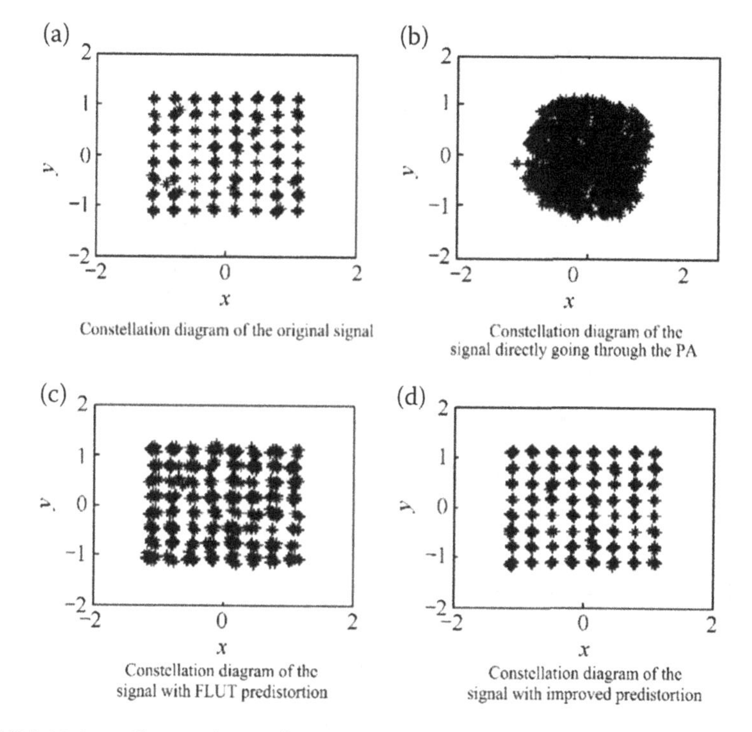

FIGURE 10.37 Comparison of output signal constellation diagrams.

As indicated in Figure 10.37, both methods effectively improve the constellation deflection and diffusion problem due to the nonlinear distortion of PA. Compared with the FLUT method, the improved method yields a better convergence of constellation points, based on which the constellation diagram basically appears very close to the constellation diagram of the original signal.

Figure 10.38 shows the comparison of bit error rates. As indicated in the figure, the bit error rate curves produced by the FLUT method and improved predistortion method both approach the ideal situation, although the improved predistortion method yields a better bit error rate compared with the FLUT method. When the bit error rate is 10^{-3}, the improved predistortion method yields a gain of approximately 1 dB against the FLUT method.

Figures 10.39(a) and (b) depict the amplitude characteristic curves of PA in absence of predistortion, in presence of FLUT predistortion and improved predistortion, respectively. As indicated in the figures,

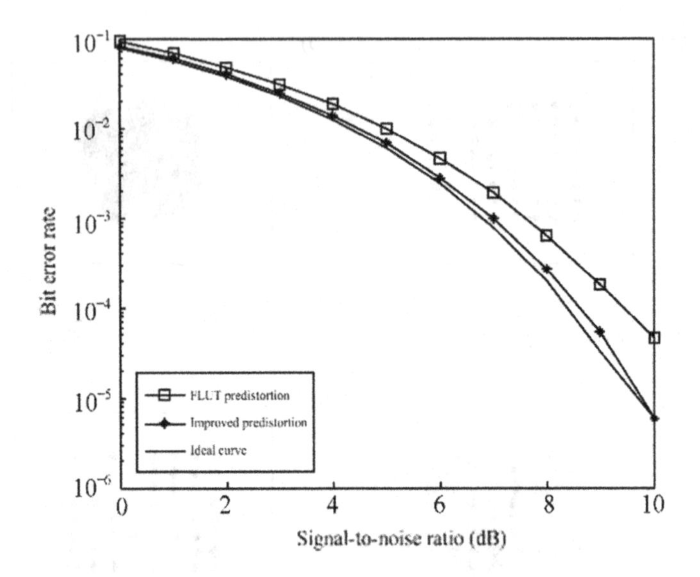

FIGURE 10.38 Comparison of bit error rates.

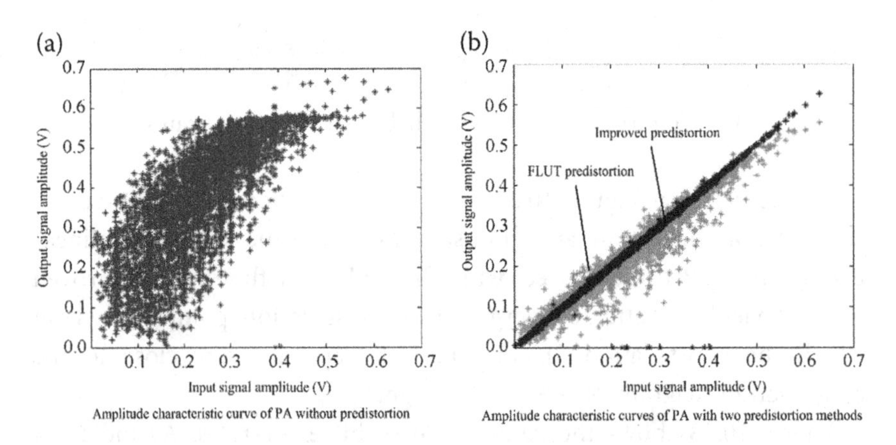

FIGURE 10.39 Comparison of amplitude characteristic curves of PA with and without predistortion.

the nonlinearities of PA are notably improved in presence of predistortion, changing from the original divergence to convergence. In addition, the improved predistortion method produces the amplitude characteristic curve of output signal appearing more like a straight line, which is superior to the amplitude characteristic curve produced by the FLUT method.

FLUT is a favorable method proposed in recent years to eliminate the distortion due to memory effects of PA, but this method is structurally complicated and fails to sufficiently compensate for memory effects of PA. Based on FLUT, this section improves the FLUT predistortion structure and simplifies the adaptive update part. Firstly, the improved method uses a narrowband training sequence to compensate for the nonlinearities of PA through a neural network subsystem. Then, a two-dimensional FLUT is used to include the impact of previous inputs on the current input, thereby accurately compensating for memory effects of PA. Compared with the original FLUT method, the improved method effectively reduces the complexity of the feedback branch, more accurately characterizes the inverse characteristics of nonlinearities of PA having memory effects and improves the overall system performance. In conclusion, the improved method proposed in this section can effectively solve the distortion problem of amplifiers with memory effects in wideband systems.

10.7 ADAPTIVE PREDISTORTION METHOD WITH OFFLINE TRAINING BASED ON BP INVERSE MODEL

In the direct adaptive predistortion learning structure, the adaptive algorithm is easy to fall into the local minimum. Literature [33] proposes an indirect adaptive learning structure using a neural network predistorter to address the low predistortion accuracy of nonlinear PA. However, the indirect structure is easily disturbed by noise, resulting in an unstable system. Based on Literatures [33] and [34], Literature [35] proposes a direct-indirect predistortion learning structure, as shown in Figure 10.40.

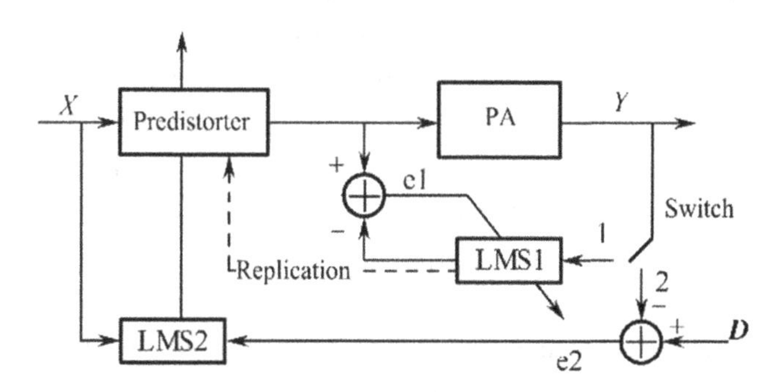

FIGURE 10.40 Direct-indirect predistortion learning structure.

Firstly, the switch connects to 1, so that the PA is modeled using Saleh and inverted using the LMS1 algorithm. Then, the switch connects to 2, so that the coefficients of the established inverse model are copied to the pre-distorter. As the initial value of predistorter in the direct-structure adaptive system, the adaptive algorithm still uses LMS. The method proposed in Literature [35] firstly adopts an indirect learning structure to make the predistorter's initial value close to the best value and jump out of the local optimal solution, thereby effectively preventing the adaptive algorithm used in Literature [33] from easily falling into the local minimum. Then the method adopts a direct structure to enable adaptive predistortion, thereby addressing the vulnerability to noise interference and system instability identified in Literature [34]. However, the Saleh model used in Literatures [33] and [35] requires a rectangular-to-polar coordinate conversion for the signal in engineering implementation. In addition, the Saleh model and LMS algorithm are both required to establish the inverse model of PA in Literature [35], resulting in a slower and more complicated process as well as an unfavorable accuracy of the inverse model. In this situation, the predistorter's initial value significantly differs from the best value, increasing the difficulty in adaptive adjustment of the predistorter.

To further improve the linearity of the nonlinear PA system, this section proposes an adaptive predistortion method with offline training based on inverse modeling of the BP neural network. Firstly, this method uses the BP neural network to implement inverse modeling of PA, and takes the established inverse model parameters as the predistorter's initial values. Secondly, this method performs offline training on the predistorter by using the BP inverse model prior to the establishment of the adaptive predistortion system to improve the predistorter's linearization effect in the initial predistortion system and accelerate the adaptive process of the predistortion system. This processing improves the accuracy and stability of the system. Finally, the method adjusts the weights of the neural network predistorter by using a direct structure and LMS algorithm, thereby eliminating the nonlinear perturbations of the amplifier. As indicated by simulation results, the proposed method reduces the adjacent channel intermodulation power by approximately 18 dB, compared with a mere decline of 8 dB yielded by the classical direct-indirect predistortion learning structure.

10.7.1 Adaptive Predistortion Method with Offline Training Based on BP Neural Network

BP neural network is a combination of multilayer perceptron (MLP) neural network and BP algorithm. BP neural network training includes two procedures: forward propagation of information and backward propagation of error. Since a two-layer BP neural network with adjustable parameters can approach any nonlinear function by means of learning and its computation amount basically increases linearly with the growth of function complexity, the two-layer BP neural network is adopted with the hidden layer containing 10 neurons. The transfer function is a tangent S-function and the output layer uses a linear transfer function.

The process for inverse modeling of PA using a BP neural network is shown in Figure 10.41. In the figure, PA refers to the amplifier's forward model, which simply exchanges the input and output of the inverse model. The forward model and inverse model have exactly the same structure. $\boldsymbol{X}(t)$ and $\boldsymbol{Y}(t)$ denote the input and output data of PA, respectively. A real-valued BP neural network is applied. That is, the input $\boldsymbol{X}(t)$ includes input signals $\boldsymbol{I}_1(t)$ and $\boldsymbol{Q}_1(t)$, and the output $\boldsymbol{Y}(t)$ includes output signals $\boldsymbol{I}_2(t)$ and $\boldsymbol{Q}_2(t)$. Consequently, the network structure is 2–10-2, as shown in Figure 10.42. This network structure reduces the computational complexity. In Figure 10.42, $\boldsymbol{Y}(t) = [\boldsymbol{I}_2(t), \boldsymbol{Q}_2(t)]$ is the input to the inverse model of the neural network, and its output is $\boldsymbol{H}(t)$ consisting of $\boldsymbol{I}'_1(t)$ and $\boldsymbol{Q}'_1(t)$. $\boldsymbol{E}(t)$ denotes the difference between $\boldsymbol{H}(t)$

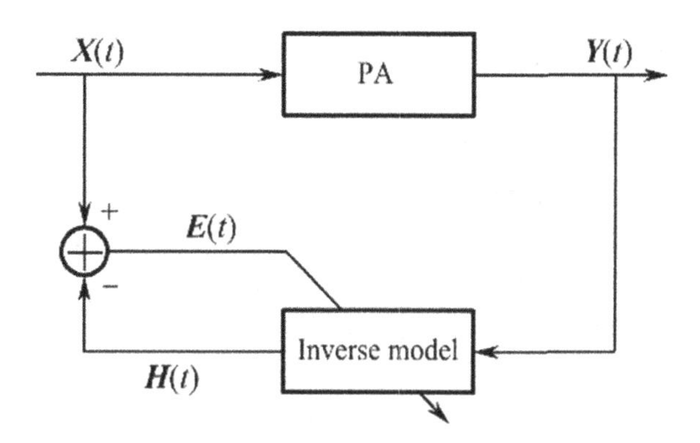

FIGURE 10.41 Inverse modeling process of PA.

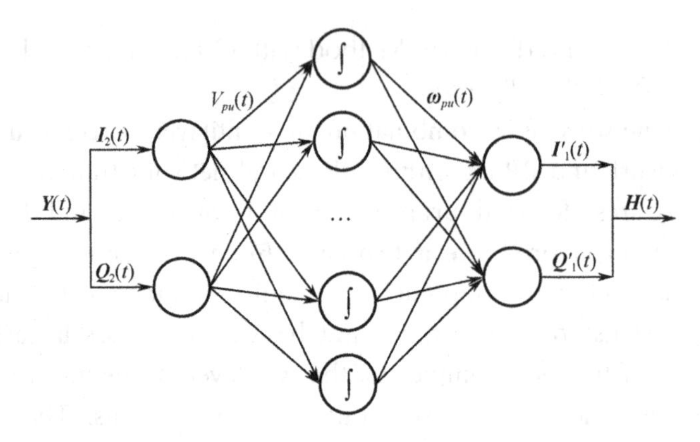

FIGURE 10.42 Network structure diagram of BP inverse model.

and $\boldsymbol{X}(t)$, and the difference is used to train the inverse model until the value of $\boldsymbol{E}(t)$ reaches the desired accuracy or the iterative process ends.

As indicated by Figure 10.42, the BP network output is represented as

$$\boldsymbol{I'}_1(t) = \sum_u \omega_{1u}(t)\boldsymbol{Z}_u(t) - \boldsymbol{\theta}(t) \tag{10.43}$$

$$\boldsymbol{Q'}_1(t) = \sum_u \omega_{2u}(t)\boldsymbol{Z}_u(t) - \boldsymbol{\theta}(t) \tag{10.44}$$

where $\omega_{pu}(t)$ $(p = 1, 2)$ is the weight value from the uth node in the hidden layer to the pth node in the output layer; $\boldsymbol{\theta}(t)$ is the threshold value of the output layer; $\boldsymbol{Z}_u(t)$ is the output of the uth node in the hidden layer, which is represented as

$$\boldsymbol{Z}_u(t) = \tan\text{sig}[v_{u1}(t)\boldsymbol{I}_2(t) + v_{u2}(t)\boldsymbol{Q}_2(t) - \boldsymbol{\theta}'(t)] \tag{10.45}$$

where $\boldsymbol{\theta}'(t)$ is the threshold value of the hidden layer; $v_{up}(t)$ $(p = 1, 2)$ is the weight value of the pth input to the uth node in the hidden layer. To overcome the weaknesses of BP algorithm (such as slow convergence and falling into the local minimum easily), the LM algorithm is used to calculate the update of $v_{up}(t)$ and $\omega_{pu}(t)$ by the minimum mean square error of $E(t)$.

The inverse model of PA describes the inverse characteristics of PA. Therefore, if the amplifier's ideal output is transferred to the inverse model, the inverse model will yield the amplifier's ideal input.

In a direct learning structure, the predistorter parameters are immune to the noise at the output of PA but the involved computation amount is extremely large. In addition, the direct use of adaptive algorithm results in falling into the local minimum. Although the indirect learning structure is simple and easy in engineering implementation, the additive noise of PA in the indirect structure makes the inverse model parameters deviate from the best values, thus deteriorating the linearization effect of predistortion. Therefore, this subsection proposes a predistortion method that combines the two structures plus an improvement.

If the input of the amplifier is directly sent to the predistorter that uses the inverse model coefficients as its initial values, the predistortion system cannot realize linear amplification of the signal, since the inverse model exchanges the input and output of the forward model of PA. Consequently, offline training is applied to the predistorter prior to direct adaptation. Figure 10.43 shows the offline training process.

The coefficients of the neural network inverse model established in the previous section are taken as the predistorter's initial values. In Figure 10.43, $\boldsymbol{D}(t)$ denotes the ideal output of the amplifier at moment t, that is, the amplifier's linear output $\boldsymbol{D}(t) = k\boldsymbol{X}(t)$ yielded when the amplifier input is determined (k is the amplifier's linear gain). In the figure, $\boldsymbol{X}'(t)$ is the predistorter-processed output of $\boldsymbol{X}(t)$, and $\boldsymbol{Y}'(t)$ is the amplifier output with $\boldsymbol{X}'(t)$ as the input. Assume that the transfer functions of the inverse model and amplifier are $G_{xy}^{-1}(\ )$ and $G_{xy}(\ )$, respectively. With $\boldsymbol{D}(t)$ representing the input to the inverse model, the inverse model output $\boldsymbol{D}'(t)$ can be expressed as $\boldsymbol{D}'(t) = G_{xy}^{-1}(\boldsymbol{D}(t))$.

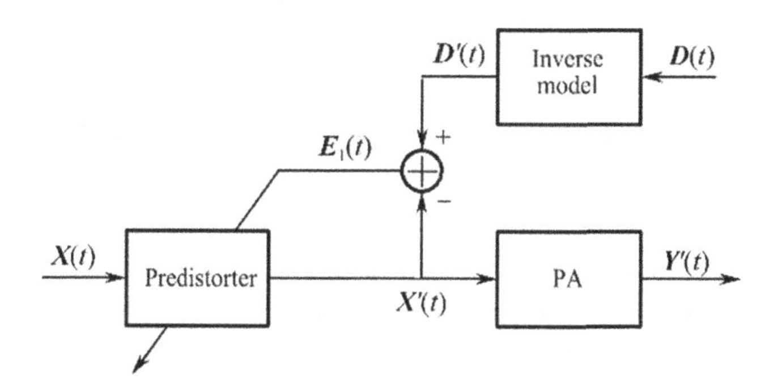

FIGURE 10.43 Offline training on predistorter.

Assuming that the predistorter module is absent in Figure 10.43 and $\boldsymbol{D}'(t)$ directly serves as the amplifier input, the amplifier output $\boldsymbol{Y}'(t)$ in this situation can be expressed as

$$\boldsymbol{Y}'(t) = G_{xy}(\boldsymbol{D}'(t)) = G_{xy}(G_{xy}^{-1}(\boldsymbol{D}(t))) = \boldsymbol{D}(t) \tag{10.46}$$

That is, the amplifier output is the ideal amplifier output $\boldsymbol{D}(t)$. Therefore, the PA can achieve an ideal linear output in the presence of the predistorter module (as shown in Figure 10.43), only if the predistorter output $\boldsymbol{X}'(t)$ equals the inverse model output $\boldsymbol{D}'(t)$. Specifically, the weight coefficient of the predistorter is fine-tuned to minimize the difference $\boldsymbol{E}_1(t)$ between $\boldsymbol{X}'(t)$ and $\boldsymbol{D}'(t)$. The objective of the BP neural network training process is to yield the minimum mean square error of $\boldsymbol{E}_1(t)$. The mean square error of $\boldsymbol{E}_1(t)$ is given by

$$E\{(\boldsymbol{E}_1(t))^2\} = E(\boldsymbol{D}'(t) - \boldsymbol{X}'(t))^2 \tag{10.47}$$

In Equation (10.47), $\boldsymbol{D}'(t)$ is the inverse model output of $\boldsymbol{D}(t)$, which is known data; $\boldsymbol{X}'(t)$ is the predistorter output of $\boldsymbol{X}(t)$. Therefore, $\boldsymbol{X}'(t)$ is a function of the weight coefficient of the neural network predistorter, and the mean square error of $\boldsymbol{E}_1(t)$ is a quadratic function of the weight coefficient of the predistorter. The weight coefficient is fine-tuned to minimize the mean square error, eventually realizing the linear output of the amplifier.

Predistorter and amplifier use the same BP neural network structure for inverse modeling, and the neural network processes data in a parallel and distributed manner, featuring simple computation and fast convergence. Therefore, offline training immediately improves the predistorter's accuracy, making the subsequent direct adaptive process faster and better.

Offline training is designed to make the predistorter's characteristics infinitely approximate to the amplifier's inverse characteristics. However, the real PA system is vulnerable to temperature variation, device aging or noise at the output and may produce perturbations. Therefore, adaptive control must be performed to correct the nonlinearities of the amplifier in real time, thereby further improving the system performance to deliver real practical value.

The above offline training process of predistorter based on the inverse model is an indirect learning structure. Beyond that, adaptive predistortion of the direct learning structure is also implemented to realize online PA tracking by the predistorter. Figure 10.44 depicts the specific process for adaptive predistortion of the direct learning structure. The offline-trained predistorter is used as the nonlinear controller of the amplifier system and placed in front of the D/A converter. Since the modulator, demodulator, A/D converter, D/A converter and oscillator have nothing to do with predistortion, these parts can be omitted from Figure 10.44. In the direct learning structure, the amplifier's output $\boldsymbol{Y}'(t)$ is divided by the linear gain k and then compared with the expected downlink signal $\boldsymbol{X}(t)$; the deviation $\boldsymbol{E}_2(t)$ is input into the adaptive algorithm and the parameters of the neural network predistortion controller are fine-tuned to achieve the adaptive linearization of the amplifier. If there is noise $\boldsymbol{n}(t)$ at the output of the amplifier, then the predistorter's parameters are fine-tuned in an online manner by the adaptive algorithm through the feedback loop to eliminate the noise interference.

Self-adaptation means that the adaptive algorithm automatically tunes the weight coefficient $\omega_{pu}(t)$ of the neural network predistorter based on the magnitude of the error $\boldsymbol{E}_2(t)$ in Figure 10.44, thereby minimizing the mean square error. The most widely used adaptive algorithm is the

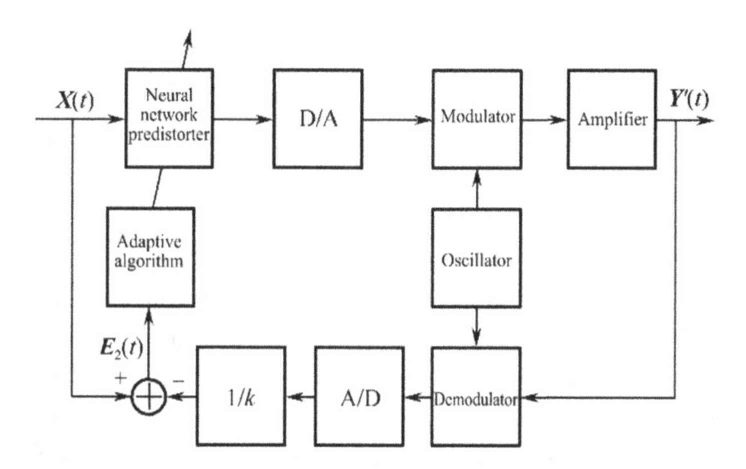

FIGURE 10.44 Specific adaptive process of direct structure.

descent algorithm, which has two implementations, that is, adaptive gradient algorithm and adaptive Gauss-Newton algorithm. The former includes the LMS algorithm and various improved types, and the latter includes the RLS algorithm and various improved algorithms. LMS is simpler in structure and less in computational complexity compared with RLS, so LMS is used as the adaptive algorithm in the improved structure.

Both the improved structure and Literature [35] use a combination of the direct structure and the indirect structure. Additionally, offline training on the predistorter in the improved structure makes the predistorter's initial value close to the best value, thereby overcoming the problem of the LMS algorithm falling into the local optimal solution and also preventing the system from noise pollution. In the adaptive control system, the feedback signal $\boldsymbol{Y}'(t)$ at moment t divided by the gain k yields the signal $\boldsymbol{U}(t)$, and then the deviation $\boldsymbol{E}_2(t)$ between $\boldsymbol{U}(t)$ and $\boldsymbol{X}(t)$ is input to the adaptive algorithm. The LMS algorithm fine-tunes the weight values $\omega_{pu}(t)$ in the output layer of the neural network predistorter by minimizing the mean square error of $\boldsymbol{E}_2(t)$. The weight update formula is given by

$$\omega_{pu}(t+1) = \omega_{pu}(t) + 2\mu \boldsymbol{E}_2(t)\boldsymbol{Z}_u(t) \tag{10.48}$$

where $\boldsymbol{Z}_u(t)$ is the output signal of the uth neuron in the hidden layer of the BP neural network; $\omega_{pu}(t+1)$ and $\omega_{pu}(t)$ are the weights from the uth neuron in the hidden layer to the pth neuron in the output layer of the BP neural network at the $t+1$th and tth iterations, respectively; μ is the learning step size determining the convergence speed of the algorithm. The system will be stable only when μ meets the convergence criterion $0 < \mu < \frac{1}{\lambda_{\max}}$, where $\lambda_{\max}$ is the maximum eigenvalue of the autocorrelation matrix $\boldsymbol{R}_{ZZ}$ for $\boldsymbol{Z}_u(t)$.

10.7.2 Experiment and Comparative Analysis

The experiment uses 2,000 sets of real-valued output data of the amplifier as the BP neural network input and 2,000 sets of real-valued input data of the amplifier as the BP neural network output to train the neural network. The amount of training data should not be too large; otherwise, the modeled network and training will be extremely complicated.

Note, however, that the amount of training data should not be too small either, lest the training accuracy declines. In addition, approximately 500 sets of real-valued I/O data are collected to test the accuracy of the trained BP network. Figure 10.45 shows the data training results plus a local zoom-in part and Figure 10.46 shows a zoom-in of the result error, for a clearer view of the accuracy of the neural network.

In Figure 10.45, the upper solid line represents the measured output, the line of triangles represents the calculated output, the lower solid line represents the error curve, and the rectangular box at the top left provides a local zoom-in view. Figure 10.46 shows a zoom-in of the result error. As indicated in simulation results, the BP neural network produces a very high fitting degree to the inverse characteristics of PA. The error between training samples and test samples is basically maintained within 0.1 and the root mean square error (RMSE) is 0.0425.

This subsection provides the comparison results of the improved direct-indirect structure against the classical direct-indirect structure proposed in Literature [35] by means of MATLAB® software simulation, thereby validating the high efficiency of the improved structure. The power spectrum comparison is shown in Figure 10.47 and constellation diagram comparison is shown in Figure 10.48.

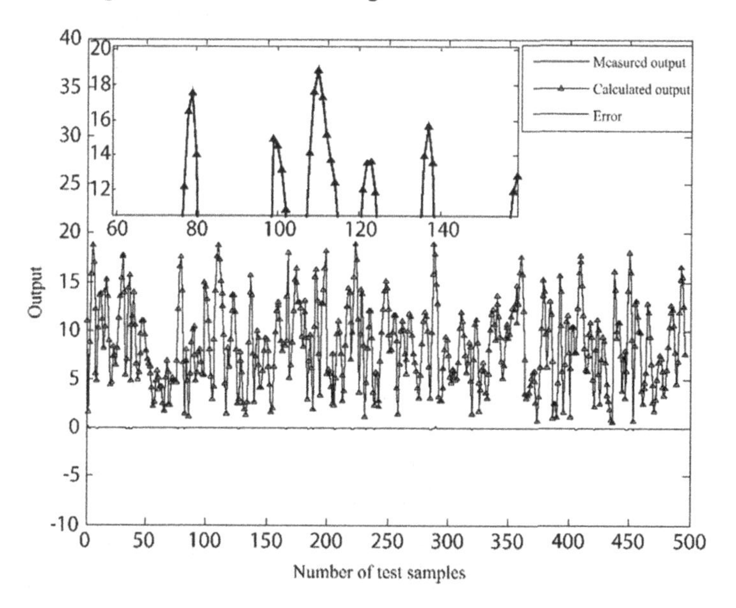

FIGURE 10.45 Data training results plus a local zoom-in part.

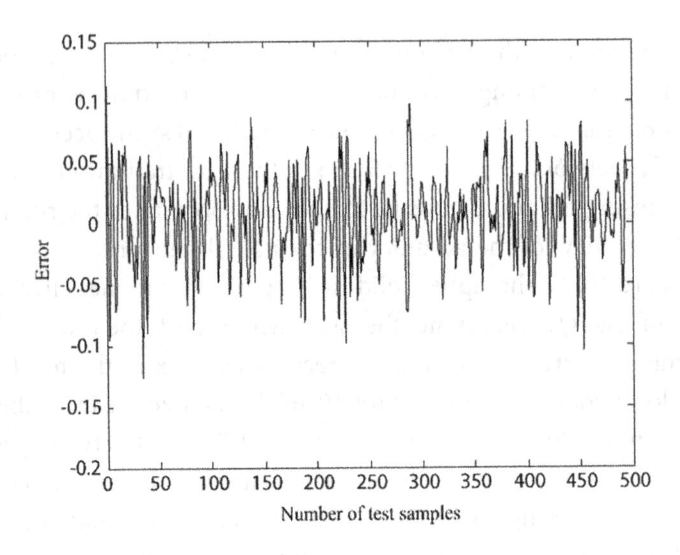

FIGURE 10.46 Result error zoom-in.

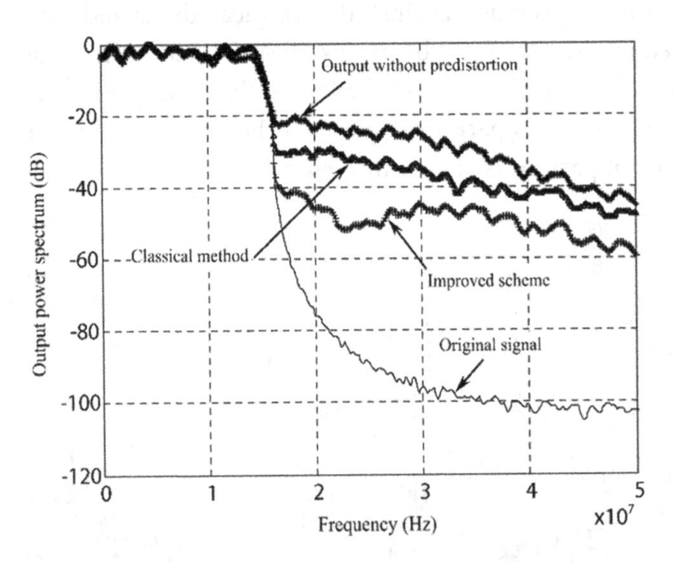

FIGURE 10.47 Power spectrum comparison in adaptive predistortion.

When there is additive white Gaussian noise in the transport channel, the input signal is modulated by 16QAM. Figure 10.47 shows the power spectrum comparison before and after predistortion with approximately 4 dB power back-off. The top curve of asterisks represents the power spectral density of PA without predistortion; the

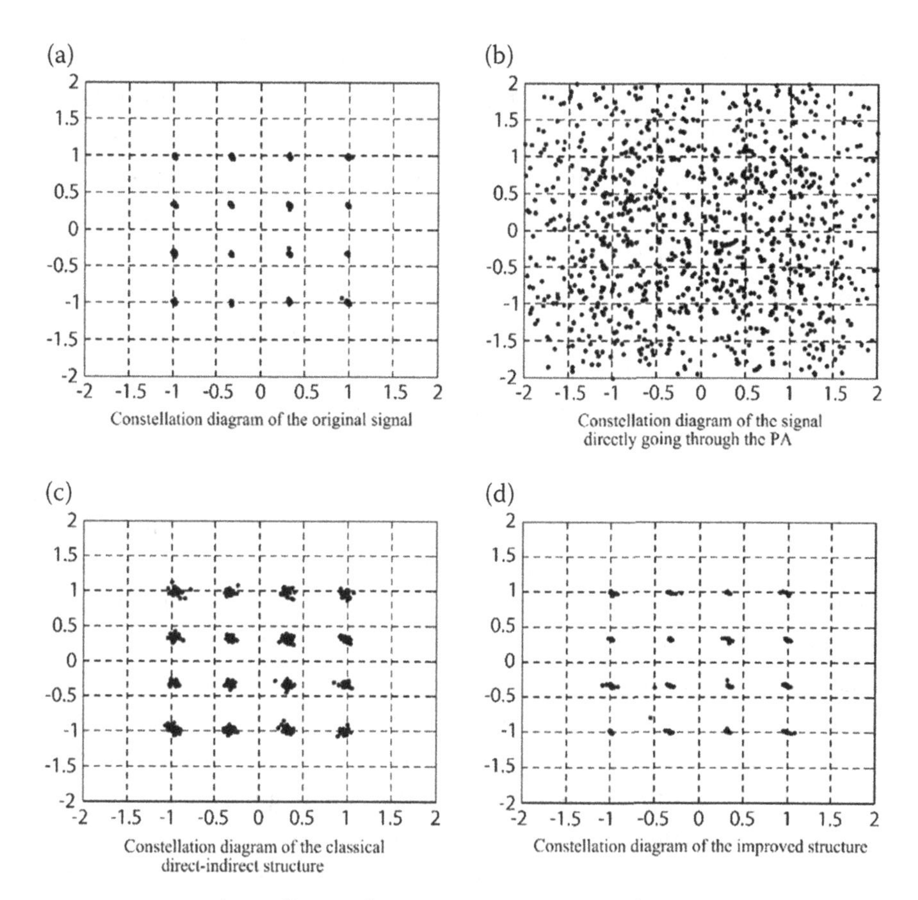

FIGURE 10.48 Constellation diagram comparison in adaptive predistortion.

second-lower curve of triangles represents the power spectral density of the classical direct-indirect structure; the third-lower curve of crosses represents the power spectral density of the improved scheme; the bottom curve represents the power spectral density of the original signal. As seen from the figure, the improved scheme reduces the adjacent channel intermodulation power by approximately 18 dB, compared with a decline of approximately 8 dB yielded by the classical direct-indirect structure. Therefore, the improved adaptive predistortion method proposed in this section delivers better performance in suppression of the out-of-band spectrum spreading, compared with the classical direct-indirect learning structure. Figure 10.48(a) shows the constellation diagram of the original signal. Figure 10.48(b) shows the constellation diagram when the signal directly passes through the

PA. Figure 10.48(c) shows the constellation diagram produced by the classical direct-indirect structure proposed in Literature [35]. Figure 10.48(d) shows the constellation diagram produced by the improved structure. As indicated in simulation results, Literature [35] and the improved structure have well corrected the constellation diagram of the signal directly passing through the PA. Particularly, the constellation diagram of the improved structure appears clearer and closer to the constellation diagram of the original signal, indicating that the improved structure better compensates the impact of additive noise and the predistorter's characteristics are closer to the amplifier's inverse characteristics. The improved structure has proven to produce a better linearization effect.

The characteristics of PA may vary due to ambient temperature, device aging or noise at the output of the amplifier. It is necessary to add adaptive adjustment techniques to the predistortion system. This section proposes an improved scheme designed to address the accuracy problem of the adaptive predistortion system. The improved scheme applies offline training on the predistorter to make it approximate to the ideal inverse characteristic of the amplifier, and then the trained predistorter is added to the adaptive predistortion system. The offline training process is based on the inverse modeling of PA with the BP neural network, which features fast modeling and high accuracy. The adaptive process uses the LMS algorithm to fine-tune the predistorter coefficients, thereby eliminating the nonlinear perturbations of the amplifier. Finally, the improved scheme and the classical direct-indirect learning structure are compared by means of MATLAB® software simulation. Simulation results have validated the feasibility and high efficiency of the improved scheme.

10.8 POWER AMPLIFIER PREDISTORTION METHOD BASED ON ADAPTIVE FUZZY NEURAL NETWORK

This section proposes an adaptive predistortion method using a double-loop learning structure based on fuzzy neural network model identification to address the low accuracy of predistortion structure for memory nonlinear PA in a wireless communication system. The proposed method combines the indirect learning structure and direct learning structure based on model identification, using the double-loop learning structure predistortion theory and fuzzy neural network model proposed in Literature [29]. On the basis of the real-valued time-delayed Adaptive Neural Fuzzy Inference System (ANFIS) model, the improved method firstly uses the simplified PSO algorithm to implement offline training on the ANFIS model for a coarse extraction of model parameters, which are used as the initial values of the predistorter. The improved method sufficiently considers the fuzzy inference ability of ANFIS and simplifies the model structure. Also, global optimization of the improved simplified PSO algorithm is utilized to accelerate the convergence speed and avoid falling into the local optimum. Then, the improved method uses the LMS algorithm to implement online training of a direct learning structure on the coarse extracted model, enabling the model parameters to be adaptively fine-tuned. In this way, the model has improved accuracy and anti-noise interference ability. As indicated by simulation results, the improved method produces a much better linearization effect with fast convergence speed and high accuracy, compared with the method proposed in Literature [29]. In addition, the improved method has a wider application scope.

10.8.1 Fuzzy Neural Network Model Structure

Fuzzy neural network has been widely applied to nonlinear system modeling. It combines the superiorities of neural network and fuzzy logic. As a fuzzy neural network, ANFIS has been used to construct the behavioral model of RF PA. In this subsection, ANFIS is proposed to be used in the model design of predistortion system. Since the PA model and the predistorter model are inverse to each other, the fuzzy neural network structure is modeled by using the real-valued time-delayed PA model.

Nonlinearities of PA with memory effects can be expressed as

$$I_{out}(n) = f\,[I(n), I(n-1), \cdots, I(n-m), Q(n), Q(n-1), \cdots, Q(n-k)] \tag{10.49}$$

$$Q_{out}(n) = g\,[I(n), I(n-1), \cdots, I(n-m), Q(n), Q(n-1), \cdots, Q(n-k)] \tag{10.50}$$

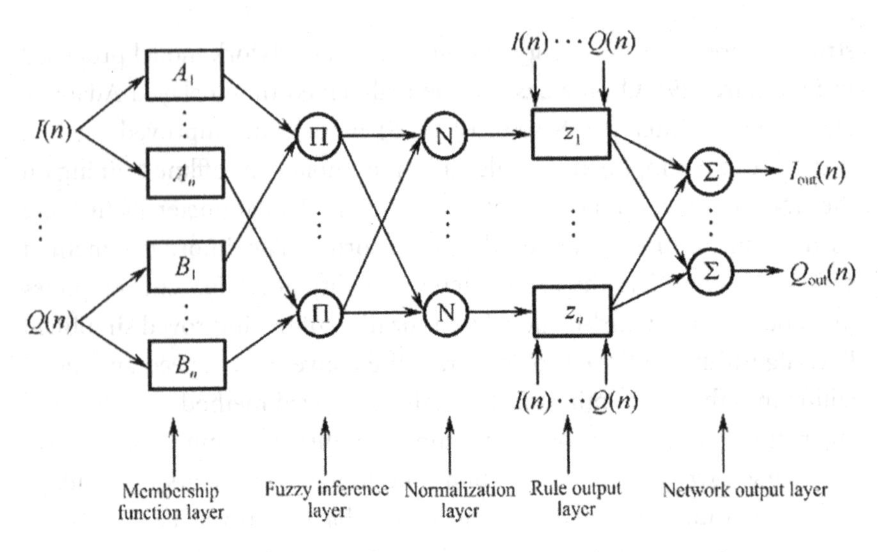

FIGURE 10.49 Simplified ANFIS structure.

where $I(n)$ and $Q(n)$ are in-phase and quadrature components of the input signal at moment n, respectively; $I_{\text{out}}(n)$ and $Q_{\text{out}}(n)$ are in-phase and quadrature components of the output signal at moment n, respectively; $f(\cdot)$ and $g(\cdot)$ are nonlinear distortion functions of PA; m and k are memory depths.

The first-order TS-type adaptive fuzzy inference system model structure is utilized to fit the nonlinear function. The ANFIS model is further simplified to reduce the model complexity. Figure 10.49 outlines the ANFIS model structure.

The structure of a traditional feedforward neural network is determined artificially, subject to arbitrariness and irrationality. That is, the structure is determined by the prior knowledge of experts and therefore an irrational structure is likely to occur. In contrast, the subtractive clustering algorithm [36] does not require the predefined number of clusters, and automatically clusters the samples into several categories based on the sample characteristics and quickly determines the cluster center, so that the number of fuzzy rules and the network structure are more rational. Therefore, this subsection adopts subtractive clustering to determine the number of rules in a fuzzy neural network for a simplified model structure and utilizes the cluster center to determine the membership function's initial parameters. Firstly, the cluster center and the number of clusters are dynamically adjusted to yield better input space division. Then, the number of fuzzy rules is determined by the number of clusters, the initial value of membership function center is determined by each cluster center, and the initial value of membership function width is determined by the Euclidean distance formula. In this way, dimension curse is prevented and the network structure is made simple and reliable.

For establishment of a predistorter model based on the real-valued time-delayed fuzzy neural network, the in-phase and quadrature components of the PA's output signal directly serve as the input, and the in-phase and quadrature components of the PA's input signal serve as the output, without considering the mutual conversion between the in-phase/quadrature component and amplitude/phase of the complex signal. In addition, the coupling effect between the in-phase and quadrature components is considered to overcome the distortion due to modulation-demodulation imbalance. The fuzzy neural network model proposed in this subsection can simultaneously characterize the nonlinearities and memory distortion of PA.

10.8.2 New Method for Adaptive Predistortion

Generally, the digital predistortion technique adopts either a direct learning structure or an indirect learning structure to yield the predistorter's model parameters. Considering the advantages and disadvantages of direct and indirect learning structures, Literature [29] proposes a classical double-loop predistortion structure combining the indirect and direct learning structures based on the direct inverse. Firstly, the indirect learning structure is used for a coarse extraction of the predistorter's model parameters; then the direct learning structure is used to fine-tune the parameters for an improvement of model accuracy. The model parameters are eventually determined by using the least square method twice. Although this method outperforms the predistortion method with an indirect learning structure, the modeling process of this method requires a large computation amount and the predistorter model is complicated. Consequently, the predistorter parameters are quite different from the best parameters after coarse extraction, thus increasing the difficulty in adaptive parameter fine-tuning. This method has a low convergence speed and an unfavorable model accuracy, although it prevents the local optimum problem.

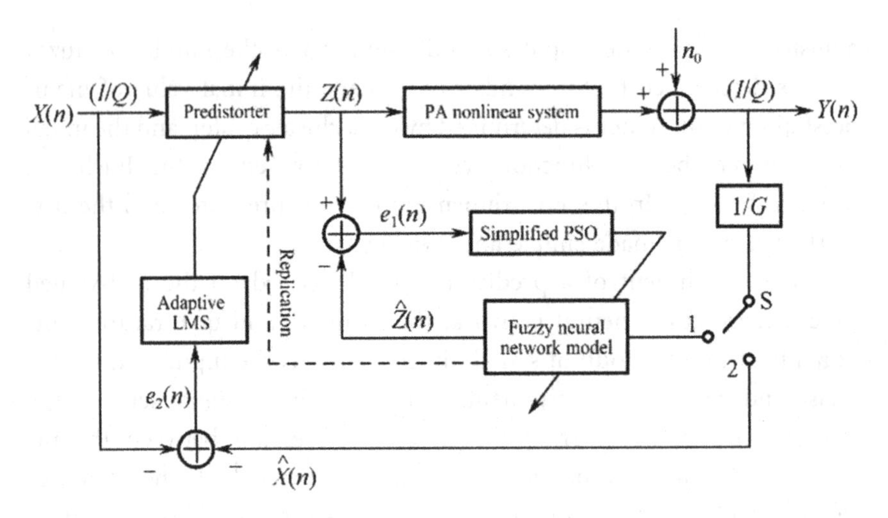

FIGURE 10.50 New double-loop adaptive predistortion system structure.

To address these weaknesses, the subsection proposes a double-loop adaptive predistortion method combining the indirect and direct learning structures based on model identification. The proposed method adopts an improved simplified PSO algorithm in the indirect learning structure to implement offline identification of the mentioned fuzzy neural network model for a coarse extraction of initial values of the model. Next, the proposed method uses a direct learning structure to enable online adaptive fine-tuning of the model coefficients, thereby improving the model accuracy. Figure 10.50 depicts the structure of the new double-loop adaptive predistortion system.

The PA nonlinear system in the structure consists of D/A converter, modulator, power amplifier, demodulator and A/D converter. Only the nonlinearities of PA are considered in the PA nonlinear system. The predistortion system contains an additive noise n_0. Firstly, switch S connects to 1, and the simplified PSO algorithm is used in the indirect learning structure to implement offline identification of the fuzzy neural network model. Then, switch S connects to 2, and online adaptive fine-tuning of the model coefficients is enabled in the direct learning structure.

Indirect learning structure includes direct inverse indirect learning structure and indirect learning structure based on model identification. The predistorter model accuracy of indirect learning structure based on model identification depends on the PA model accuracy and the error of its inverse model algorithm [38]. Due to slow convergence and low accuracy of the least square method, this subsection proposes to use the improved simplified PSO algorithm to offline train the fuzzy neural network model for identification of the model parameters. The extracted model parameters are taken as the predistorter's initial values. The improved simplified PSO algorithm features a global search ability, fast convergence, small number of iterations and high accuracy. Consequently, the initial predistorter model yielded by the indirect learning structure has sufficiently approached the ideal inverse model of PA.

Literature [37] proposes a simplified PSO algorithm from the model's perspective. The simplified algorithm removes the particle velocity term and is able to find the optimal solution iteratively with only the particle position term. Consequently, the algorithm becomes simpler and more efficient. On this basis, this subsection introduces an improved simplified PSO algorithm, which enriches the diversity of populations, accelerates the convergence speed and avoids the "premature phenomenon". Considering information sharing among particles, the improved simplified PSO algorithm introduces the term of optimal candidate solution for random individuals, and also uses the inertia factor, learning factor and Laplace coefficient to regulate the impact of "previous moment value", "self-learning part" and "mutual learning part", thereby diversifying the search directions and improving the training efficiency.

The improved simplified PSO algorithm performs the following steps to optimize the fuzzy neural network model.

Step 1: Input the sample data of PA. Determine network structure and initial parameters by means of subtractive clustering.

Step 2: Randomly initialize the particle swarm according to the constraints. Determine the position vector of each particle in the swarm.

Step 3: Calculate the fitness of each particle in the swarm. $E = f(x_i^t)$ serves as the fitness function. Set the ith particle's p_i^t as the current position of the particle. Set p_g^t as the position of the optimal particle in the swarm.

Step 4: Update the individual extremum and global extremum. In the current iteration, if the fitness E of particle i is better than the fitness of the individual extremum p_i^t, p_i^t equals the particle's current position x_i^t. If the fitness E of particle i is better than the fitness of the global extremum p_g^t, p_i^t equals the particle's current position x_i^t.

Step 5: Update the variables such as inertia factor, learning factor and Laplace coefficient based on Equations (6–24), (6–26), (6–27) and (6–28). Update the position of each particle based on Equation (6–29).

Step 6: Determine whether the termination criteria are satisfied. If so, the algorithm terminates. The global optimal solution p_g^t and the corresponding global optimal value E are yielded. Otherwise, return to Step 3 to continue the search.

The predistorter model and PA model can adopt the same fuzzy neural network structure, and the fuzzy neural network has the ability to process data in a parallel and distributed manner. Therefore, the indirect learning structure is able to establish the predistorter model and PA model at the same time. Offline training makes the predistorter's characteristics sufficiently approach the amplifier's inverse characteristics. It simplifies the complexity of system structure, accelerates the modeling process and improves the model accuracy.

Although the indirect learning structure based on model identification makes the predistorter's initial model sufficiently approach the ideal model, the predistortion effect cannot be guaranteed because the real PA is easily disturbed by temperature variation, device aging and additive noise (such as noise at the output). This weakness can be addressed by the direct learning structure, so the adaptive algorithm is used for online fine-tuning of model parameters to further improve the system performance.

To compensate for the impact of noise interference, the offline-identified model parameters are taken as the predistorter's initial values in the direct learning structure, and the LMS algorithm is used for online training to fine-tune the model parameters. In addition to offline extraction of the high-accuracy model, the proposed method enables adaptive fine-tuning of model parameters using the LMS algorithm based on gradient search, which makes the adaptive convergence process of direct learning structure faster and better.

The LMS algorithm fine-tunes the weight values of the fuzzy neural network predistorter by minimizing the mean square error of $e_2(n)$ indicated in Figure 10.50. The update formula is given by

$$\omega(n+1) = \omega(n) + 2\mu e_2(n) p(n) \tag{10.51}$$

where ω denotes the weight; μ denotes the learning step size; e_2 denotes the difference between the real input and ideal input; p denotes the input to each layer of the network.

As described above, the adaptive predistortion process of double-loop learning structure based on the identification of the fuzzy neural network model is shown in Figure 10.51.

10.8.3 Experimental Validation Analysis

This subsection describes a simulation validation for the predistortion system of nonlinear PA with memory effects on the MATLAB® platform, thereby validating the correctness of the proposed predistortion method.

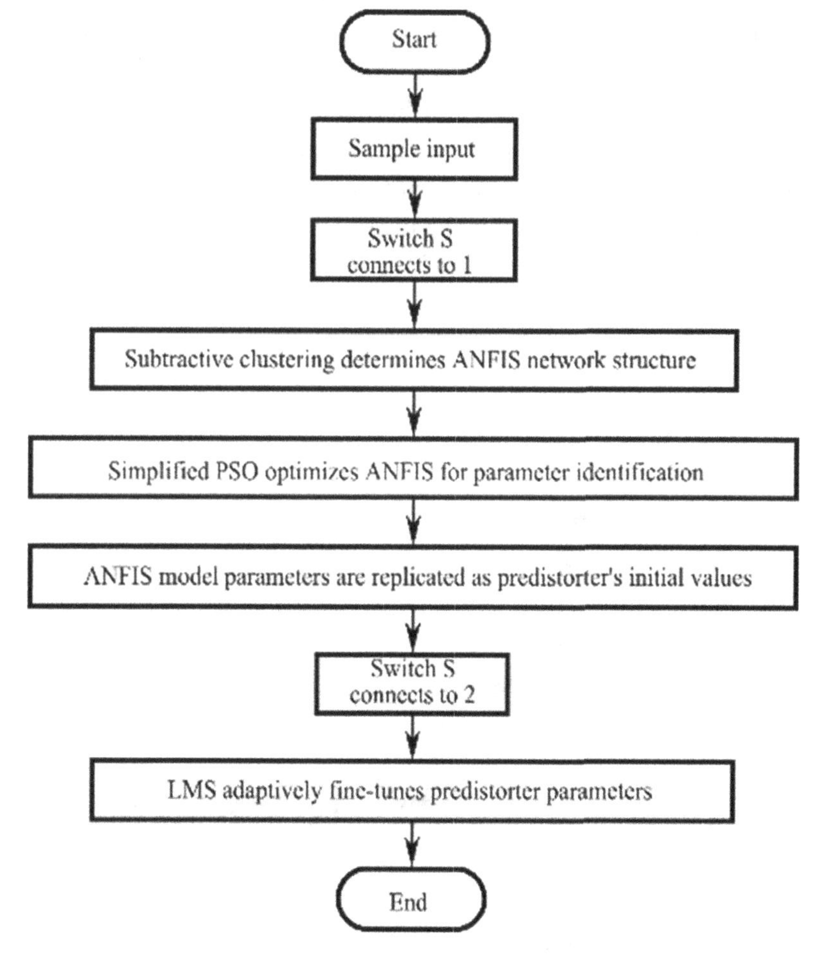

FIGURE 10.51 Adaptive predistortion process of double-loop learning structure.

Firstly, 2,000 sets of input and output data of PA are collected for behavioral model identification, of which 1,000 sets are used to identify model parameters and the other 1,000 sets are used to validate the model performance. The model's memory depth is 2. The input signal is a 64QAM baseband signal with a bandwidth of 15 MHz. An LDMOS Doherty PA is driven with a gain of 50 dB and a center frequency of 1.96 GHz. Then the yielded model is used to simulate and validate the predistortion system.

Figure 10.52 shows the power spectral density of PA model using the fuzzy neural network. The line of crosses represents the real output. The line of circles represents the model output. The solid line represents the error curve between the real output and model output. As indicated by simulation results, the fuzzy neural network optimized by the improved simplified PSO algorithm produces a very high fitting degree to the PA model. The error between the real output and model output is extremely small, with an RMSE approximately −37 dB. High model accuracy and fast convergence are observed in the simulation. Since the predistorter model is the inverse model of the PA model, this modeling method is able to yield a predistorter model with high accuracy.

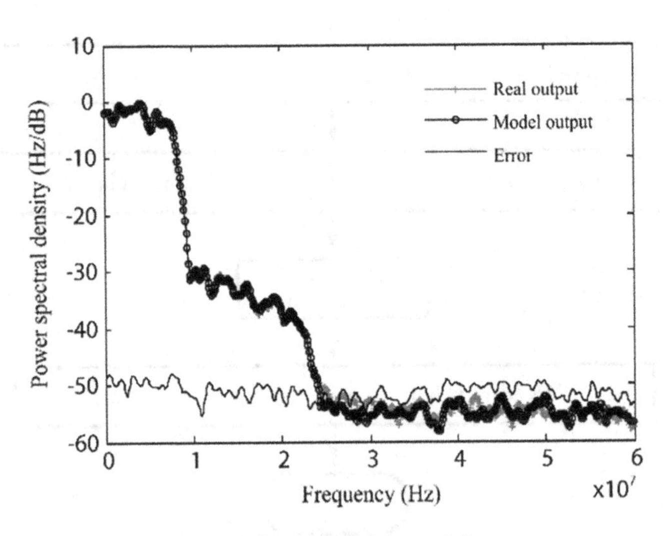

FIGURE 10.52 Power spectral density of PA model.

Simulation and comparison are performed between the proposed method (double-loop structure predistortion based on model identification) and the classical double-loop structure predistortion method, thereby validating the high efficiency of the proposed method. Figure 10.53 shows the power spectral densities of the PA nonlinear system in presence of additive noise with and without predistortion. Curves a and b in the figure denote the power spectral densities of the input signal and the output of PA without predistortion, respectively. Curves c and d denote the power spectral densities of the output with predistortion using the method proposed in Literature [29] and the method proposed in this section, respectively. As indicated from simulation results, predistortion has suppressed out-of-band spectrum spreading. Compared with the method proposed in Literature [29], the predistortion method proposed in this section improves the adjacent channel power ratio by approximately 7 dB, providing a better predistortion effect and stronger anti-noise ability.

Figure 10.54 depicts the adaptive convergence characteristic curve in the direct learning structure. It is seen in the figure that the method proposed in this section notably outperforms the adaptive fine-tuning method proposed in Literature [29] in terms of online parameter fine-tuning. Since the predistorter's initial values are pretty close to the ideal

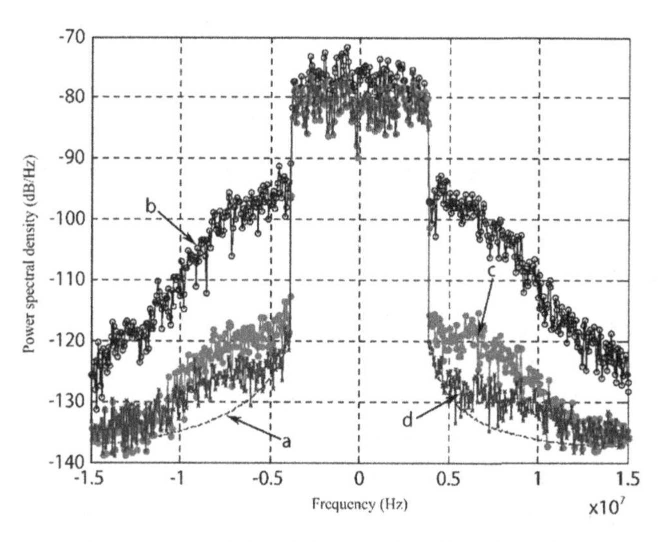

FIGURE 10.53 Power spectral densities of PA with and without predistortion.

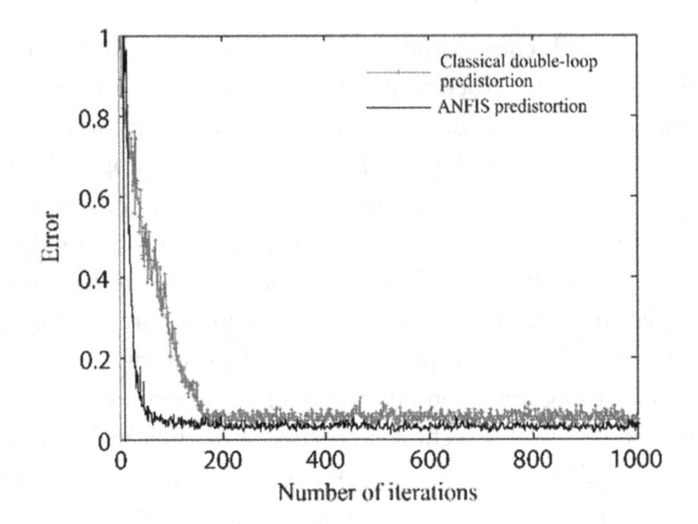

FIGURE 10.54 Adaptive convergence characteristic curve.

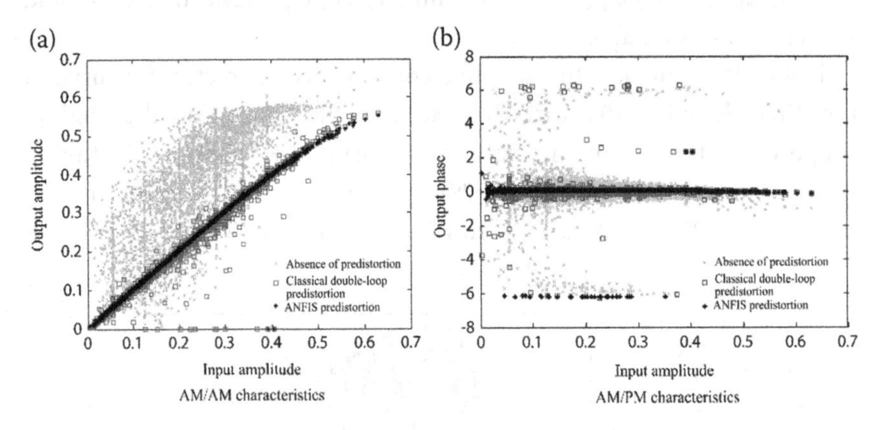

FIGURE 10.55 AM/AM and AM/PM characteristic curves of PA predistortion.

parameter values, the fine-tuning process requires less iterations and higher model accuracy is yielded.

Figure 10.55 depicts a comparison of AM/AM and AM/PM characteristic curves of PA with and without predistortion. As indicated in the figure, the memory amplifier's characteristics appear as a cluster of curves with hysteresis, instead of just a curve as produced by the memoryless amplifier. In presence of predistortion, the predistortion system proposed in this section has produced a favorable linearization effect, which enables the characteristic curve of PA to appear more like a straight line with little divergence. It is noted that the proposed predistortion system outperforms the method proposed in Literature [29]. In addition, the correctness of the memory nonlinear behavioral model proposed in this section has also been validated.

This section proposes a double-loop adaptive predistortion method that combines the indirect and direct learning structures based on fuzzy neural network model identification. The proposed method utilizes the improved simplified PSO algorithm to offline train the adaptive fuzzy inference system for identification of the model parameters, which are used as the predistorter's initial values. Then, the proposed method uses the LMS algorithm for online training of the direct learning structure to fine-tune the model parameters. In this way, the predistorter model is eventually yielded. As indicated in simulation results, the proposed method fully considers the nonlinearities and memory effects of PA and improves the accuracy and anti-interference ability of the model. Furthermore, the proposed method features fast convergence, high accuracy and notably improved linearization effect. In conclusion, the proposed method is able to significantly improve the performance of the digital predistortion system.

REFERENCES

1. STEVE C CRIPPS. Advanced Techniques in RF Power Amplifier Design [M]. Norwood: Artech House, 2002.
2. POTHECARY N. Feedforward Linear Power Amplifier[M]. Norwood, MA: Artech House, 2000.
3. KENINGTON P. High-Linearity RF Amplifier Design[M]. Norwood, MA: Artech House, 2000.
4. SEUNG YUP LEE, YONG SUB, SEUNG HO HONG. An Adaptive Predistortion RF Power Amplifier with a Spectrum Monitor for Multicarrier WCDMA Applications[J]. IEEE Transactions on Microwave Theory and Techniques (S0018-9480), 2005, 53(2):786–793.
5. WAN JONG KIM, SHAWN P STAPLETON, JONG HEON KIM, et al. Digital Predistortion Linearizes Wireless Power Amplifiers[J]. IEEE Microwave Magazine (S1527-3342), 2005, 6(3):54–61.
6. WANGMYONG WOO, MARVIN D MILLER, STEVENSON KENNEY J. A Hybrid Digital/RF Envelope Predistortion Linearization System for Power Amplifiers[J]. IEEE Transactions on Microwave Theory and Techniques (S0018-9480), 2005, 53(1):229–237.

7. FREDERICK H RAAB, PETER ASBECK, STEVE CRIPPS. Power Amplifiers and Transmitters for RF and Microwave[J]. IEEE Transactions on Microwave Theory and Techniques (S0018-9480), 2002, 50(3):814–825.
8. KATHLEEN J MUHONEN, MOHSEN KAVEHRAD, RAJEEV KRISHNAMOORTHY. Look-Up Table Techniques for Adaptive Digital Predistortion: A Development and Comparison[J]. IEEE Transactions on Vehicular Technology (S0018-9545), 2000, 49(5):1995–2002.
9. 赵洪新，陈忆元，洪伟．一种基带预失真RF功率放大器线性化技术的模型仿真与实验．通信学报[J]2000, 21(5):41–47.
ZHAO H, CHEN Y, HONG W. A Baseband Predistorting Linearizer for RF Power Amplifier Prototype Simulation and Experimentation. Journal on Communications[J]. 2000, 21(5):41–47.
10. 龚珉杰．数字预失真技术在ROF系统中应用研究[D]．北京：北京邮电大学，2010.
GONG M. Study on Applications of Digital Predistortion Technology in ROF System[D]. Beijing: Beijing University of Posts and Telecommunications, 2010.
11. 刘明，吴椿烽，张慧，等．RoF技术在光通信领域中的研究与应用[J]．光通信技术，2009,6:43–46.
LIU M, WU C, ZHANG H, et al. Study and Application of RoF Technology in the Optical Communication[J]. Optical Communication Technology, 2009, 6:43–46.
12. FERNANDO X N， SESAY A B. Adaptive Asymmetric Linearization of Radio Over Fiber Links for Wireless Access[J]. IEEE Transactions on Vehicular Technology, 2002, 51(6):1576–1586.
13. 林倩，郭里婷．OFDM系统中峰均比抑制与预失真联合技术研究[J]．电视技术，2011，35(2):17–19.
LIN Q, GUO L. Research on Combined Technology of PAPR Reduction and Pre-distortion in OFDM Systems[J]. Video Engineering, 2011, 35(2):17–19.
14. 张福洪，孔庆浩，栾慎吉．宽带系统中有记忆大功率功放的数字预失真器研究[J]．电子器件，2008，31(6):1903–1906.
ZHANG F, KONG Q, LUAN S. Study of the Digital Pre-Distorter for High Power Amplifier with Memory Effects in Wide Band System[J]. Chinese Journal of Electron Devices, 2008, 31(6):1903–1906.
15. JAYALATH A D S, TELLAMBURA C. SLM and PTS Peak-Power Reduction of OFDM Signals without Side Information[J]. IEEE Transactions on Wireless Communications, 2005, 4(5):2006–2013.
16. ALVAREZ M, PAN A, RAPOSO J, et al. Extraction Lists of Data Records from Semi-Structured Web Pages[J]. Data&Knowledge Engineering, 2008, 64(2):491–509.
17. 周明建，高济，李飞．基于本体论的Web信息抽取[J]．计算机辅助设计与图形学学报，2004，16(4):535–541.
ZHOU M, GAO J, LI F. Ontology-Based Information Extraction from Web Sources[J]. Journal of Computer-Aided Design & Computer Graphics, 2004, 16(4):535–541.

18. DU T C，LI F，KING I. Managing Knowledge on the Web-Extracting Ontology from HTML Web[J]. Decision Support Systems, 2009，47(4):319–331.
19. 贾赛，乔鸿．基于本体的Web信息抽取及本体的构建实现研究[J]．图书馆学研究，2011(5):31–33.
JIA S, QIAO H. Research on Ontology-Based Web Information Extraction and Ontology Construction Implementation[J]. Researches on Library Science, 2011(5):31–33.
20. 廖涛，刘宗田，孙荣．Web表格定位技术的研究与实现[J]．计算机科学，2009，36(9):227–230.
LIAO T, LIU Z, SUN R. Research and Implementation of Web Table Positioning Technology[J]. Computer Science, 2009, 36(9):227–230.
21. BI LEI, SHEN JIE, XU FAYAN. Extracting Web Business Information Using Domain-Specific Ontology[J]. Computer Engineering and Design, 2008, 29(24):6393–6396.
22. 刘建成，赵宏志，全厚德，等．迭代变步长LMS算法及性能分析[J]．电子与信息学报，2015，37(07):1674–1680.
LIU J, ZHAO H, QUAN H, et al. Iteration-Based Variable Step-Size LMS Algorithm and Its Performance Analysis[J]. Journal of Electronics and Information Technology, 2015, 37(07):1674–1680.
23. AI B, YANG Z Y, PAN C Y. Improved LUT Technique for HPA Nonlinear Pre-distortion in OFDM Systems[J]. Wireless Personal Communications, 2006, 38(4):495–507.
24. 柳佳刚，陈山，黄樱．一种改进的基于本体的Web信息抽取[J]．计算机工程，2010(2):39–42.
LIU J, CHEN S, HUANG Y. Improved Ontology-Based Web Information Extraction[J]. Computer Engineering, 2010(2):39–42.
25. KEKATOS V，GIANNAKIS G B. Sparse Volterra and Polynomial Regression Models：Recoverability and Estimation[J]. IEEE Transactions on Signal Processing, December 2011, 59(12):5907–5920.
26. REINA TOSINA J，ALLEGUE MARTINEZ M，MADEROAYORA M J，et al. Digital Predistortion Based on a Compressed-Sensing Approach. 43rd Eur. Microw. Conf (EuMC 2013)，Nuremberg，Germany，October 6–11，2013[C]. Piscataway：IEEE，408–411.
27. ABDELHAFIZ A, KWAN A, HAMMI O，et al. Digital Predistortion of LTE-A Power Amplifiers Using Compressed-Sampling Based Unstructured Pruning of Volterra Series[J]. IEEE Transactions on Microwave Theory and Techniques, 2014, 62(11):2583–2593.
28. GUAN L，ZHU A. Simplified Dynamic Deviation Reduction-Based Volterra Model for Doherty Power Amplifiers. Workshop Integr. Nonlinear Microw. Millimetre Wave Circuits (INMMIC 2011)，Vienna，Austria，April 18–19，2011[C]. Piscataway：IEEE，1–4.
29. 曲昀，南敬昌，毛陆虹，等．射频功放数字预失真新方法研究[J]．微电子学，2013，43(3):440–444.

QU Y, NAN J, MAO L, et al. Study on New Digital Pre-distortion Method for RF Power Amplifier[J]. Microelectronics, 2013, 43(3):440–444.

30. JARDIN P, BAUDOIN G. Filter Look-up Table Method for Power Amplifier Linearization[J]. IEEE Transactions on Vehicular Technology, 2007, 56(3):1076–1087.
31. CHEN H H, LIN C H, CHEN J T, et al. Joint Polynomial and Look-up Table Power Amplifier Linearization Scheme[J]. IEEE Transactions on Circuits and Systems，2006，53(8):612–616.
32. 张玉梅，南敬昌．基于Saleh函数的功放行为模型研究[J]．微电子学与计算机，2010，49(12):35–39.
ZHANG Y, NAN J. Study of Behavior Models Based on the Saleh Function for Power Amplifier[J]. Microelectronics & Computer, 2010, 49(12):35–39.
33. 沈英杰，刘郁林，胡中豫．自适应数字预失真方法在功放线性化中的应用[J]．重庆邮电学院学报（自然科学版），2005，(06):687–690.
SHEN Y, LIU Y, HU Z. Application of an Adaptive Digital Predistortion Technique to HPA Linearization[J]. Journal of Chongqing University of Posts and Telecommunications (Natural Science Edition), 2005, 17(06):687–690.
34. 崔华，赵祥模，艾渤．记忆功放的BP神经网络分离预失真方法[J]．西安电子科技大学学报，2010，37(3):565–569.
CUI H, ZHAO X, AI B. Separate BPNN-Predistortion Method for Nonlinear HPA with Memory[J]. Journal of Xidian University (Natural Science), 2010, 37(3):565–569.
35. 詹鹏，秦开宇，蔡顺燕．新的射频功放预失真线性化方法[J]．电子科技大学学报，2011，40(5):676–681．40.
ZHAN P, QIN K, CAI S. New Predistortion Method for RF Power Amplifier Linearization[J]. Journal of University of Electronic Science and Technology of China, 2011, 40(5):676–681. 40.
36. 孙伟，张小瑞，唐慧强，等．基于自适应遗传粒子群优化模糊神经网络的疲劳驾驶预测模型[J]．汽车工程，2013，35(3):219–228.
SUN W, ZHANG X, TANG H, et al. Fatigue Driving Prediction Model Based on FNN Optimized by Adaptive GA-PSO[J]. Automotive Engineering, 2013, 35(3):219–228.
37. 胡旺，李志蜀．一种更简化而高效的粒子群优化算法[J]．软件学报，2007，18(4):861–868.
HU W, LI Z. A Simpler and More Effective Particle Swarm Optimization Algorithm[J]. Journal of Software, 2007, 18(4):861–868.
38. 李成法，陈贵海，叶懋，等.一种基于非均匀分簇的无线传感器网络路由协议[J].计算机学报，2007，30(1):27–36.
LI C, CHEN G, YE M, et al. An Uneven Cluster-Based Routing Protocol for Wireless Sensor Networks [J]. Chinese Journal of Computers, 2007, 30(1):27–36.

Printed in the United States
by Baker & Taylor Publisher Services